The Big Disconnect

The Big Disconnect

Protecting Childhood
and Family Relationships in the Digital Age

Catherine Steiner-Adair, EdD,

WITH TERESA H. BARKER

HARPER

www.harpercollins.com

THE BIG DISCONNECT. Copyright © 2013 by Catherine Steiner-Adair. All rights reserved. Printed in the United States of America. No part of this book may be used or reproduced in any manner whatsoever without written permission except in the case of brief quotations embodied in critical articles and reviews. For information, address HarperCollins Publishers, 10 East 53rd Street, New York, NY 10022.

HarperCollins books may be purchased for educational, business, or sales promotional use. For information, please e-mail the Special Markets Department at SPsales@harpercollins.com.

FIRST EDITION

Designed by Jennifer Daddio / Bookmark Design & Media Inc.

Library of Congress Cataloging-in-Publication Data has been applied for.

ISBN: 978-0-06-208242-8

13 14 15 16 17 ov/RRD 10 9 8 7 6 5 4 3 2 1

To Fred, Daniel, and Lily,
my beloved home page

Contents

Introduction: The Revolution in the Living Room   1

Chapter 1   Lost in Connection: How the Tech
Effect Puts Children's Development at Risk   33

Chapter 2   The Brilliant Baby Brain: No Apps or
Upgrades Needed   66

Chapter 3   Mary Had a Little iPad: The Wisdom
of Tradition, the Wonder of Tech:
Ages Three to Five   99

Chapter 4   Fast-Forward Childhood: When to
Push Pause, Delete, and Play: Ages
Six to Ten   129

Chapter 5   Going, Going, Gone: Tweens, Screens, and
the Perils of Independence: Ages Eleven
to Thirteen   162

Chapter 6   Teens, Tech, Temptation, and Trouble:
Acting Out on the Big (and Little) Screen   193

Chapter 7   Scary, Crazy, and Clueless: Teens Talk
about How to Be a Go-To Parent in
the Digital Age   226

Chapter 8   The Sustainable Family: Turning Tech
              into an Ally for Closeness, Creativity,
              and Community   260

Acknowledgments   297

Notes   305

Bibliography   335

Index   361

The Revolution
in the Living Room

All the wisdom in the world about child-rearing cannot, by it-self, replace intimate human ties, family ties, as the center of human development . . . the point of departure for all sound psychological thinking.

—SELMA FRAIBERG, *THE MAGIC YEARS*

Alarm travels fast through the human brain. Awareness dawns more slowly. It is a cognitive process, a matter of comprehension that involves a different part of the thinking brain. It's how we connect the dots. Parenting has always included both. But parenting in the digital age challenges us in ways the human brain—and heart—can hardly process fast enough. The dots rush past in a blur: flashes of insight with no time for reflection, downloads of details with too little time to both feel and respond thoughtfully. Once in a while a moment seems to hang for a fraction in time—OMG moments—when we sense we are glimpsing a piece of some larger picture taking shape around us—something important—but it's hard to tell exactly what. As a psychologist, a child and family therapist, a school consultant, and a mother of two, I see a lot of dots.

Sally, the mother of four school-age children, comes to a parent coffee I'm facilitating. She is at her wit's end and goes on to describe a domestic scene from the night before. She had been folding laundry while keeping an eye on her children—ages four, eight, eleven, and fifteen—immersed in their various activities in the adjoining family room. Their family room is off the laundry nook but not quite visible, so Sally was multitasking, folding clothes and checking her e-mail on her smartphone, as many of us do, alternately stepping in and out of the family room for a few moments at a time to keep an eye on the kids. Finally she needed a solid five minutes to search for socks without partners, and she asked the two older children to keep watch on their four-year-old brother. The fifteen-year-old was on her iPad, and the eleven- and eight-year-old brothers were playing a two-player racing game on the Wii. The four-year-old was romping around them as four-year-olds do. As usual, the older siblings nodded their agreement without looking up; *Yes*, they indicated, *got it.* So mom left them all together and took the laundry upstairs, tending a few e-mails while there. When she popped back into the room five minutes later everyone was precisely as they had been—except the four-year-old. He was putting the flourishes on a colorful wide-tipped purple Magic Marker trail that looped and snaked across the hardwood floor and beautiful wood cabinetry, weaving throughout the entire room. It was *Harold and the Purple Crayon* come to life.

When Sally saw this she went ballistic, she tells me. "How could you let him do that?!" she shrieked at her older children. And in that nanosecond before her children could even look up to see what she was talking about, she saw the answer to her question. All three of them were so transfixed in their screen activities that they hadn't noticed the four-year-old running wild, decorating the floor and walls. This might just as easily have occurred if they'd had the TV on or if they'd had their noses buried in a book, reading the afternoon away like generations before them. But in that instant Sally saw something more worrisome, felt something deeper and more ominous. First, she realized that when she had turned her attention to e-mail, however briefly, her intuitive "third eye"—the parental antenna—had instantly lost its signal. Honestly, she'd lost track of time, too; had it really been "just five minutes" or had it stretched beyond that? She had no idea; it was as if she'd been lost in another world. And then there was the specter of her three older children, each lost in a mind meld with screens. "I know this could have happened without the iPad or the Wii," she says, "but I feel there's something about those screens that sucks my children away from me. Something that I don't think watching *The Brady Bunch* did to me and my siblings when we were growing up."

The other mothers nod in agreement. She is right. Our screens are sucking us in—all of us—in ways that shows like *The Brady Bunch* never did circa 1970. But it wasn't nostalgia for *The Brady Bunch* that made Sally wistful; it was remembering what it was like to be a kid and watch a TV show. You could love watching it but then you would leave the screen and get on with your real life: homework, outdoor play and after-school hours with friends, a solid dinner hour with your parents and siblings, and unhurried telephone conversations that arrived on your family's landline. Whoever answered the phone might ask "Who's calling?" and then yell the name out for all to hear as they called you to the phone. TV had its allure—we have a few generations of couch potatoes now to prove it—but it was essentially a passive activity, contained and time limited, and it was something a family often did together; not the stimulating, interactive, mesmerizing 24/7 experience we can access solo today through the media and technology.

Tech Is Our New Home Page

We live in the glow of the digital age and we're hooked on tech—an ever-expanding array of electronic devices and capabilities designed to enable our every wish. Few of us would even consider trading our wired (or Wi-Fi) ways for the unplugged and bucolic *Brady Bunch* life. We've adapted happily, for the most part, to the convenience and connectivity of the computer age. Our children, born into the digital culture, are natives; they speak the language, tech is their frame of reference, their mind-set. Kids between the ages of eight and eighteen, according to a 2010 Kaiser Family report, are spending more time on their electronic devices than any other activity besides (maybe) sleeping—an average of more than seven and a half hours a day, seven days a week (i.e., *every day*). Not only that, they typically multitask on computer, simultaneously instant messaging (IM), uploading YouTube videos, posting updates on Facebook, and continually searching the Web for fresh diversions. The so-called downtime they spend on computers is neurologically, psychologically, and often emotionally action packed. Stimulation, hyperconnectivity, and interactivity are, as the psychiatrist and creativity expert Gene Cohen put it, "like chocolate to the brain." We crave it.

Designed to serve us, please us, inform us, entertain us, and connect us, over time our digital devices have finally come to define us. We step in and out of our various roles throughout the day as co-workers, family, and friends. But with our phones in our pockets, our laptops handy, and our panoramic screens, game systems, and online lives just a click away, for many of us our relationship with technology is our single most consistent domain. It is our digital backdrop and theme music. In any given moment, with a buzz or a ping, our devices summon us and we are likely to respond, allowing ourselves to be pulled away from our immediate surroundings and anyone in them, into the waiting world of elsewhere and others. Whether we use it for work, shopping, or socializing, for communicating with our children or their teachers, for wonderful reasons or sometimes for meaningless and addictive stuff, the effect is the same: We turn our attention away from those present.

We use the language of addiction to joke about our texting and online habits. We compulsively check our "crackberry," lament not-really-joking that we're "addicted" to e-mail, or complain that we need our YouTube or Facebook "fix." We mean to step away from the screen and call it quits for the day, but we sit down for one last check of one last thing—and then one more. We take our phones to bed. We take them to the bathroom. We don't leave home without them. "I thought when my kids got to be a certain age I'd stop feeling like I *had* to have it with me, but now I keep it with me anyway," says the mother of three children in their twenties. She has become a little self-conscious about her habit, she says, "but when I start to leave it behind, it's like a little voice inside says, *better to keep it handy just in case—you never know.*"

Pregnant women who regularly use cell phones may be more likely to have children with behavioral problems, "particularly if those children start using mobile phones early themselves," according to a study that appeared in *The Journal of Epidemiology and Community Health* in 2010. Another study suggests that the quality of sperm may be compromised in men who carry a cell phone in their pants pocket. The American Academy of Pediatrics (AAP) has called on the federal government for a review of cell-phone-radiation-emission standards out of concern for children's health. Have these risks been enough to motivate us to change our tech habit?

We know that some aspects of tech are addictive and that different types of brains are more vulnerable than others. We joke, but the truth is that research shows we are in fact enjoying neurochemical hits and fixes—the neurotransmitter dopamine most notably—in the brain's pleasure centers when we're "on" our devices. Talk of addiction is not hyperbole; it is a clinical reality in some users' lives today. As adults we may choose to mess with our minds and gamble with our own neurology, but I have never met a caring parent who would knowingly risk his or her child's future this way. And yet, we are handing these devices—that we use the language of addiction to describe—over to our children, who are even more vulnerable to problems of use and abuse and the impact of everyday use on their developing brains. Are we blindly cultivating a generation of "crackberry" kids? In our enthusiasm to be early adapters, and to give our kids every advantage, are we putting our children in harm's way?

It appears that our distraction with tech is in fact putting our children in harm's way quite literally—physically. Doctors and researchers, from local emergency rooms to the Centers for Disease Control (CDC), link the growing use of handheld electronic devices to an alarming increase in injuries to children, especially when parents or caregivers are distracted and fail to properly supervise young children in the moment. The *Wall Street Journal*, in a roundup of research and interviews with experts on the subject, noted that injuries to children under age five rose 12 percent between 2007 and 2010, after falling for much of the prior decade, according to the most recent data from the CDC. That 14 percent of adults—and 22 percent of adults who send text messages—reported being so distracted by their devices that they have physically bumped into an object or person. Barbara Morrongiello, a psychology professor at the University of Guelph in Canada who has studied the relationship between child supervision and injury, told the *Journal* that most people don't realize how much they are distracted by devices. If you ask a parent or caregiver sending a text message "if they are paying attention, they would say, 'of course,'" she said. People "often underestimate how much time they're taking to do something."

The tech effect has transformed every facet of our lives—from work to home to vacation time away—emerging, dot by dot, to reveal a new and unsettling family picture. While parents and children are enjoying swift and constant access to everything and everyone on the Internet, they are simultaneously struggling to maintain a meaningful personal connection with each other in their own homes. It is the parental paradox of our time: never before has there been so much opportunity for families to plug in and at the same time disconnect.

John is distraught because his marriage and family are disintegrating. A flashpoint is the invasive presence that online time, social networking, texting, and the like have in their lives. His wife has a long commute to her job and she routinely communicates by text with him and their two children, eight and fourteen years old. This includes when she is home and they are all in the house. She's a tech enthusiast; she enjoys bringing the newest gadgets home and isn't bothered by the time the children spend on them. John prefers face-to-face conversation, or phone calls—voice contact. They fight over whether access to all this tech is hurting their children. The emotional disconnect between the two of them is acted out and amplified through their battles over technology.

John worries about his children and the way their online diversions, texting, and social networking have become the default mode for their attention. Car rides used to include conversation or quiet time with them. No more. Wherever they go his children are either fighting over who has the digital tablet or they are plugged in and oblivious. On their last trip, he recalls, their son texted constantly on his phone; the family also had a handheld digital movie player. "So they're watching a movie, but all their heads are down like this [he looks down at his lap], and they're playing games, too, doing some idiotic thing like putting cubes into a box or making a car jump over and over." Back home, what used to be old-fashioned downtime before dinner to hang out as a family, swap stories from the day, and make small talk has vanished. The kids stay in their rooms with their laptops until dinner is ready. Or they come to the table with screens to continue "looking up stuff—you know, looking up shoes or looking up music."

"You know, it's almost like nervously filling the time with a passive activity, like biting your nails," he says. "Instead of biting their nails, people are texting people and Googling them. And young people being the way they are, I mean, what information are they really exchanging? *Hey? Huh? Hey, yeah. What you doing? So bored."* How can that be meaningful? he wonders. What is the quality of all this connection?

John is also disturbed by what he sees in the broader social context and the reflection of that in his children's friendships. "There's a social weirdness to it," he says of the way people continue to text or take calls even when they're engaged in face-to-face conversations. "That's what I see in my children, in their relationships. . . . My son has had a friend come over for a playdate, and my son is, like, skateboarding, and his friend is just standing there texting, right?" He feels defeated in his efforts to rein in his kids' media usage or to reconcile with the digital makeover of his family. "Yeah, I mean, we're sort of stuck with it, is the problem. I guess that's really the very crux of it. Because the society's all wired around it now."

Everywhere I go, from parent conferences to kitchen table conversations, a similar mix of confusion and clarity characterizes the concern about tech and family.

"When you have very busy lives, your relationships become completely utilitarian and nagging," says Helene, reflecting on life with her husband and two teenage children. She rattles off the to-do list of deadlines and scheduling that dominates their conversations: homework, camp application deadlines, games, sports, concerts, practice, the family social calendar. "It's like we're this little business, and we just interact, so if you want to have any kind of connection otherwise, you send the YouTube video, send the text . . . we never talk directly, we never look each other in the eye anymore."

Another mother tells me of her dawning awareness that something fundamental is changing about family life, echoing the overwhelming sense of unease so many parents share: "I have this feeling that we're all just camped out in the house on these different screens, and I feel like there's been this deterioration of connection, and I don't know how," she says.

I know that feeling as a parent, too. I remember this eerie feeling that our family's way of being a family felt like it was slipping away as our home computer began to dominate our children's time and attention. The dynamic shifted beyond what I had anticipated as normal adolescent behavior, especially during our son Daniel's teen years, as he spent more and more time on computer games and online socializing. He was in that first generation of digital natives, those children of the early 1980s who grew up around bulky home computers, floppy disks, and loaf-size mobile phones used almost exclusively by their parents and mostly just for work. These were the kids who were in elementary school when the first wave of computers hit the classroom. It was easy for us to set limits on TV (none during the school week) and to set clear limits on the weekends. But like so many of his peers, Daniel entered adolescence fully computer literate, his intellect wired for upgrades, his appetite eager for mastering games. Suddenly his after-school social life shifted to a new playing field: computer games. Homework called for screen time, too, and exploratory research on the Internet. So we said yes to a computer, then yes, however hesitantly at times, to new tech, updates, and more computer games when it seemed they were the new requirement for a teenager to stay connected with his friends. On the one hand, his home computer transformed Daniel's academic life in ways that excited all of us. His success in the screen world of wizardry and warfare transformed his social life in ways that thrilled him and us. One wave after another, tech swept through the home front. Our son's feelings about it were predictable: he loved it. But we were conflicted. It was fun watching him have great experiences. Not so much having to argue or negotiate constantly with him over screen time or game choices. And then feeling: *this is invading my family and eating my son alive and I don't like it.* Even when his grades and behavior were fine, and he was fine, something about that computer's gravitational pull just didn't feel right.

Another experience that focused the evolving and unsettling picture for me arose not at home but one afternoon in Washington, D.C. My husband, Fred, and I were visiting Arlington National Cemetery across the river in Virginia. Walking through the grounds that still vibrate with the pain and sacrifice of war, we noticed that no matter where we went people were talking on their cell phones with no regard for the people around them or for their own privacy. A family of several generations hovered at one stone and right next to them a man was arguing and swearing to someone on the phone. Parents knelt quietly at a specific gravesite while their kids were texting and laughing on their phones. The solemn surroundings invited quiet reflection, or at least quiet respect for the place where so many are buried. But quiet respect was no longer the norm.

We eventually stopped for a bite to eat at a bustling street café where, at the table across from us, sat a teenage girl, maybe sixteen or so, with a middle-aged woman who appeared to be her mother. Through the entire meal the girl said only about three sentences to the woman just inches away and her comments were brief and perfunctory. The rest of the time she looked steadily down, texting on the smartphone on her lap under the table. A noticeable silence defined the space between the two, the woman glancing at the girl periodically with a hopeful look, while the girl maintained eye contact only with the tiny screen on her lap, continuing her lively texting. It was disturbing to see. There they sat together, but not really. It felt sadly familiar. It was the quintessential family portrait of the Digital Age of Disconnection.

In my work, both as a therapist and school consultant, in the context of our relationship with tech I am always struck by the one eternal and incontrovertible truth about families: children need their parents' time and attention and families thrive when parents have strong, healthy relationships with their children and children are attuned to the family milieu. But this reality can be so easily lost when we are lured away by the siren call of the virtual world.

What Children See: Parents Missing in Action

We read so much about kids tuning out and living online, but that's only half of the problem. More worrisome to me are the ways in which parents are checking out of family time, disappearing themselves and offering that behavior as a model for their children. Every day there are twenty Google alerts about kids and tech. Where are the Google alerts with critical concerns about parents and tech? We complain about kids' love affair with tech, but children—even those who love their screens and smartphones—describe in almost identical ways a sense that their parents are virtually missing in action, routinely either engaged in cell phone conversation and texting or basking in the glow of the computer screen with work or online pastimes. A group of seventh- and eighth-grade boys and girls describe their parents' tech habits:

> There are definitely some parts that make it seem like
> they're really addicted to their phones [and] I feel like my dad
> is with his computer definitely. I mean if he's like awake at one,
> he'll check on the dog—you know, make sure he's OK. And
> then he'll look at his computer and be on it, and end up going
> to bed at five thinking it was five more minutes. Then the next
> day he's really tired. (Collin, twelve)

> What I wish my parents understood about technology is
> that technology isn't the whole world. It's sort of a new thing for
> them because they're sort of really old. [Laughter.] And so it's

like, "Oh my gosh, this new thing came out," but it's annoying because it's like *you also have a family! How about we just spend some time together,* and they're like, "Wait, I just want to check something on my phone. I need to call work and see what's going on." (Angela, thirteen)

When they're on the computer or something, it's always like they're following it, like they're *entranced* by it . . . they are addicted to it, and they just do it all of the time. First they got their phones, and then computers, and then all these other extra things. They didn't have that as a child, so maybe that's why they're doing it. We're growing up in like a new technology era, and maybe they're doing their best to try to fit in. (Carlos, thirteen)

The younger ones speak candidly about having to compete for their parents' attention, not always successfully. A group of elementary school children describe the tech scene at home:

My mom is almost always on the iPad at dinner. She's always "just checking." (Tyler, seven)

I always keep on asking her let's play let's play and she's always texting on her phone. (Penny, seven)

Once my dad was ignoring my mom so bad for over thirty minutes, so I sat on his keyboard. I got sent to my room. I was just trying to help my mother. I got in trouble, though. (Owen, nine)

When my mother and father are really mad at each other on the phone, and I am in the car and it is on the blue phone [Bluetooth] I hear things that make me scared that something bad is going to happen. I worry they are going to get divorced. (Samantha, ten)

A lot of time at home when my parents are home and on their computers, I feel like I'm not there, because they pretend like I'm not there . . . they're like not even talking to me, they just are ignoring me. I feel like, ughhh, sad [sigh]. (Ava, seven)

In a play therapy session, seven-year-old Annabel talks about the loneliness and distress she feels when she is unable to get her parents' attention. "My parents are always on their computers and on their cell phones," she tells me. "It's very, very frustrating and I get lonely inside."

"What do you do when that happens?" I ask.

She then acts it out for me, with the expressive eyes, face, and voice that break my heart:

When my dad is on the phone I have this conversation in my head: "Hello! Remember me? Remember who I am? I am your *daughter*! You had me cuz you wanted me. Only it doesn't feel like that right now. Right now it feels like all—you—care—about—is your phone!"

Then she adds:

But I don't say that, because they'll get mad at me. It doesn't help. It feels worse. So it's just the conversation I have with myself.

Young children's drawings of their families and typical scenes from family life tell a similar story. We often look to children's drawings for what they can reveal about a child's environment or about the child's inner life—things they can't articulate or perhaps don't want to say out loud. Once upon a time they would draw family members with a few distinguishing characteristics—funny hair, eyeglasses, shoes. Sometimes the drawings included the family pets because that's what children thought of when they thought about family. A few years ago in therapy with young children, I began to notice a change in their family drawings. In addition to brown eyes or curly hair or glasses or blue jeans, moms and dads often were talking on cell phones (some with voice bubbles that said, "blah, blah, blah!"). Dads, in particular, might have a hands-free headset hugging their ear. Sisters and brothers wore dangling earbuds tethered to iPods, or they sat at screens "playing computer." Children often depicted themselves at the computer, too, or with handheld devices. Not always. I remember one eight-year-old girl whose picture showed her family members all happily plugged in, with a lone little figure off to the side with no gadgets. "That's me," she said.

An elementary school counselor tells me about a session with a taciturn little boy, in which he used a sand tray with plastic figurines to show his family at dinnertime. "It was one of the saddest I've ever seen," the counselor says. He continues:

> There was all the sand in front, and then we had the mom in this corner, the dad in this corner, the younger brother in this corner, and he and the dog in the middle. There was no interaction. When he was finished, I asked him to tell me about it. You know, every one of them—mom was on the phone, dad was on his computer, and the younger brother was on his computer eating dinner. The boy told me, "My dog keeps me company. He sits on my feet while I eat."

In play therapy, young children talk about "daddy's stupid computer," as if they are jealous rivals. They tell their parents, "I hate your computer!" "Your phone is a *bad* phone!" Toddlers will throw their parents' cell phones in the trash, vying for their parents' attention. Or babies want to hold the phone, want to handle it, drool on it. The phone is almost a transitional object like the old-fashioned blankie—a soothing stand-in for the mom or dad.

I was playing pretend grocery store with my friend's nearly-three-year-old in the park when our game took a sudden emotional shift. Young Anders was the "man at the store" underneath the climbing structure and I was buying groceries and books with my pretend money. He wanted to switch roles, so he became the buyer and he wanted a cell phone. I pretended to hand him one and he got upset. "No, Cafrin! I want a REAL phone!" I was unable to calm him down or bring him back into pretend play. Wendy, his grandmother, tried to calm him down, as we both realized that the word and symbol of the phone triggered something deeper, more emotional and powerful than "groceries." Anders desperately wanted the real phone "right now!" Like so many young children I see, by two or three years old they don't want a toy phone anymore; they want the real thing. They can sense the difference and at some deep level the real thing connects them to their parents.

To parents, multitasking via screens and cells may seem a reasonable work-life compromise, a way to feel available to the children while still tending to work and other interests or commitments. To children, the feeling is often one of endless frustration, fatigue, and loss, not compromise. In the old days the phone would ring and you might be on a call for a bit, but the phone didn't travel with you all day in your purse or your pocket, with the power to pull you away instantly, anywhere, anytime.

"You become distracted and you're not part of the conversation," says the sand-table counselor. "As soon as you put your head down, you know, the so-called BlackBerry nod or whatever you want to call it when people are looking at their devices, then you're not in the conversation. You've just gone into another world. As soon as you look at your device, you see, oh, there's an e-mail from so-and-so, and then you want to look at that. And then, all of a sudden you've just decoupled from the conversation."

You versus U: Why Your Child Needs the Real Thing

Parents' chronic distraction can have deep and lasting effects on their children. Psychologists know this from work with children who grow up with unavailable, disconnected, or narcissistic parents; they struggle and often don't do very well. In addition to the issue of distracted supervision putting children at risk for injury, at some point distracted, tech-centered parenting can look and feel to a child like having a narcissistic parent or an emotionally absent, psychologically neglectful one.

In nonclinical settings, most notably in focus groups in schools around the country, the take-home message I am hearing from children of all ages is this: They feel the disconnect. They can tell when their parents' attention is on screens or calls and increasingly they are feeling that all the time. It feels "bad and sad" to be ignored. And they are tired of being the "call waiting" in their parents' lives. The word "hypocrite" comes up a lot from middle school and high school kids. It is confusing when you know it is time to leave the house and your father is on his laptop or tablet or cell phone and gets mad at you for telling him that you will be late for school. It is defeating when you try to do the right thing (get to school on time) and get in trouble at both ends (at home and at school). To a child, getting mixed messages from parents undermines trust and security. Inconsistency undermines a sense of safety and stability.

A lot of the disappointment involves kids giving up. Little moments of promises broken, of feeling let down. *Dad was supposed to read with me. Mom said she'd play a board game.* Teens offer an older version of the same yearning: *I don't see why Mom can't just not take a call when we're talking—she's always telling me that's what I'm supposed to do. I know Dad's busy, but it's like nothing I do is important enough to really matter to him.*

Children don't need us constantly, but they do need to experience our being there for them, genuinely connected with them, at times when our presence matters to them. That will be different for an infant from what it is for a toddler, a tween, or a teen, but at any age there are daily moments when they initiate connection and they need us to respond attentively. The message we communicate with our preoccupation and responsiveness to calls and e-mail is: *Everybody else matters more than you. Everything else matters more than you. Whatever the caller may say is more important than what you are telling me now.* Meanwhile, a child is waiting to connect. Maybe she's waiting to play or waiting to ask for help with homework. Or they are waiting to tell you about a test, or a crush, or a quandary. As one teenage girl told me, "I think in the olden days families mattered more. It feels like we're losing the idea that family matters."

Apple creator Steve Jobs famously said about the unpredictable steps that led him from college dropout to mastermind of the technology that transformed our world: "It was impossible to connect the dots looking forward . . . but it was very, very clear looking backwards." That may be acceptable if you're tracking the evolution of modern technology, but it won't do for those of us living it, those of us with children and families we cherish and whose lives are in some ways increasingly remote to us, defined more by tech's pervasive presence than the unmediated heartfelt personal one. We can't afford to wait and we don't need to wait to see this much of the picture clearly: Technology, social media, and the digital age have converged on the American family, first transforming it and now threatening to replace the deepest and most vital human connections that children need to grow and thrive. Parents who have long prided themselves on protecting, providing, and promoting a values-rich childhood for their children are feeling increasingly irrelevant in their children's technology-driven lives. And they are right. Parents have lost their job—sometimes unwittingly abdicated it—at a time when they are most desperately needed by kids who are not only growing up faster but growing into a world that no longer protects childhood.

As computers, smartphones, digital tablets, e-readers, and the Internet have become integral components of everyday life, tech has not only sucked us in; it has gained a de facto coparenting role: continuously engaging, informing, entertaining, and modeling its digital version of the connected life. If we are too busy to spend time with our kids, too busy to listen to them, or too intimidated or overwhelmed by tech to respond to them, an online world of "intimate strangers" or diversions is ever ready to welcome them. Tech is undeniably allowing us to connect in new and wondrous ways. Unfortunately, in ways that matter it is often a model of connection that favors quantity over quality, breadth over depth, and image over intimacy. The tech culture is conditioning us to accept that as an unquestioned norm. And it is training our children to think that way, too.

The trouble with tech as a coparent lies in what it can and cannot do. In some ways, to our dismay, it can replace us as the source of values, information, context, community, and coaching in our children's lives. But despite marketing claims to the contrary, it cannot provide the direct, nourishing, and uniquely human dimension of relationship essential for healthy neurological and psychological development in human children.

Our relationship with tech is redefining the way humans have interacted since time began. As we scramble to keep up with tech and to make the most of it, with each step we must consider its impact not only on our time and attention, but on the irreplaceable substance of relationships with those who matter most to us.

Families at the Intersection Where Cultures Converge, Collide, or Collapse

We live today in what the media theorist Henry Jenkins has called the "convergence culture," which he describes as "an age when changes in communications, storytelling and information technologies are reshaping almost every aspect of contemporary life—including how we create, consume, learn, and interact with each other." Where I work, on the family front, in the trenches, the impact of the convergence culture is taking its toll on the inner life, the social and emotional infrastructure of families.

This historic cultural shift has looked something like a huge flip book; the pages whiz past and with each page tech takes over more and more and more of the picture. The invasion has been astonishing, exciting, entertaining—and now troubling. The swift takeover has transformed tech from the merely helpful tool or harmless diversion that it once was, to a dominant role as the hub and hearth of family life. The transformation has been nothing short of a revolution.

Beyond the technical and societal challenges the convergence culture presents, the psychological dimensions are dicey for all of us. We're confronted with a mix of communication conundrums. The content and tone of an e-mail may be unclear or off-putting, making an appropriate reply hard to craft. The sheer number of e-mails, texts, voice mails, and other communications can be exhausting. Facebook, which invites us to carefully craft online personas and commentary, has over one billion active users; it has redefined the meaning of "friend" and added "unfriending" to the relational vernacular. Intentional or not (and you never know), gaps in response time can add to confusion, or worse, be misconstrued in negative ways. Technology is redefining the fundamental cues, content, and cadence of our communication and the improvisational, uniquely human dimension of connection.

Tech has altered our social discourse so rapidly that we've had no time to thoughtfully decide whether this is what we want. Time-honored expressions of basic courtesy and civil conduct have simply disappeared. Absent that thoughtful standard of established etiquette, it has become acceptable to have a highly personal conversation in a highly public space, at full volume, completely disregarding strangers next to us on the street, at the airport, or in a store. Beyond the annoyance factor, in that digital context the other individuals around us are now irrelevant, dismissed, tuned out. We authorize ourselves, when on our phones, to behave in ways that were just a generation ago commonly held as discourteous, uncivilized.

All of this matters. Our children watch us at home, on our cell phones, and at every other opportunity, for cues to help them navigate life. As parents, our relationship with tech and our patterns of behavior around it become a training ground for these impressionable youngsters as they forge their own relationships with tech. It is all too easy for us to complain when our children favor screen diversions and ignore us, when that is what we teach them by example.

Our Relationship with Technology versus Our Relationships with One Another

What I see and hear a lot in therapy now with children, young adults, and equally if not more so with grown-ups is how tech is playing a destructive role in relationships. Everything that is vulnerable and challenging for people and human nature can be played out and amplified through tech. Many of these human interactions and relational challenges are not new. Friendship, courtship, intimacy, and casual interactions with strangers are, after all, timeless elements of human relationships. But the new context in which they occur is different, particularly when the conversation plays out online or in snippets of text on screens, whether that is texting, Facebook, Instagram, instant messaging, or longer e-mail or online exchanges. Tech not only accelerates the speed of exchanges, it changes the way we express ourselves.

It can ramp up the emotional discharge, encouraging us to emote without accountability for the impact of our communication because we can't see the other person's face, note the impact, and adjust our tone. When we are upset, we often say things in texts, e-mail, or online that we wouldn't say in person. The likelihood of misunderstanding grows without the more nuanced sensory feedback that a face-to-face conversation affords. Even a heartfelt apology, or an expression of sympathy, empathy, understanding, or support, lacks the unique resonance that is communicated through the human voice whether it is on the phone or in person. Reading the words *I get it* or *been there* may be an attempt to communicate a caring thought, but to the inner ear of the psyche, it can sound thin. Voice delivers something more; even to sit in the silent pauses of conversation and hold that space together has meaning beyond words.

Texting, in particular, is often inadequate to the task of important interpersonal communication, but its availability and our pressured pace prompt us to use it more and more at times when we'd be better served by talking together, if only by phone or video chat. When real face time is impossible, a real voice is next best. But evolving social convention has made a quick text response more routine than a call, more routine even than the simple thought that if it isn't an emergency then we might push our mental pause button and wait to call when we're in a good place to have a good talk. Experts suggest that much of our tech connection taps into the part of our brain that seeks order, the left hemisphere where the impulse is to organize and accomplish. In the flow of those just-do-it impulses and the stimulating neural feedback when we do, we unconsciously—and neurologically—recast conversations as tasks. They pop onto the left brain's to-do list, where they join the multitasking queue. So although it may be unnecessary, thoughtless, or even reckless to make that call or text while you are otherwise engaged, the left-brain is exultant: *impulse-action, think it– do it—mission accomplished!* The impulse to respond instantly begins to *feel* necessary. Never mind the consequences.

For parents arguing or for those fighting over divorce or custody issues, all this offers a fresh arena in which people can reach new lows, constantly berating, texting, and showing such utter lack of self-control because it is so easy to do—the immediacy of the medium encourages impulsive action. What you used to think privately while in the shower, you now can text at a stoplight, and that is often not a good thing. When people are angry and upset, tech plays to reactivity, for children and for any of us. That has dramatically changed the dynamics of relationships. We need to understand our own internal wiring as it relates to our use of tech so that we can think clearly about when it is okay and when it really is not okay to, for instance, communicate intense upset, anger, hurt, loss, or fury, through texting. (Hint: It is almost never okay to communicate intense upset, anger, hurt, loss, fury, or any other emotionally laden message through texting or e-mail, since the speed of your response can fan reactivity.)

Some say that people haven't changed, that technology has just changed the way people communicate. They suggest that the essence of human relationship hasn't changed either, that relationships are just taking place in a different context in text or online. This might be true from a sociological perspective, but not from a psychological one. Psychologically we are indeed in new territory. Technology has changed the basic construct of our relationships. It has triangulated our connections with each other, becoming the ubiquitous third party in our conversations, sometimes connecting us, but often interrupting us and ultimately disconnecting us.

What originated as a mechanism for communication is now driving, demanding, and sometimes distorting our communication. In addition to communicating impulsively, we find ourselves driven to communicate more than we want or more than is healthy for us. We have never had the expectation before that we should be available to anybody, *everybody*, anytime anywhere. "Big Brother" used to be synonymous with an unsavory "other" presence that shadows our every move. Now it is tech itself that is a constant presence, sometimes useful, sometimes annoying, but always, it seems, commanding our attention. We have never lived with that level of on-demand presence, and it is wearing us out. Children and parents are showing signs of relational fatigue—tech burnout—from the pressure of constant communication, the endless competition with screens for each other's attention, or trying to "be there" for all people all the time.

Finally, there is the new expectation that we accept being routinely ignored or treated as unimportant, that we ignore feelings of emotional disconnect as no big thing, when in fact those things do, and should, affect us. In earlier days, a personal relationship in which we were treated this way, or we treated someone this way, would have been considered an unhealthy one. A normal reaction to such treatment would have been, *that's inconsiderate, it's rude, it hurts.*

The emotions of feeling like you don't matter, feeling invisible or unloved have certainly been around forever. But we've never had a lifestyle that made screen communications the priority and made it acceptable behavior to ignore others, or to accept when someone drops our conversation midsentence to turn to another. Research confirms what we sense is true: the time we spend online or on electronic devices is eroding the time we spend with the people around us. Researchers at the University of California's Annenberg Center for the Digital Future report that the percentage of people who say they spend less time with family members because of the Internet nearly tripled from 11 percent in 2006 to 28 percent in 2011. Similarly, the Digital Future report showed that in the same time period, the time family members spend together each month dropped from an average of 26 hours to 17.9 hours a month. The university's study, which surveyed about two thousand American households, also showed that the number of people who reported feeling ignored, at least sometimes, by family members using the Internet grew by 40 percent over the same period.

The unlimited possibility of the World Wide Web speaks to all of us with its potential for creativity, new knowledge, and new relationships. It offers us an immediate escape from our daily world where so much of what we do is tedious. But it also imprisons us in its world of infinite access. It has added a new workplace boss—not a person but the expanded job demands made of us round the clock with no regard to our need to be separate from work and connected to our self and our families. In a much larger sense, the medium itself has become an even more demanding boss, the digital boss composed of any and all people and pitches, distractions and demands. This composite enchantress/slave driver/sorcerer knows no boundaries. It calls us to social banter and to business conversation, family news and drama, entertainment and education, gaming, bargain hunting, browsing, current events as they happen, and commentary that flashes past. We feel compelled to connect and we struggle with the guilt, the lost time, and the exhaustion that come with it.

As adults we are free to choose unhealthy habits for ourselves. But this is no way to treat our children. However sophisticated our children may seem in this tech-driven culture, their street savvy is deceptively naive. Still on the learning curve in their interface with the adult world and in so many ways in relationships, their capacity to connect through tech often outstrips their capacity to manage the emotional voltage it can deliver. Or to recognize the potential danger of exposing themselves to others who would use social networking to do them harm.

In my work with families, I see battle lines drawn at ever earlier ages over texting, screen time, and freedom to roam the Internet. I hear disturbing reports from the school front where teachers report adverse effects on children's play, their intellectual curiosity, their learning, and their social and emotional development. Crisis calls come from home and school when children's online or social media missteps lead to damaging consequences. There is nothing here we cannot figure out and turn around to better serve us, but it requires some thoughtful recalculation of old assumptions and new choices.

Digitalized Childhood: Deciding What We Want for Our Children

I am not anti-tech. To the contrary, I use all kinds of tech, to connect to my family and friends, for my research and work, for my love of painting and gardening, and so on. I work in schools and see the use of tech in educational settings and am continually amazed by its potential. And its power. And its pull. Used well, it is an exciting tool to engage and expand learning. But I believe the digitalized life we now take for granted is taking a far greater toll on family cohesion and childhood itself than we imagine, or perhaps than we want to allow ourselves to imagine. Research already points to serious concerns for infant and child development, neurological effects that are narrowing the way the baby brain organizes for lifelong learning. Yet we place tech in the hands of infants and young children and encourage them to play. So I worry about tech's addictive qualities and the way we've accepted our own compulsive behaviors about it even as we lecture our children about excessive screen time and texting. I worry about what it's doing to our culture, where it's leading us. I want us to be thoughtful about technology, more in charge of the way we integrate it into our lives. I want us to adapt technology to serve us well, rather than surrender ourselves unquestioningly to adapt to technology.

In 1982, the educator, theorist, and media culture critic Neil Postman wrote in *The Disappearance of Childhood* that television and the TV-saturated culture had collapsed the protective barriers between children and the adult world, robbing children of the time and place to develop at the pace nature intended. At the time his book was published the personal computer was just coming into popularity for home use. That year *Time* magazine celebrated "The Computer" in place of a person as "Machine of the Year." Facebook creator Mark Zuckerberg wasn't even born yet; his social networking brainchild was still twenty-two years away.

Postman targeted the TV culture but traced the disappearance of a protected childhood back much farther, to the advent of publishing and early telegraphy that allowed for swift, broad transmission of ideas. He argued that through centuries of technological advances, "What appears to have happened is that the certainty of opinion about the nature of childhood began to be questioned."

Today we are at that crossroads anew. More than at any other time in history, we have ample evidence to evaluate the risks and benefits tech poses for our children. We have enough information to reach "a certainty of opinion about the nature of childhood." Now we must decide what that is, and what kind of childhood we want for our children.

I would add that the experiment is not only influencing children's ventures into the wider world but affecting their social, emotional, and intellectual development, the most fundamental internal processes of childhood, as well. More than twenty-plus years into this vast mind/media experiment, scientific research is now upending some assumptions we made in the honeymoon period. New research offers unprecedented views into "the celestial openness of the child's mind," says scientist Patricia Kuhl. With the modern tools of neuroscience, she says, "we are embarking on a grand and golden age of knowledge about child's brain development."

As scientists work to expand our knowledge about the brain, studies are shedding light on the intersection of tech and neurobiology. One thing is clear: the brain processes mediated interaction differently from the way it processes direct human-to-human interaction. Most of all, research keeps coming back to *the role of parents* in knowing their children, knowing what they need, and staying connected to our children IRL—in real life.

Parenthood: The Ultimate Role-Play Game

One morning I sit with ten mothers, some with professional careers and some home-based, all parents at a Manhattan preschool where I'd been hired to talk about raising kids of character in our rapidly changing digital culture and the shifting role of parents within it. When I ask these women how tech is changing their family lives, there isn't even a moment of contemplative silence. They leap into the conversation, brimming with anxieties, insights, and stories of raising their children surrounded by the silvery shine of so many electronic gadgets.

"Our household is eerily silent at night now because everyone is on their own machine, their own little screen," one mom says, describing a now familiar scene in homes across the country. "And even after we've put our kids to bed, my husband and I don't spend real time together. We sit at the dining room table, facing each other but staring into our laptops."

"I read something the other day that said more four-year-olds know how to download an app than tie their own shoes," adds another mother. She'd seen a news item about a 2010 international survey of 2,200 mothers, which found that more children aged two through five can play with a smartphone application than tie their shoelaces.

"I have a six-year-old daughter," a soft-spoken young woman at the corner of the table says. "She doesn't have too much of an obsession with the Internet yet but I worry that I'll blink and suddenly I'll have a thirteen-year-old daughter who is ignoring me for Facebook. And it's going to come back to haunt me that I spent so much time during her early years texting and e-mailing my friends. It makes me think of that song about the father who never really stops to appreciate his son and then, when his son is grown, the same thing happens in reverse."

She is recalling "Cat's in the Cradle," Harry Chapin's song from the 1970s that tells a poignant story of remorse about lost time between father and son. It is a song about a young boy's longing for connection with his father, who is always too busy and preoccupied with other priorities to connect with his son. Years later, the father is old and wishes for more time with his adult son, but the young man has no time for him: the boy has grown up to be just like him.

It is a haunting song, nearly forty years old now, but only more relevant today as a cautionary tale about the fleeting opportunity we have to truly *be* with our children when they are young—to be present and focused on what counts most—and the lifelong lessons they learn from watching us.

Here, as in so many focus groups, therapy sessions, and conversations with parents across the country, parents tell me about their varied concerns and their common quandary: they want to be good parents, want what's best for their children—including tech—but they can feel what's slipping away and they feel helpless to do anything about it. A friend of mine calls parenting "the ultimate RPG"—role-play game— and I agree that at times the new digital landscape can feel foreign and threatening. I imagine a cast of parental avatars armed with good intentions but overmatched. One moment the action is fun and exciting and the next, with one wrong step, we or our children have wandered into troubling territory and we can't press the escape key to restart. Tech is hardly a villain as characters go. But it does have Jekyll-and-Hyde potential, as insidious as it is engaging. Our ongoing quest: to develop a healthy relationship with tech that includes taking charge and unplugging from it, to model that for our children, and to find ways to use tech to our genuine advantage as families.

Winifred Gallagher, in her book *Rapt: Attention and the Focused Life*, describes the "grand unifying theory of psychology" as simply this: *Your life is the sum of what you focus on.* Our expanded ability to be technologically connected on screens to the world almost anytime, anyplace, is unquestionably pulling us away from making families primary in children's lives and in grown-ups' lives. Without being aware of it, we've shifted our attention from the primacy of nourishing family connections in the ordinary ways children need it, and turned instead to self-interests, work, and other sources of fulfillment. Left to their own devices, our children will do the same.

Over time, our family, like most, has been steadily swept along with each new wave of tech, from desktop computers to laptops, from flip-phones to smartphones, to Skype, texting, Facebook, and Instagram. And, with each new experience, I have alternately felt amazed and fatigued, alarmed and nostalgic. One minute I am worrying that my kids are disappearing into their screens. The next I am marveling at the new and creative, efficient ways we can connect to each other as a family, not to mention the world. Beneath it all, I am profoundly aware that family life, and the capacity for connection, is forever changed and will continue to change in ways we cannot imagine.

So many parents, educators, counselors, pediatricians, thoughtful public policymakers, and caring communities share these concerns. We are in a panic at the loss of boundaries, confused about how to reestablish them, overwhelmed by what feels like a tsunami of crass cultural norms.

We don't need to resign ourselves to obsolescence or wish ourselves back to a predigital past to empower ourselves. And we don't need to outthink the electronics industry geniuses or their marketing teams. Just as we are, we have all we need to set right our relationship with tech and incorporate it wisely to create vibrant, sustainable families.

No matter how tech shapes and changes the world around us, one thing remains universal and unchanged: the connection that begins in the family shapes a child's brain, mind, body, and soul in uniquely human ways that tech cannot replace. So much in this digital culture is beyond our control, but as the research shows, our locus of power as parents, educators, and advocates for children remains essentially unchanged and it resides in the space of our relationship with them. Nothing can match the power of our attention and our capacity to connect in affirming, loving, nourishing ways. Screens and tech cannot match it, but they can replace it—if we let it happen.

As a psychologist, I believe that the better we understand our relationship with tech and with each other, the better equipped we are to make wise choices about how we and our children use tech as our options grow, without sacrificing what many believe is the single most important human thing we can do: create loving, sustaining families. Parents, teachers, grandparents, nurses, pediatricians, counselors— all of us who work with and care about children are already in this conversation. This book is written in the spirit of the teachable moment so that with fresh eyes we can further unpack the psychological dynamics emerging as tech becomes an integral part of how we relate and connect. In particular, we must seek to understand the potential psychological fallout for our children and the implications of that fallout for us as parents.

In the chapters ahead, I share stories and observations from my clinical practice with children and families, from my work as a consultant to schools, and from extensive interviews with students and educators for the purposes of this book, to examine ways in which tech and media are putting our children at risk at every stage of development. With fresh eyes, an open mind, and the will to act on what we see and learn, we have the opportunity now to nourish our families and protect and prepare our children for a meaningful life in the digital age that is transforming family life, human life, in situ.

Chapter 1

Lost in Connection

How the Tech Effect Puts Children's Development at Risk

Stimulation has replaced connection, and I think that's what you need to watch out for.

—NED HALLOWELL, PSYCHIATRIST AND
AUTHOR OF *DRIVEN TO DISTRACTION*

Tom is one of the most thoughtful, attentive, engaged dads I know, and his sons—four, seven, twelve, and thirteen years old—are good kids. He takes his boys hiking, coaches their sports teams, helps them raise animals, and travels with them to study the world. He is not a man absorbed in tech and he has been mindful about the TV and electronic games he allows the boys to use. This is especially challenging because their range of ages and development means he has to set different limits for his two older boys ("the bigs") from those he sets for their younger brothers ("the littles").

Some of the things he might permit the bigs to play—gory war games, for instance—he would never want the littles exposed to. And then there is Grand Theft Auto, a game so vile he never intended to let it in the door. In the game, the player is a sociopathic getaway driver who racks up points for running down pedestrians and other innocents, basically killing anyone who gets in his way just for sport.

Between kills he is immersed in a criminal world—not as a cop or heroic good guy, but as a really, really bad guy.

"The guy can go into a strip club. There's pornography. There's drugs. There's shooting police. It's horrendous," Tom tells me. "So I kept saying, 'No, we're not going to get it. I put my foot down.'" And he stood firm. Then for eldest son Sam's thirteenth birthday, one of his best friends gave him Grand Theft Auto.

Tom was exasperated but wanted to be reasonable; rather than force Sam to surrender the gift, he modified his strategy. He banned all but Sam from playing it and only when the others were not around. That didn't work. He made new rules; they got bent. He hid the game; it got found. Finally he thought he had it under control. The two bigs obeyed the limited-play rule and the game stayed out of sight and inaccessible to the littles—four-year-old Ben and seven-year-old Teddy.

One day soon after, Tom was driving "blissfully along," his younger boys quietly engaged in the back, when he discovered Teddy, the seven-year-old, was playing Grand Theft Auto on his iPhone. He had used the touch-screen web browser to access it. It had never occurred to Tom that the boys could do that—or would. The only reason Teddy even had a phone at his age was because his mother traveled extensively for work, and this allowed them to communicate with each other if need be. Tom had never thought of it as a potential security breach in his protective parental firewall.

Tom could have thrown up his hands, given up, surrendered to tech's incursion on his parental authority, but instead he immediately set about plotting his next strategic move: downsize the tech capabilities on the phones and update parental controls so he could more closely monitor the boys' activity. He banned Grand Theft Auto in the house altogether, knowing that his oldest son could and would play it at his friend's house, but at least his message was back on course and clear.

It is an endless challenge, this role of IT parent in the digital age. Tom wants his children to have the benefits and enjoyment that tech offers, he wants them to be tech savvy and media literate, and his home is outfitted for work and fun on screens. What he does not want to do, he says, is leave his kids on their own with it to "roam and learn" and hope for the best. There is just too much at stake.

About fifteen years ago I began getting calls to visit schools and talk with parent groups about raising children of character. Parents were deeply worried that a breakdown in cultural norms, growing consumerism, cynicism and crass entertainment, overscheduled lives, and an increasing pressure for competitive success in school were shortchanging their children's moral development. Since screens, tech, and online access became commonplace in children's lives, the calls now come from parents and teachers of children as young as preschool age, alarmed at the extent to which children's screen play or online lives are affecting their learning, their social and emotional development, their family interactions, and their school communities. Teachers share their concerns about the subtle but pervasive ways they see tech impinging on the school experience: four-year-olds who want to imitate computer games on the playground and hesitate to play with blocks or peruse books; elementary school children who struggle to problem-solve and who depend on adults to help them with the simplest tasks; high school students who struggle with any assignment requiring more than shallow attention and prefer a virtual tour of a museum to a field trip to see the real thing.

Parents call in a panic. A child is showing signs of gaming addiction or has been caught watching porn on a pal's laptop. A mom snooping on her fifteen-year-old daughter's Facebook page has learned that she plans to sneak away to the movies to meet a forty-something man she friended there. Or a twelve-year-old has posted pictures of herself online at "Am I Ugly" inviting anonymous critics to rate her looks.

I also hear much more anxiety from parents about sleepovers than ever before. One of the biggest concerns is that there will be older kids in the home who have been influenced by sexual content or YouTube videos and other sites and will put their children in harm's way by exposing them to inappropriate content. Or that the sleepover kids will have unsupervised access to computers and go looking for trouble— or find it by accident. "I used to just assume these things couldn't happen," says one mother, "and now I have to assume that they can." That is not only a safe assumption but a wise and durable one for the years ahead. Technological innovation by definition takes us into unknown territory and will continue to alter the landscape of everyday life in ways that hold us in thrall. Research into the effects of those practical innovations on human life, from brain synapses to sleepover experiences, will necessarily lag behind. Every new thing, every upgrade, takes us farther along the slippery slope of the cyber culture where we must expect to be continually challenged in new ways.

The tech paradox we all confront as parents is that the very thing that can get our kids in deep trouble can also deepen and enrich their lives in unimaginable ways. Technology has transformed the ways we can connect with family and friends at a distance and manage the traffic flow of work and family commitments. Our children can access extraordinary resources to explore their healthy interests and connect with others who share those passions. Tech has transformed what it means to be a student, the opportunity for lifelong learning, and the very process of education itself. Tech has transformed what it means to be a global citizen and our capacity to empathize, understand, and truly see the world from the perspective of people we may never meet. The possibilities are nothing short of thrilling, often inspiring.

Yet we know the darker side is there. Research already shows detrimental effects on the developing brain, early learning, and emotional development. We know that the entertainment and online culture is in many ways antisocial, crass, and demeaning and that kids have such easy access to it. In an era when children need adult supervision the most, parents say they feel more ineffectual than ever. They cannot control the landscape and they cannot control their children's journey through it. They want to trust their children and believe that their children will know how to navigate, protect themselves, and respect others in the chaos and moral indifference of the cyber culture. But as much as parents want to trust their kids to make the right choices, it's not a matter of trust but a question of whether they are prepared to make their way safely and wisely through what is for all of us new territory. As for trust, at best all you can trust is that they are good kids who will inevitably roam into bad tech terrain. But unlike grown-ups, whose fully matured brain should be able to tell right from wrong, a joke from bullying, and tasteful content from trash, and should be able to exercise impulse control and mature judgment in how we use tech, our children are not there yet. They are still children.

Our species is notable for the amount of growth and length of time required for the brain to mature after birth. Too often, conversations about child development focus on what a child can do and how to make it happen faster, when instead we should be talking about how a child can think, how the developing young brain is prepared to process experience, and how we can support that growth in healthy ways. We know now that it takes twenty-five-plus years for the prefrontal cortex, the part of our brain that enables us to link consequences to behavior (called executive functioning) to fully develop. In the adolescent brain executive functioning is still a work in progress, neurologically not yet a fully functioning piece of a teen's decision-making process. So it falls to us. Older and ostensibly neurologically wiser, we are the ones equipped to think of consequences. At times, though, our own love affair with tech clouds our view of the serious consequences the same habits hold for our children.

We all work so hard, juggling life's big worries and ordinary demands, trying to stay afloat and feel we are competent, doing no harm, being our *almost* best selves as much as possible despite the one hundred interruptions that splinter our attention and ability to accomplish something. It is easy to slip into denial about the downside, reassuring ourselves: *It's got to be fine, everyone else is doing it. It's not really that bad—I know other parents are letting their kids do much worse. They're going to see it sooner or later anyway, there's nothing I can do.*

We delete from memory the steady flow of news stories about the known dangers of texting and driving, or research showing likely links between children's media habits and health concerns like anxiety, aggression, addiction, attention deficit and hyperactive disorder (ADHD), developmental delays, obesity, and eating disorders. Or the stories that end tragically. If it hasn't happened to us yet—the crash or the crisis, the diagnosis, or the call from the school or whatever worrisome thing is next—it doesn't seem likely it will happen at all. But it is happening. Research and behavioral trends already show that when tech becomes an early and continuing presence in children's lives, it can undermine family and child development. In the struggle to preserve our families and protect our children we are losing ground on some critical fronts. Psychologically, these losses in fundamental aspects of child development and well-being can set our children up for trouble in school and in life.

Tech Replaces Family Primacy: What's @ Stake as Peers and Pop Culture Delete Parents

What is family? When I ask children that question, their responses reflect the very different things that family represents for children at different ages and stages of development. Four-year-old Amber describes family as "my mommy and my daddy and my sister and . . . what makes me happy."

"Your family is who loves you," says Max, five.

"They're the people who matter most," says Emory, eight.

Naomi, ten, describes family as "where you learn about your values and love."

Andrew, thirteen, concludes: "Sometimes they can be pretty annoying but you know, they're always there for you. Yeah. Good stuff."

However we describe family in everyday terms, the primacy of family has special meaning from a developmental perspective. The infant's experience of itself and its environment (everyone and everything included) is undifferentiated; there is no *me*, only a *we*. "Family creates our first experience of ourselves in the world, and it becomes the foundation of our view of the world," writes the psychoanalyst Harvey Rich in his book *In the Moment: Celebrating the Everyday.* "Family is the organizing theme around which our consciousness grows. . . . It is where we begin to define ourselves relative to others, and as part of the larger story of family, community, history, and humankind. At the deepest level, it is where we first discover ourselves."

A couple's relationship and expectations, the constellation of personalities and circumstances into which a child is born, all of that constitutes the so-called nurturing surround that shapes the way a child thinks, grows, and engages the world from birth. In child development, when we talk about "the primacy of family" we are not simply suggesting that family is very important to a child as a home base. We are referring to the family's role as the deepest, most profoundly defining influence in a child's formation of self—her neurological, psychological, and physiological growth and development.

The psychologist Selma Fraiberg, in her classic book *The Magic Years*, wrote that family is "how a child becomes humanized." Ultimately, she was referring to the way in which family serves as our first and most significant teacher in what it means to be fully human in the best way, from cognitive capacities to qualities of character. We have reason to be concerned, then, when the intimate nurturing surround of family is breached, and media and tech displace family as the context for defining values, modeling relationship, mentoring, and meaning making for our children.

We might think that the transient, insubstantial content of so much of the media and tech culture would be relatively harmless against the deep, primal influence of family. The opposite is true. The content is more powerful. The psychoanalytic theorist Harry Stack Sullivan, whose seminal work in the 1930s defined psychological theory about family primacy, observed zones of interaction between parents and children at every age. These "relatively enduring patterns" of interaction give children a sturdy relational base from which trust, empathy, optimism, and resilience can grow.

Shared rituals or conversations you create with your child around meals, bath and bedtime, playground, and drive times together—all these are zones of interaction. You create them in the repeated resonance of consistent responses to everyday moments: *It's time to brush your teeth . . . I love you so . . . Yes, you have to take a nap now . . . It's your sister's turn on the swing . . . I see you are sad but it's time to leave the playground.* Or setting limits: *If you can't take turns with the Wii, off it goes . . . Yes, you have to stay at the dinner table till we are all done.* Or when you remind your teen for the umpteenth time: *Drive safely . . . Call if you're not comfortable . . . Curfew means curfew or there will be consequences.* And mean it.

It is the talks in the car, the grunts at the breakfast table, the conversations at dinner, over and over, day after day, that weave us together. And perhaps most of all, it is parents teaching their children: what is okay and not okay; what is rude and insensitive; when they have crossed the line from a joke to teasing; the meaning of limits; and the fact that misbehavior has consequences and consequences are enforced. Kids don't tend to thank us in the moment, and often it is disruptive and painstaking to enforce rules or discipline at home. Yet that is our job as parents. So when we set limits, when we say no, we also teach them. Our children develop their deepest sense of self and internal stability from this pattern of intimate connections. Every time our child's texting, TV, electronic games, and social networking take the place of family, and every time our tech habits interrupt our time with them, that pattern is broken and the primacy of family takes another hit. "There are a lot of minimoments of disconnect that are cumulative in the lives of children today," says Liz Perle, cofounder and editor in chief of Common Sense Media. "Kids are bright and they sense when their parents are present and they know when they are not."

Too Much, Too Soon: The Premature Loss of Childhood Innocence

Children come to life innocent, unaware of the harsh aspects of pain and suffering and how cruel people can be. Part of the job of parenting is to protect them from that harsh truth long enough for them to develop a sense of goodness and core values of optimism, trust, internal curiosity, and a hunger for learning. If they see too much too soon—before they're neurologically and emotionally ready to process it—it can short-circuit that natural curiosity. Boys and girls alike are easily traumatized by premature exposure to the media-based adult culture that cultivates cynicism and cynical values, treats sex and violence as entertainment, routinely sexualizes perceptions of girls and women, and encourages aggression in boys.

Today's kids are growing up in this culture that normalizes lying, cheating, crass sexuality, and violence. These things are nothing new, of course, but prior to the Internet and personal tech that put it within reach, children generally did not have access to that world without parental permission. We have lost a protective barrier, individually as parents and collectively as a culture. When you go into a drugstore and see the *Playboy* magazine discreetly displayed under partial cover behind the counter it harks back to a sweeter, more innocent time when grown-ups at home and in the community together would protect children from premature exposure to unhealthy values and behaviors.

Children are no longer sheltered in this way. Further, the adult culture "adultifies" children. Some of it is incidental: kids tap into general content that is intended for grown-ups—graphic coverage of disturbing news events and rants and rhetoric that adults understand with a maturity and a context that children lack. But plenty of it is intentional, backed by big-bucks research and product marketing interests that target children as consumers. These marketers cultivate kids' "must have" consumer appetites, right down to click-'n'-buy options on free app games for toddlers—linked to the parent's credit account. Much of so-called children's programming isn't tailored to protect children from those exploitive cultural messages; it is simply tailored to package them in language and content that appeal to kids and gets the okay from parents who assume it is okay for their children. The same cynicism, cruel humor, destructive gender stereotypes, and disrespect that distinguish so much of popular adult TV and online commentary is shrunk to fit for children's viewing in the likes of *Bratz Dolls*, *Power Rangers*, and *SpongeBob SquarePants*, and later on, *Gossip Girl* and the like.

With TV tutors like that for hours a day, it shouldn't surprise anyone when girls approach adolescence with a jaundiced view of themselves and what it means to be pretty, popular, and powerful. Or when boys act on the behavioral scripts they find in computer and digital games, media, and porn that promote the "boy code" of sarcastic, aggressive, humiliating one-upmanship of each other and of girls.

Early exposure to sexualized images and gender messages goes straight from screen to real life and often into my office. A well-behaved, sweet third-grade boy explains why he recruited a girl friend (not a *girlfriend*) to go into a school closet, pull up their shirts, and kiss each other's nipples. Cruising the Internet on the family computer in the kitchen, he had stumbled onto a YouTube vein of videos depicting adults doing that, suggested it to his friend, and she went along with it. At a middle school, a sexting incident shocks everyone: a naive and obliging eleven-year-old girl sexted a topless photo of her barely developed breasts to a twelve-year-old boy who had asked her for it and then showed a friend, and the friend e-mailed it to the entire school. At another school a nine-year-old girl sent a ten-year-old boy several graphic e-mails alluding to the size of his genitals and other sexual innuendo that she clearly did not even understand but found online. Sadly, these are no longer unusual calls for me to receive.

One afternoon I get a call from Sarah, a distraught mother, who is unable to calm her usually sturdy daughter. Amy, age ten, had come home from the after-school program hysterical and had been on-and-off crying for hours. It was now 9 p.m. That afternoon Ginny, a thirteen-year-old friend with whom she often hung out happily at after school had showed her a YouTube video that Ginny thought was "hilarious." It was a promo trailer for a horror film depicting sexual violence and sadistic torture. Amy had been mortified. When her mom called the school to see how this could have been allowed to happen, she learned that while the school has a no-Internet policy during the day, including after school, students with smartphones and tablets are on the honor system after 4 p.m. Ginny had been genuinely surprised that Amy was so upset by something she herself had found laughable. After meeting with the school principal, Ginny willingly wrote Amy a sincere apology; she truly had meant no harm. Ginny also did detention, where she had to research more about inappropriate online content for kids. But Amy couldn't shake the images for days. Any humor was lost on her young brain, and the shocking, sadistic, gory details became pop-ups in her head when she tried to fall asleep. I gave Sarah some ideas about how to help Amy shed the disturbing images and in a few days Amy was able to concentrate again. But she lost a piece of innocence that afternoon, and it crushed her heart as well as her mother's.

In each case, this behavior clearly ran counter to the family's values. What it means to be a good person or a good friend, how boys and girls act and treat one another: all those messages that used to come from family and friends are now being challenged by outside sources—programs, people, and profiteers with their own agendas. Protecting your child's innocence is not one of them.

Tech Trades Away Family and Personal Privacy and Exposes Vulnerabilities

Privacy is fundamentally a way we have of protecting ourselves. As media and social networking have erased the boundary between protected childhood and the adult world, they have also blurred the distinction between the public and private dimensions of life, especially in our children's minds.

Family traditionally was a private, protective realm for children. It provided a safe place to be yourself, at times your most unflattering self, your scared or sullen or angry self. You whined and aired petty grievances and social revenge fantasies to your mom or dad (or perhaps your diary, secreted away). You teased your siblings, sometimes cruelly. After a long day at school, you brought your woes home and debriefed with your parents or the family dog. Your parents comforted you or corrected you or talked things through with you, however clumsily at times. You experienced consequences, learned important lessons about love, life, and relationships. You were forgiven and you grew up, your early missteps relegated to family lore. In the privacy of family you shared your hurts and failures. If you were shaken up, they were the people you would turn to first—not a public chat room or online stranger.

You also learned that you were responsible for protecting others—that family loyalty meant protecting your family, recognizing boundaries, and thinking carefully before you revealed family information or troubles to people outside the family. No one thought to post complaints about family members or embarrassing pictures of them in a public place.

Now, instead of coming home to a snack and telling us the stories of their day as they unpack their backpacks, kids mark the transition from school to home by plugging into screens. Social networking has switched out the private family space for a public square that promotes freewheeling communication, impulsive sharing, and uncensored feedback. It has turned what was once a child's private life into a universe where personal disclosure plays out to that infinite audience of intimate strangers. It has also replaced the old family teaching that some matters involving others aren't for public sharing. Now everyone is fair game and the inner voice of restraint is lost in the impulse to join the carping chorus in texts and posts.

Not all families are safe places for children to share their vulnerability, and at times a young person's online disclosures—about traumatic events, social cruelties, sexuality, depression, and suicidal thoughts, for instance—have connected him or her to needed acceptance and support not found at home. For some kids going online is a lifeline in this way, even lifesaving. As is often the case with discussions of our relationship with tech, the issue is not only whether there is potential for good, but how certain patterns of use are diminishing

our children's experiences in ways that put healthy development—
and basic safety—at risk.

The Indelible Digital Footprint
Makes Errors Costly

The family album or a child's scrapbook or diary used to be the extent of the physical evidence of our younger childish days. Kids hated
it when Mom pulled out the snapshots of them as bare-bottomed
babies or awkward broody teens; when you are still in the process
of growing up, it is embarrassing to see the younger you whom you
are trying so hard to leave behind. Today, the private scrapbooks
and diaries aren't private anymore. Children (and many parents, too)
are sharing everything online: photos, video clips, rants, reflections,
gossip, and secrets—theirs and others'. Never before have children
grown up with such a public and permanent record of their daily
lives. This includes their immaturity and poor judgment.

Jake Strong, a middle school principal in the Midwest, sits with
me one afternoon sharing stories from the front lines where kids'
texting, Facebook, and Internet activity have transformed incidents
that might once have been disciplinary misdemeanors—literally,
mistakes in demeanor—into unforgivable offenses. In a more serious
case he recalls, a boy posted a "racist, homophobic, and exceedingly
explicitly violent post" on a classmate's Facebook wall. Other students saw it and, understandably, "were horrified." They reported
him, and when Strong got a transcript of the post, he felt he had no
choice but to expel the student.

> What I ultimately told him was, there are lots of opportu-
> nities for making mistakes and for recovering from it, but this
> was one you can't come back from a year from now. Because

here, in this school, you're going to always be the person who said this. . . . There are very, very few things which are not fixable—especially for kids—but the explicitness of this, and the horrifying choices of language and imagery were unforgettable, and in some ways, unforgivable. I couldn't justify allowing the student to continue in the community having said things that would make it so that there would be many people who would not ever feel safe.

The parents argued that their son deserved another chance, that his comments did not reflect the family's values, and that he clearly had picked those ideas and that language up elsewhere and had tried them out impulsively—like scrawling graffiti—on the Facebook wall.

In pre-Facebook days, had this boy spoken this way in the lunchroom or at recess, or even scrawled it as graffiti on the side of the building or in the restroom, his actions would have deserved serious consequences. But the impact of his comments could have been contained. The graffiti could be erased and the offender required to atone for his comments. His parents and teachers might have been able to make more of the teachable moment. Given the boy's general history of behavior, Strong says, it is doubtful he would even have expressed those ideas and acted so recklessly in a personal setting. But once he posted on Facebook, the deed was done. He couldn't take it back. And he couldn't control its life online: who saw it, who shared it with others, how far and how long his hate-filled message would travel on the Internet.

None of us can know the long-term effects of public self-disclosure in online posts and photos, saved texts, tweets, and YouTube videos. Children are especially vulnerable both as instigators and victims as they develop online personas even as their own identity development is a work in progress. Every day there are adolescents posturing online as haters, hookers, and hell-raisers. There are shy kids trying out bold personas and ambitious kids testing the limits. They may use language from a popular song that they think will make them look cool, only to land them in the principal's office. They may post, as some seventh-grade girls did, photos of themselves in bikinis, pouty-lipped and posturing seductively as if for the *Sports Illustrated* swimsuit issue. Or they may join the ranks of college kids posting photos of themselves drunk and puking, naively thinking that their privacy settings will protect their privacy, when in fact those settings offer only scant protection. Many employers now routinely review applicants' online lives before making a hire. More important, we know that this kind of online socializing can spiral out of control and hurt our kids.

We are often absent, distracted, or unaware when our children take missteps online; we miss the teachable moments and they miss the needed guidance. Whatever the reasons for our absence, it is amazing how little all of us, as parents, know about our children's lives online and the identities they are creating for themselves there. It is where they regularly hang out and socialize, and they are wildly unsupervised. Kids want to—and need to—own their identity in a way that's not accessible to parents and the online world has given them a place for that. But the sad truth is that it is not as protected a place as they need.

Empathy Is the Missing "E" in Our E-Culture

Think back to the kinds of notes you passed between friends at school when you were sitting quietly working on an assignment. Recently I asked high school students attending a conference for girl leaders to write down memorable ugly posts or texts they saw or received while doing homework at school. Here are a few of the milder ones:

"You are a lonely disgusting human being."

"Someone ought to kill that bitch"

"Everyone knows that your friends hate you, why are you even
    trying?"

"You r an ice cold bitch"

"U just want to get your hands dirty"

Our children are growing up immersed in a culture where it is
cool to be cruel, where media influences encourage it and social networking facilitates it. In my work as a school consultant, I spend a
lot of time helping schools get to the heart of what is amiss in their
school culture. Parents and teachers describe a disturbing new presence of sarcasm and meanness across age groups. I see the effects in
my patients, often children struggling through situations in which
their own or others' lack of empathy is problematic.

Eight-year-old Ellie is refusing to go to school and can't sleep at
night. She tells me:

I was on Google Buzz and these girls who I thought were
my friends were really mean. They told me that I was fat and
that I wasn't going to be invited to any of their birthday parties.
I keep seeing those words "u r FAT" when I try to go to sleep.

In a suburban school, a seven-year-old boy taunts another boy,
telling him that he is going to go to the boy's house and shoot his dog
and his brother. His teachers have been concerned because he plays
video games for hours after school and now he is quoting them, using
the same threatening game language with other students.

A seventh-grade teacher asks me to help her deal with a "culture
of fear" that has permeated their entire grade and its counterpart at
another school. Two boys called a girl "a fucking bitch" online, adding "you know what is ugly/your face is ugly," she says. Others sent

online Valentine Day cards designated "for sluts." They are falsely
outing kids online in the other school, spreading rumors about sexual
activity using innuendo and language that many of the girls don't
even understand. I get similar calls from so many schools and I know
this is not just about "that school" or "that child." It is about us,
about our cultural crisis.

Incident by incident, we put out the fires, but the damage remains.
Child victims of this kind of cruelty, and perpetrators of it, too, are
psychologically affected. The effects of being bullied by peers as a
child or adolescent are direct and long-lasting, according to a 2013
study in the *Journal of the American Medical Association*. The study
focused on 1,420 people who reported having been bullied and/or
having bullied others multiple times between the ages of nine and
sixteen. Victims and perpetrators were at higher risk of psychiatric
problems as children, although it was unclear whether the risk continued into young adulthood. In addition to an increased risk of young
adult and child psychiatric disorders that might also be associated
with other life circumstances, the researchers found that victims had
a higher prevalence of agoraphobia (feeling unsafe in public places),
generalized anxiety, and panic disorder. Bullies were at higher risk for
antisocial personality disorder. And those who were both victims and
had bullied others were at increased risk of young adult depression and
panic disorder. Of that group, only girls were at increased risk of agoraphobia. Boys were at greater risk for suicidal thoughts and behavior.

Empathy might seem a "soft" skill for children when compared
to reading, writing, and math, but it is actually a neurological phenomenon as well as a soulful one. The development of empathy
comes from direct experience that lays down neural pathways in
both the left and right hemispheres of the brain and through the
body: to say we "feel" empathy for someone is biologically accu-
rate. Those pathways expand and deepen with experience, creating
what the noted child psychiatrist and author Dan Siegel has called

a "neural map" of our interdependent selves, a system designed by nature to enable us to pick up cues about other people's feelings and intentions and to be moved by their experience. When those neurons fire "they dissolve the border between you and others," he says. That's empathy.

The development of empathy is a critical step in early childhood and over a lifetime. Empathy is the caring glue that creates our humanity, our compassion. It has been identified as one of the key markers for success in school, family, and work life. Empathy also enables us to have compassion for ourselves, a vital component of mental health. Absent experiences that cultivate emotional awareness and empathy, the neural map takes shape differently.

"The brain is what it does," says Duke University professor Cathy Davidson in her book on brain science and learning. What it does in this regard has changed dramatically in the past twenty-five years as the media themes in children's and young adult entertainment have made teasing, sarcasm, bullying, and physical aggression the norm. On your child's neural map, an afternoon spent with TV's mean-spirited Angelica on *Rugrats* is a far cry from Mister Rogers and his kindly neighborhood.

Neurologically speaking, empathy takes time and practice to sink in. The neuroscientist Maryanne Wolf writes extensively about the tech effect on cognitive processes in the young brain in her book *Proust and the Squid: The Story and Science of the Reading Brain.* She explains that the speed and superficiality of the tech experience have thinned the neural experiences that create empathy. In contrast, activities such as reading books or other substantive content create complex arrays of neural pathways, the necessary rich weave of interconnectedness that develops empathy and allows it to deepen.

We each have what Wolf describes as a "beautiful embroidered circuit" of interconnected neural networks that are "connected to everything." However, she says, "if you're just going super-fast, you're

not making those connections. . . . All these things take extra time."
The deeper connections that children make through reading, reflection, and conversation are what teach them "this is what it means to
be good, this is what it means to be callous, this is what it means to
be evil," Wolf says.

Common sense tells us that communicating via e-mail, texting,
and other media does not make us uncaring. The fact that social networking and online activity have been used so successfully to rally
support for altruistic initiatives suggests there is something in that
collective connection that can nudge our neurons toward compassion. At the same time, a Stanford University review of research findings showed that increased dependence on tech has resulted in the
diminishing of empathy by reducing the amount of direct human
interaction involved. One analysis of seventy-two studies performed
on nearly 14,000 college students between 1979 and 2009 showed a
sharp decline in the empathy trait over the past ten years. That is an
outcome; the groundwork for it likely began a decade or so earlier in
children's day-to-day lives.

In his wonderful description of family's role in empathy development, Ron Taffel in *Parenting by Heart* describes family as an "empathic envelope . . . like a container around your kids in your family . . .
made up of your values, your expectations, and your ways of being with
your children." In that environment, he says, a child learns empathy
in the experiences of compassion, consequences, and communication
through meaningful time together and in conversation.

Through playing and fighting and sharing and pushing, through
put ups and put downs, through all the messiness that is sibling relationships and family dynamics, we learn to empathize, to say I'm
sorry, to understand when we've crossed the line. We learn to be accountable for our words and actions and their impact on others. We
learn to live with each other. When those experiences go missing, so
does a child's training ground for empathy.

Tech Takes Away Time for Independent, Self-Generated, Creative Play at Every Age

The American Academy of Pediatrics has for years urged parents to eliminate or at least minimize TV and tech time for young children and monitor it closely for older ones. Based on an extensive review of three decades of research, the AAP concluded that unstructured, unplugged play is the best way for young children to learn to think creatively, to problem solve, and to develop reasoning, communication, and motor skills. Free play also teaches them how to entertain themselves, the report said, concluding: "In today's 'achievement culture,' the best thing you can do for your young child is to give her a chance to have unstructured play—both with you and independently. Children need this in order to figure out how the world works."

Nonetheless, recent research shows a mass exodus of children from outdoor free play and free time to the glowing screens of video and computer games, TV, and, increasingly, apps and games for handheld devices. At the same time, childhood obesity has more than tripled in the past thirty years, with more than one-third of children and adolescents overweight or obese. It's not just that they are sedentary, it's that marketing of junk food has become so pervasive—and successful. When we connect those dots, the unhealthy picture only gets worse with the expansion of digital products and junk marketing to infants and toddlers.

It's bad enough that time for unstructured, imaginative play is shrinking for children of all ages as they become as overscheduled as everyone else, their days packed with school and organized activities or tutoring after school, squeezing out downtime. But adding hours of TV and computer play to the mix (65 percent of families have TVs in their kids' rooms) means that technology-focused play has taken the place of unplugged play. With more than half of families with TVs in kids' rooms, it is no wonder.

Computer play in moderation has merit for the school-age crowd. But they rarely play in moderation, and for children on the whole, TV and screen time has replaced the kind of old-fashioned free play that develops children's capacity for creativity, deep thinking, social interaction, emotional self-regulation, and reflection.

On the bright side, Pew research says games can have positive effects. They can connect children with their peers, promote social interaction, create opportunities for strategic thinking, and teach collaborative play, as well as develop the capacity to think quickly when solving problems. Pew found that prosocial games encourage prosocial behavior off-line. Tech and games that have positive themes can be a good source of family and peer fun and promote some of the values—collaboration, optimism, stick-to-itiveness—that we want for our children. When the tech or games include voice-to-voice or text chats, kids do socialize in the tech sense. What and how kids play makes a difference.

So let me be clear: there are excellent TV programs and computer games for children. *Mr. Rogers*, LittleBigPlanet, *Planet Earth*, SimFarm, and Minecraft quickly come to mind. But we are talking here about losses, and the math is simple: For every minute or hour your child spends on screens or other digital diversions, he or she is *not* engaged in healthful, unstructured, creative play. When they're engaged on screens, as social as it may be in one sense, they are not outside with other kids, taking in the day, relaxing and chatting, inventing games, and interacting directly—or arguing face-to-face, debating fairness directly, not via a game or headset. They are not running around, shooting hoops, and skateboarding, developing coordination and physical strength. Yes, they may be learning some computer skills and online etiquette (such as it is), but the issue is what they are not learning, the loss of which undermines healthy development. They are not learning how to deal with the frustration of real forts crumbling and block towers falling, of having to rethink and start over again. They are not alone with themselves, learning to be comfortable with solitude, with their own thoughts, with no alternative but to let their mind wander and drift, explore, discover, feel.

This is necessary throughout childhood—not just for young children. When little children make their own worlds with sticks or figurines, they are discovering the joy of making meaning, making story, creating drama, and problem solving. At every age, children use play and the realm of imagination to understand their selves and the world, to integrate learning, process conflict and loss, to replay again and again moments of anguish or joy. Eventually with that comes a capacity to think more deeply, to reflect, and to muse. Interactive games do not invite a child to daydream or ponder life's big questions. If you give them too many programmed games or if they become addicted to playing on screens, children will not know how to move through that fugue state they call boredom, which is often a necessary prelude to creativity. We talk so much to children about finding their passion, but the capacity to fall in love with a subject or a sport or an instrument begins with the capacity to cultivate a deep connection, a drive that comes from the inner self.

Michael Rich, a pediatrician and director of the Center on Media and Child Health, says that unplugged downtime for kids doesn't get the respect it deserves from adults. "We have to be very careful, particularly around the brain development period, but I think throughout childhood, that we're not so busy stimulating these kids that they don't have any downtime in their brains. The mind needs time to wander, the mind needs boredom, basically, to work well. Both to do maintenance, but also to do the kind of free-form creative thinking that none of these activities let you do."

"Computers are the new playground," laments Michael Thompson, a psychologist and the author of numerous books on child development. In his most recent one, *Homesick and Happy: How Time Away from Parents Can Help a Child Grow*, Thompson suggests that one of the most valuable aspects of sleep-away camp is that it typically provides what has become almost impossible for parents to provide: a tech-free environment and immersion in outdoor play. "It is so unusual to see a group of twelve-year-olds without handheld electronic devices that seeing them that way at camp is startling," he writes. "At most camps, the children have turned in their cell phones and there are no computers for them to use. And they thrive. They are happy and they are proud of themselves."

Tech Is Eroding the Capacity for Sustained Attention

Marina, age fourteen, was sent to see me when her parents grew alarmed at how long it routinely took her to complete relatively simple homework assignments and how agitated she became. She would take three or four hours to do assignments that her teachers claimed should take no more than an hour and a half, two hours at most. Her parents worry she has attention deficit disorder (ADD). They know she labors over her homework each night; she isn't out partying.

As we sort out what might be ADD from what might be caused by other factors, her online activity quickly stands out as a contributing factor. When I ask her to keep behavioral charts of every time she does something different while doing her homework, it becomes clear that her multitasking is the culprit, not her neural wiring.

She would start an assignment, then video-chat with a friend working on the same assignment; she would receive from six to more than a dozen texts during those three hours, have many IM conversations, and "have to check Facebook" two or more times. She often received e-mails from her teachers, too. There were so many distractions that she could rarely sustain focused attention for as little as five minutes. The longest time she focused on any one assignment was fifteen minutes. In addition to those interruptions, she'd be distracted by online pop-ups and find herself checking out new UGGs, YouTube videos, and favorite *Glee* songs—all part of a night's work.

As Marina identifies each distraction that was getting in her way, I ask her what she could do to get rid of it so she could achieve her goal of getting her homework done more efficiently and effectively. The result is that she decides to do her homework in the kitchen and to keep herself honest adds a self-imposed ban on the POS (Parents over Shoulder) screen switch kids routinely use so their parents won't see what they're doing when it's not the homework. She also decides that, in order to eliminate temptations, she would do homework for periods of half an hour to an hour and then reward herself with twenty to thirty minutes' play time. She would also put up an away message if she was anxious about her friends being mad or upset at her absence online, telling them when she'd be available. She tells me later that, although at times she regrets the no-POS plan, she knows it is working, and the best part was that she told her parents what she was going to do, not vice versa. She is in charge and it is working.

A 2006 survey by the Kaiser Family Foundation found that middle and high school students spend an average of 6.5 hours a day hooked up to computers or otherwise using electronic devices, and more than a quarter of them are routinely using several types of media at once. It also found that when teens are "studying" at the computer, two-thirds of the time they are also doing something else.

"Children's rooms are now almost *pathogenic* because they have so many distractions," says Martha Bridge Denckla, a neuroscientist at Kennedy Krieger Institute and Johns Hopkins University who studies attention deficit disorders in kids. "I think the most devastating thing that has happened is giving a child a room with a computer in it—you think you're being a good parent by doing so. Well, a funny thing can happen on the way to the homework." If we are going to give children computers to keep in their rooms, we can set controls when they are young, but ultimately, they need to learn—we have to teach them—to self-regulate, self-monitor, and stay on course.

Scientists don't yet know how screens and media use may shape the brain in ways that contribute to serious attention disorders, but research has shown negative effects in terms of distractibility and reading comprehension. In a Pew Internet/Elon University survey, tech experts acknowledged mixed feelings about the hyperconnected generation, dubbed AO for "always-on." On the plus side, they said, teens and young adults brought up with a continuous connection to each other and to information will be nimble, quick-acting multitaskers who count on the Internet as their external brain and who approach problems in a different way from their elders. However, the experts also predicted this generation will have "a thirst for instant gratification and quick fixes, a loss of patience, and a lack of deep-thinking ability" due to this "fast-twitch wiring."

Multitasking has become an asset and for many a requirement in our pressure-cooker culture, so much so that we've come to think of it in positive terms. It suggests a quickness, clarity, and efficiency of thought. But those who study the brain describe it differently. "It's not really that you multitask, it's that your brain oscillates between two activities," says the noted pediatrician and researcher Dimitri Christakis, director of the Center for Child Health, Behavior, and Development at Seattle Children's Hospital. Christakis's research aims to identify optimal media exposure for children and his findings thus far suggest that multitasking undermines the capacity for sustained attention and deep thinking.

"The young, nimble minds that have trained themselves to do it do so quite well," Christakis says. "But they nevertheless are oscillating and paying a price for it. They're not focusing as well."

A brilliant, beloved high school teacher, Steven Fine, tells me about the growing number of students for whom even the most engaged, creative teaching "just can't hold their attention." His greater concern is that screen-based learning is training the brain to process information more superficially through screens rather than through the printed page or human interaction. Once the brain has become accustomed to the more superficial screen-based learning, then this is what it is best prepared to do. Superficial thought processing becomes its default mode. Each year Fine sees the tech effect eroding students' capacity for sustained attention, reflection, and deep thinking.

"The Internet promotes a hunt-and-bump dynamic—you hunt and bump around looking for information," Fine says. "When you go on a web site you jump around a lot through very small, compartmentalized and very manageable pieces of information. Our students have more difficulty going more deeply into content." He now assigns long magazine articles as a way to counteract the hunt-and-bump mentality.

Based on studies of highway accidents involving drivers who were texting and reports of pedestrians injured while reading or texting, we know that the distraction factor is real and significant, even when our life and the lives of others depend on our staying focused. A sixteen-year-old boy told me about getting hit by a car while crossing the street at a major intersection because he failed to heed the "don't walk" pedestrian signal. He hadn't gotten a text or heard from anybody in what felt "like forever," and thought he'd take a quick look to see if he had missed something. With his attention half-focused on his phone, he heard the warning buzzer on the pedestrian signal but didn't look up and stepped into the path of an oncoming car.

We have to be concerned about tech habits that train our children for compulsive connection and fast-twitch wiring. We may feel that by the time a kid is sixteen, he is old enough to know he has to pay attention when he crosses the street. But the fact is that when our tech habits condition the brain for distraction and dopamine hits, then that is what the brain seeks. We know it from our own experience. We can hardly resist checking e-mails or texting while driving, so self-assured are we that we are in *total control* of our car and fully present to anticipate any oncoming car. This despite public health warnings that texting and driving is equivalent to driving drunk and that the risk of an accident increases by 50 percent. What but an uncontrolled compulsion would lead us to risk our lives and the lives of others, often those we love the most, to "just check" what could easily wait for a safer moment? With digital tablets and apps in the hands of infants and preschoolers, we are fast-twitch wiring them younger than ever.

Can You Hear Me Now?: The Art of Conversation Is Getting Lost in Tech Translation

My friend Martha remembers when she was young and visits with her grandparents and extended family included cooking together, eating together, and relaxing together after the meal. In the migration from kitchen to dinner table to family room, the intergenerational conversation was continuous. The young cousins might convene a board game or maybe watch a TV show together, but family conversation wove through it all.

"We used to sit around with my aunts and my uncles and they would just talk about anything," she says. "There was no point to it, it just went, but it had this great sort of humanizing, literary kind of impact on you."

The contrast between that rich conversational flow and the relatively shallow, staccato one that dominates her own household today is dramatic—and discouraging, she says. Absent the texture and cadence of old-fashioned family conversations, she believes her children are growing up in "an extremely isolated little bubble" of digital dialogue, a kind of conversational Muzak that fills the space, mimicking style but lacking substance. She and her husband routinely text with their college-age kids, more than she ever communicated with her parents at that age. Yet something is missing in all the quick quips and updates. She has lamented this loss to her children, who don't have a clue what she is talking about.

A generation ago a mother like Martha might have complained about her kids' lack of interest in family conversation, but at least she could have assumed that their "extremely isolated little bubble" included direct face-to-face visits and conversations with their friends. That is how children forever have learned and practiced the skills of dialogue: mutual listening, talking in sentences, extracting meaning from events and feelings, and sharing feelings in the give-and-take of conversation. Not so much anymore. Children have turned instead to texting and posting online, eliminating the need to participate in a full-fledged conversation, and losing those skills or failing to develop them at all. Many report an *inability* to sit, listen, and participate in a family conversation without texting. "It's too slow, so boring," they complain.

In focus groups with more than six hundred teens over the past four years they described texting as their primary way of communication. Teens don't even consider calling anyone anymore, and if you want to call your teen, you probably know by now that you are better off texting first to set up the call.

As our kids have grown accustomed to the detached and superficial quality of texting and online messaging, they have become averse to spontaneous conversation. They'd rather avoid it: with you, with their friends, with anyone. They describe a phone conversation, even with a friend, as "too intense," or "so intrusive." To call someone and have direct personal contact with your voice feels too needy, too forceful, too "in your face," they tell me. A conversation by phone or in person feels too risky. You can never be sure what the other person is going to say, and the immediacy of it puts you on the spot to respond. With texting, the other person can't see your emotional response so you are not nearly as vulnerable as in a live conversation. You can think before you text or not respond at all.

Absent conversation, kids are developing a seriously disordered understanding of what it means to truly communicate: to hear a voice, process the incoming and outgoing messages, engage directly with someone that way. The result is a new cautiousness, a tentativeness about the art of talking and the psychological capability of being direct or intimate with another person. This extends to e-mail, which they have long since abandoned for similar reasons. To e-mail a friend would be, as one older teen told me, "too weird, too personal."

This disconnect is most concerning when we see it eroding the foundations of communication on which love, deep relationships, and emotional commitments are built.

Lucy, a college sophomore, tells me, "The irony is that although our generation is in touch with each other all the time texting, video chatting, on Facebook and all that, we are really bad at intimacy. It's really kinda sad." The emotional detachment of digital correspondence is now the norm.

Adolescence is an awkward time anyway, and socially anxious kids always hate making phone calls. So for many kids, texting facilitates connection because it does feel safer. But what heavy texters are losing is the opportunity to practice: first to get up the nerve and then to carry on a spontaneous conversation directly with someone else. Never before has it been possible, probable, *or preferable* for two people to sit on a couch next to each other and "talk" with each other by texting! I'm seeing it more and more with teens and adults whose texting habits have put all sorts of relationships in jeopardy as their fast-twitch communications career from high-speed emotional fights to the deliberate detachment that kills emotional intimacy.

In studies of infants' responses to their mother's voice, researchers have found that even a very young baby can discern between the sound of its mother's voice in person and the sound of her voice through a screen presence, with the baby being more responsive to the "real" embodied voice. This example of "embodied cognition" is a new area of scientific study, but we know from our own experience that in-person conversation resonates in ways that deepen our experience of one another, and that with practice we become more comfortable and confident communicators. When texting begins to take the place of substantive in-person conversations for any of us, we are training the language and speech centers of our brain for a new, unnatural, and superficial model of connection. When that training starts early, as it does now for young texters, they get so used to it at such a young age that, unlike the newborn baby who innately knows something is missing and complains about it, our older tech-trained children don't even know what they have lost.

For many kids texting and the urge to spontaneously communicate also is taking the place of conversation within themselves—the capacity for reflection—that enables them to sit alone, think about things, and come to insights. To enjoy the experience of solitude, their own company, is an unfamiliar and often uncomfortable experience.

Lisa: I Text Therefore I Am

Lisa, age fifteen, comes to the therapy session one day still fuming from a school-sponsored weekend team-building retreat with her classmates at which they were not allowed to have their phones. She and her close-knit circle of girlfriends were used to sharing their thoughts about everything instantly in their secret online journal, in online chat, and through texting. The tech-free experience was boring and unsettling, she said. She especially disliked what was called "the solitude activity" in which each of the students sat alone on a rock, separated from the others doing the same thing throughout the woods.

"Why would anyone with best friends want to be alone in nature when you can text?"

As she shares her thoughts with me, Lisa doesn't hear that her own inability to "be alone in nature" speaks to the very experience that the school was trying to create for these endlessly wired, hyperconnected kids: the capacity to have oneself be the "other" with whom you share your observations and insights. Lisa had been unable to experience any calm, creative, meaningful connection with herself, a capacity that comes with practice.

Like many children I work with, Lisa expects to be able to text her feelings as she is feeling them. For her, texting is often the medium that leads to self-awareness, much as an extravert needs to hear herself speak to know what she thinks. Lisa will say things like, "when I texted her I realized that I felt" as if her experience doesn't register inside her until she texts it to her friends. In fact, she struggles to generate this kind of insight and self-awareness off-line. (Lately she is working in therapy on texting herself—to strengthen her connection to herself and to learn how to be her own ally.)

The neurological impulse to communicate, coupled with the addictive impulse to hit the keyboard to text, can be overpowering. It overrides what in other times would have been a pause and a thought—*I need to call, what do I want to say? I need to write a letter, what do I want to say?*—in which you would first connect with your own thoughts before you would address the other person. The immediacy of connecting as we do in texting interferes with the natural neurological and psychological process of communication.

Thoughtful conversation is one way we humanize our experience in a largely dehumanized digital culture. It remains a universal way in which parents provide something children need that they can't get from tech. Kids depend on their parents for caring, candid conversation about life. Not just life with a capital *L*, and all the big issues and long-range plans that implies, but little-*l* life, not small at all, but personal and individual, teeming with the details of the day. For that kind of textured, nuanced conversation, reflection and hashing things through, tech can't deliver. That's what parents are for.

Our children may roll their eyes at required attendance for family dinners or at the stories we share and others we ply from them about their day. And while they may not miss the practice of conversation, they do feel a sense of loss when that connection is missing. "My family is sort of busy," a middle school boy tells me. "My dad works late hours, my mom works all day, and then she has to cook dinner, and wash clothes, and stuff. Then I have homework to do. When we finally get family time, it's sort of like awkward cause we don't really talk to each other much. So it doesn't really feel very close."

Closeness counts. There is no substitute for genuine felt connection. We cannot control the culture outside our homes, but as parents we can create the culture we choose inside our families and communities. Our children need us to step into that role, reclaim our parental authority to "know what's best," dig deep for resolve, and tap the resources available to help us do it. Especially in today's tech-oriented environment, it is in the humanizing qualities of family and empathy, of a protected childhood rich in play, with sheltered time for reflection and conversation that closeness grows.

The more we learn about the architecture and interior design of child development, the more we see that windows of opportunity and vulnerability exist at different ages and developmental stages. Although exciting new research is showing that the brain can continue to grow anew as we grow old, long-standing research also shows that in the earliest days and months of life the brain organizes itself for lifelong learning in deep, defining ways. The windows are wide open.

Chapter 2

The Brilliant Baby Brain

No Apps or Upgrades Needed

Look, the brain of the child is shaped by the interactions they have with parents—that's just absolutely clear. We need to be in the physical and relational world before we reduce it down to screens.

—DAN SIEGEL, CHILD PSYCHIATRIST AND AUTHOR

In the seconds after Kaitlin was born, the nurse held her, cleaning her up, and Betsy looked at her husband and said something. Twenty years later she can't remember what she said to her husband but her newborn daughter's response is still vivid: "Kaitlin turned her head—she clearly recognized my voice," Betsy says. "I was so surprised—silly, I guess, as she had been hearing me talk for months. But it was just a moment of recognition—that she knew me. It was the moment I realized that she was as bonded to me as I was to her."

In my travels and talks with school groups, I often begin by asking parents to recall that earliest touchstone memory of falling in love with their baby. Angela, thirty years old when her first daughter was born, says she had been through so many miscarriages that when little Linda arrived, the first feeling was profound relief. That night with her baby in the bedside bassinet, Angela listened to

Linda sleep and her heart filled with gratitude: "She was here, and she was alive, and she was mine." A few years later, when second daughter Lori came along, "I would hold her against my chest and she would put her tiny hand out and rest it on my chest right under my collar bone, and just leave it there," Angela says. "I was her anchor, and I felt it."

Many mothers and fathers point out that their love affair with their baby began in utero. Others, including adoptive parents, have told me their stories of love at first sight or snuggle. Some parents describe a worried wait through a rough start with medical complications or fussy feeding or colicky wailing or overwhelming exhaustion that made them think they might never bond with their child.

Blair was ill with a 104-degree fever when she delivered Claudia, who had to be in a boxy incubator on IV antibiotics for the first week. Breastfeeding was complicated and clumsy, and for the first two days both seemed mostly exhausted and distracted. On the third day, in a moment when her baby's vulnerability overwhelmed her, "I started sobbing," Blair says, sobbing anew as she recalled the moment more than five years later. "I was overwhelmed with the instinct to protect and heal. I can't put into words how much the love came down over me all at once . . . the love fell over me like a blanket."

Jared, an older first-time dad at fifty-five, took the week off when his son Brandon was born, and he vividly remembers his intense focus on caring for his new son and his wife, Anne, who had endured a difficult delivery and challenging start to breastfeeding. "I sat in a chair and read *Harry Potter* to them, hour after hour," Jared says. "I have a lump in my throat thinking about it now [eight years later] because of the overwhelming sense of being totally *in* that moment. It was definitely what Maslow referred to as a 'peak experience,'" he says, referring to the psychologist Abraham Maslow's term for a transcendent moment of life-affirming awe. "I recall changing Bran-

don's diaper for the first time and marveling how there was only joy
in the act, no nausea . . . not what I expected."

Take a moment and remember that first feeling. Perhaps it was hum-
ming to your baby in utero, gazing into his eyes for the first time,
or just watching her fall asleep on your chest, hearing her breathy
baby sighs or feeling the damp spot where her drooly little cheek was
pressed against your shirt. Maybe it was in a frightening moment,
a "dark, murky" moment, as one father put it, when the fragility of
your child's life was nearly unbearable, and into that abyss came a
powerful experience of a bond beyond words. Perhaps none of it felt
natural to you at all, but a friend, neighbor, or caregiver proved the
perfect mentor for you, or you picked up what you and your baby
needed from watching other parents you admired. Close your eyes
and connect with that feeling and you have the best possible reason to
follow the consensus of expert medical, scientific, psychological, and
other child development opinion to leave tech out of your baby's life
for the first twenty-four months. Fresh to life, open to your imprint,
your infant watches and listens for your face, your voice, your touch,
your gaze. Tech offers nothing your baby needs more than you.

Tech and Me and Baby Makes Three

Babies come equipped with their own bonding instinct, and we
know from studies of infants who thrive and those who don't how
vital the bond between parent and child is for their healthy growth
and development. From the moment your baby is born, you define
her world not only in the ordinary sense of organizing her external
environment, but also from the inside out. There is a natural, wonderful sequence in the way grown-ups have tended and played with

infants forever, and it is those simple interactions—nursing, soothing, bathing, diapering, singing, strolling, reading, and playing—that provide all a baby needs to lay the foundation for the next levels of cognitive development at three, four, and five years old, and for all their future learning.

The mirroring exchange that occurs when we return our baby's gaze or giggle or respond to his cues about hunger or comfort or curiosity allows us to communicate wordlessly to one another aspects of emotion and presence—neurological "statistics" as one researcher put it—that continue to develop, deepening with every interaction. If that connection is stable, steady, secure, and supportive, baby and parent form what we call a "secure attachment." When those qualities are weak or missing, the attachment may be compromised. When this emotional connection is strong, the young brain is open to new experiences: the sweet smell of milk, the sight of a spinning mobile, reaching for a stuffed animal. The social brain needs the same emotional foundation to develop relationships, which begin with parents and caregivers and broaden to extended family, sitters, and neighbors.

The first and continuing lesson your infant learns from you is that she exists—*Oh someone recognizes me, I am a being in the universe, there is something worthy about me—someone is noticing and paying attention to me!* This is what forms her sense of being a person and makes her feel safe and secure in her environment, in her primary attachment to her parents, and in the widening circle of trusted others as she grows. The baby brain comes hardwired for human relationship because that is the most essential connection for survival and all future learning. And the single most important relationship is the one your baby finds with you.

Consider what happens when you add a cell phone to that picture and your baby's view of you includes a handheld or headset device, or a tablet or computer screen that routinely pulls you away, interrupts your attention, or captures it completely. Perhaps your frequent e-mail or Facebook activity becomes an integral part of your baby's experience of you. What does it mean when you use screens and apps as a pacifier for your baby or toddler, or as a babysitter or teacher, or in a pinch as a stand-in for you? Or when you use them to distract yourself or pacify yourself to relieve the stress and tedium of your parenting time. Tech not only changes the iconic picture of the parent-child relationship in early childhood—it changes the relationship itself.

Ellen worries about that when she sees her six-month-old son Henry's expression as he watches her work on her iPad during playtime on the floor or sometimes when she's holding him. Like most of us, she is often preoccupied with all the things she has to do, and if Henry is engrossed in his own little world of play, she will discreetly pull her tablet out, hoping he will not notice.

> He's just lying here and playing, so I'm on the iPad and suddenly he stops playing and he is looking at me! I mean *so* many times—that happens 90 percent of the time—and I don't know at what point he stopped playing and started looking at me. It breaks my heart because I don't know how long he has been staring at me. I mean, what is he thinking? I feel so guilty that I'm not present with him and he knows it. It's one thing if I'm unloading the dishwasher and talking to him. That doesn't require brainpower, but e-mail does. It's impossible to really be doing both. I know he knows I am completely disengaged, you can just see it in his eyes. So what does that mean to him [that] we are both in the same room together and I'm not being present with him?

There is really no way to know what it means to Henry; at six months old he can't tell us. However, based on long-standing research of infants' reactions to their mother's voice and facial expressions, we know that Henry is indeed capable of detecting that his mother is disengaged. We also know that babies are often distressed when they look to their parent for a reassuring connection and discover the parent is distracted or uninterested. Studies show that they are especially distressed by a mother's "flat" or emotionless expression, something we might once have associated with a depressive caregiver but which now is eerily similar to the expressionless face we adopt when we stare down to text, stare away as we talk on our phones, or stare into a screen as we go online. More recent studies using brain imaging scans on infants show that brain centers critical for higher order learning and language development "light up" when a mother is present and fully engaged as she speaks to her baby. When the mother's proximity changes, the brain's response changes, too.

We also know that our child's fascination with the gadgets we use is a response to the cues we give; the more attention we give an object, the more they want it, notes the Yale psychiatrist Bruce Wexler, author of *Brain and Culture: Neurobiology, Ideology, and Social Change.* Wexler, who also directs a neurocognitive research laboratory, says that research from child development to anthropological fields shows that children have always been quick to adopt whatever we put within reach.

"If an infant is given a choice of playing with an object being handled by an adult or with an identical copy of the object that is closer, the infant will reach past the copy to play with the one the adult has," he explains. "By six months of age more than half the infants will follow their mother's gaze, and by the time they are a year old nearly all will do so."

Whether the behavior is inborn or conditioned, or a little of both, Wexler says, the result is that through this "instrumental parenting"—the human habit of collecting and introducing our children to material things—we influence what our infants pay attention to, what they become "most aware of, become most familiar with, and think most about." A parent's focus of attention—whether it is cookies or computer screens or books—becomes the infant's object of desire.

So we can assume that Henry's watchful gaze reflects that natural curiosity. Whatever Henry is thinking when he watches his mother sneak some screen time, he is thinking it often—90 percent of the time they are together this way, Ellen says. Whatever he is thinking, it has power in his young mind; it repeatedly interrupts his attention to his own toys and self-directed play, riveting his attention on his mother and her screen. Perhaps he is simply curious, enamored with his mother's visually enticing screen or perhaps with the sight of her so entranced by it. Perhaps he has designs on the screen for himself. Or, as Ellen worries, perhaps he is drawing meaning from it, aware how completely disengaged and "not present with him" she is and feeling unsettled by that.

Although she cannot know what Henry is thinking, Ellen's sensitivity to the emotional connection and disconnection she has with her son is a good example of what Dan Siegel calls "mindsight." In his book by that name, Siegel says parents can cultivate their ability to "see" the minds of their children through the basic signals they can perceive. For parent and baby alike, the nonverbal messages of eye contact, facial expression, tone of voice, body posture, and the timing and intensity of response may reveal our internal processes more directly than our words. For a parent, this means paying attention to your state of mind as well as your child's, so you can recognize the emotions in play without being swept up in them.

Ellen understands that she is the emotional foundation for Henry (as is her partner, Amy) and that Henry's well-being is influenced by her ability to, for meaningful periods of time, be fully present and emotionally available to him. She is tuned in to the two-way communication between them, aware of her own thoughts (*I'm just checking e-mail . . . I'm ignoring Henry but I'm anxious to read e-mail*) as well as what Henry's experience of the moment might be (*I'm a little anxious. Sometimes she cares about me, sometimes she doesn't*). Staring into the screen to read feels different—more intense and oblivious to Henry—than when she's folding laundry and chatting with Henry at her feet. And it is: no matter how brief, screen activity locks our attention, visual and otherwise, in a way that excludes everyone and everything around us. Even when we multitask, the second we engage in a screen the accompanying disconnect from those around us is palpable. You know the feeling—you've been the one left waiting when someone says, "one sec—I just need to check this."

Ellen knows she should be attentive when they are together and respond to his coos and other nonverbal cues, encouraging his attempts to communicate with her. She knows that at six months old, he needs to feel her connection, not the repeated experiences of disconnect in routinely having to wait for her to look up from her screen.

By the time children are three or four years old, they can play and communicate more independently, but from birth to two they rely on us completely and they need our engaged presence during these connecting interactions. They can tell when we are distracted. We can't fool them, yet, even knowing that, it can be tough to wean ourselves from established tech habits.

Many parents who come to see me are struggling with difficult adjustments to the demands of new parenthood or a growing family and, for all the ways tech is helpful to them, it is also problematic. Today, in addition to issues of potty training, picky eaters, sibling rivalry, and chronic waking up, our conversations inevitably turn to the presence and role of tech in a young family's life.

Donna Wick, a clinical and developmental psychologist and director of Mind to Mind Parenting, helps new parents develop what she calls "reflective parenting," a quality akin to mindsight. Very little in life prepares parents for the dramatic shift from a me-centered life, or even a couple's we-centered life, to the baby-centered life, she says. "A crucial part of parenting in that first year or two is that you realize that if you are going to be a responsible and responsive parent, then it's all about the baby—you are going to come second, and that's a hard lesson," Wick says. "Tech aids and abets the impulse we have to avoid that level of responsibility and that level of self-sacrifice. But you just can't."

This includes computer time for work and personal enjoyment. It also includes parents' TV, phone, and texting habits and their children's casual or "secondhand" exposure to all that. Then there is the matter of time their infant or toddler may be spending on screens. That often already includes TV, cell phones, tablets, and toys with electronic play features. It is rare anymore that screens are completely absent from the family scene. According to the Kaiser Family Foundation, 74 percent of children under age two watch TV, 59 percent watch TV for an average of just over two hours daily, and 30 percent of children birth-to-three years old have TVs in their bedrooms. A 2011 survey by *Parenting* magazine and BlogHer Publishing Network found that one-third of Gen Y mothers (born between the mid-seventies and mid-nineties) allow their two-year-olds to use a smartphone, presumably to watch videos and play games. Another study sponsored by the Sesame Street Workshop concluded that more than 60 percent of children under age three watch videos online. The younger the parent, the earlier kids are plugging in.

More significant is how often we find that screen time and other tech habits are contributing to tension and anxiety for the parents and disruptions in sleep patterns that can lead to irritability and other symptoms of overstimulation in their children. People think it's harmless to watch TV or be on the computer while their child plays nearby with her blocks. It's not. Children are seeing and hearing secondhand what their parents are plugged into. Parents tell me they feel both guilty and defensive about their tech use.

I can't be unavailable to my coworkers.

When am I supposed to have "me" time?

How can I keep up with all the check-ups and other scheduling for everyone?

My eighteen-month-old starts to whine as soon as I pick up my phone.

The only way I can use my iPad now is in a separate room because my toddler wants it so badly.

Is it ever okay to talk or go online with my baby in my arms?

These questions about tech and the way it can connect and disconnect us as families, how it can foster and hinder child development, used to come up as children reached preschool age and began to engage in the wider circles of school and community. Now the concerns come home with the newborn. This is especially troubling because as informative as tech is for adults, everything an infant needs to thrive happens off-line, off screens. Everything that a baby needs from its environment between birth and two years comes from people, from relationships with people and interactions with the environment— physically exploring, playing, crawling, and interacting with others. When we triangulate our relationship with our babies and tech, we compromise that essential connection.

Nature, Nurture, and Multitasking

You don't have to look far to see the way screens, cell phones, digital tablets, and other electronic devices have become the ever-present third party in our conversations and downtime with our very young children. How often have we looked askance at the sight of a parent pushing a baby in a stroller while talking intently into his or her cell phone, only to be that person the next day? Who hasn't made a promise to themselves—*today when I go to the park with my baby I'm going to turn my cell phone off and give her my undivided attention*—only to stay plugged in and let that promise slide just one more time?

Immersed as we may be in a tender, relational "right-brain" moment with our baby, our left-brain taskmaster, our neurological "wizard of connection," as Dan Siegel calls it, drives us to seek completion, mastery, order, predictability. "They're all left-brain dreams," he says, "that's what the left brain lives for." So just as mother's milk is soothing for an infant in the middle of the night, for the frazzled, sleep-deprived parent the allure of checking e-mail provides a satisfying fix, even if it takes place while nursing an infant at 2 a.m.

We all know how hard it is to resist that impulse, especially when our baby appears completely absorbed. It seems harmless enough to hold the baby on your lap as you sit at the computer to catch up on e-mail or Facebook. Or when you bring the laptop into the kitchen at mealtime, letting the baby play with her toy screen while you're on yours. It feels odd *not* to slip your phone in your pocket or your iPad in your pack as you pick your baby up to go into another room or out the door. Our right-brain relational self says *pick up the baby*; our left-brain multitasker says *let's do a two-fer—grab your phone*. But our babies need to bond with us, not our gadgets.

This early triangulation is a relational game changer for our babies, but also for us. Parenthood is a reciprocal process in which our babies and the act of parenting make us who we are. In the hyperconnected digital culture it is easy to forget that the parent-child bond is just that: a bond between two people. Just as we are shaping our baby's understanding of what it means to be a person, our babies are teaching us what it means to be a parent. They need for us to be attentive partners if we are to learn their rhythms, their temperament and personality, their moods and body language, their stimulation needs and limits.

Babies and toddlers are dependent on us in so many ways, we

sometimes overlook the brilliant teachers they are and the enormous task they set out from birth to do: develop *our* skills to be the loving, effective parents they need to prepare them to be able learners in the wider world. From the moment they are born, long before they become verbal, they are teaching us how to be parents, grooming our intuitive sense, teaching us to understand their needs, wants, and ways of being. The cranky fussing, the wide-eyed gaze, the sudden smile: in their preverbal language, they show us what counts, cheer us on when we get it right, and let us know when to try again.

These are skills we will need—and want—for the rest of our lives as parents. Years from now the quality of your relationship with your child won't be measured by the high-tech toys or apps you gave them when they were babies, but by the quality of connection you created together in those early years. It is precisely that loving human connection we call bonding, attachment, and attunement that stimulates optimal brain development.

Your interactions with your baby are the classroom of early childhood. Language, reading, play, movement, all of it begins with you and the quality of the connection you establish and continue to nourish from birth through the early years. Tech cannot do it, but tech can come between the two of you, and it can undo some of the critical brain building under way.

The Sensorium: The Baby Brain's Bandwidth for Development

When your baby locks her gaze on you or on a spot on the wall, when she smiles at you or grabs your finger for the first time, when she startles at a sound or pumps her little legs, or when she squeals with delight in a game of peek-a-boo, you are witness to the miracle of development. These are all visible expressions of the earliest sequences in developing neural pathways and networks of what we call the sensorium—the collective capacities of the brain to receive, process, and interpret sensory information. The sensorium sounds like some fantastical play palace in the neighborhood where you might take your baby for an afternoon of fun. In a way, it is. Everything your baby sees, hears, tastes, and touches, every move or sound she makes, every sensation and emotion she experiences—and all of your interactions with her—contributes to the robust development of the sensorium.

Tremendous brain growth occurs in the first two years of life—more dramatic than at any other age—such that the overall brain size doubles in the first year, reaching about 70 percent of its adult size; it reaches 85 percent of adult size by age two. During this time your baby's brain is busily building structural and functional connectivity, creating the essential neuro-architecture to support life and learning. Too much tech at any age, but especially too early an introduction to it—before age two—shortchanges a young child on the time and mix of experiences the sensorium needs for well-rounded development. Neurologically and psychologically, the tech effect becomes a wild card in your baby's development.

"The brain was designed to develop in all the areas through natural human interactions and play, and by putting kids in front of screens we are changing their brains," says the developmental psychologist JoAnn Deak. In her "baby brain boot camp" talks for parents, Deak is unequivocal about the value of a tech-free infancy and toddlerhood: "The job of early childhood and family life is to make sure that every sector of the sensorium has a chance to fully develop, that every sector gets strong enough to do the job. If every sector doesn't get to develop to its fullest potential, then the deficit is lasting." If you let the brain do just the things it likes—for instance, repetitive play on a touch screen—then that imbalance, too, can hinder the brain for a lifetime. "It's what screens take away and destroy," Deak says. "For every hour they play a computer game or watch a program that someone else has made for them, they lose the opportunity to do that for themselves."

Language development and the neurological foundations for later reading are especially prominent threads in the "embroidered circuitry" of the sensorium at this age.

Unlike speech and language development, for which the brain comes equipped with neural circuitry from day one, there are no ready circuitries in the brain for reading. Those neural pathways and networks take years to develop, years of layered learning to create a circuit that moves from being "a decoding machine"—a term Maryanne Wolf uses to explain what children are doing in the early stages of reading—to being a circuit wired for comprehension, what she calls "a place where the deepest thought can happen and be brought forward to insight and epiphany." The time you spend talking and reading with your child acts on many levels to strengthen those neural pathways. Tech can interrupt or weaken that connection.

Contrary to what commercial educational or developmental learning programs would have you believe, when we talk about whether media content, toys, or gadgets are developmentally appropriate, we are not talking about whether an infant or toddler *can* manipulate a device or sit still for a show. We are talking about what it does to them when they do. Just because your baby can tap a touch screen to change a picture does not mean that he should, that it is a developmentally useful or appropriate activity for him. In fact, research suggests that the process of tapping a screen or keypad and engaging with the screen activity may itself be rerouting brain development in ways that eliminate development of essential other neural connections your child needs to develop reading, writing, and higher-level thinking later.

Research by Patricia Kuhl and others shows that babies learn language most effectively from human interaction, not from audio or video programs or entertaining apps or TV shows, as advertisers (and sometimes our own wishful thinking) would suggest. The mediated connection does other things; it visually stimulates, which the studies show does not support language development and may negatively affect it and negatively affect attention span. Studies also show that the mediated connection fails to stimulate certain other neurological connections needed for language and cognitive development. And it appears that certain learning centers of the infant brain respond only when interacting with real people—physically present and attentive—and particularly when they are parents or caregivers with whom the baby has formed an attachment.

Specifically, it appears that babies need what we call an embodied connection to stimulate the part of the brain that governs language development. Ordinarily, these neural connections are made naturally in the face-to-face interaction between parent and child. Putting a baby in front of videotape or on an iPad does not teach language, and when a mediated connection replaces a human connection the benefits of the human connection are, neurologically speaking, lost or diminished.

Neurobiologist Wolf says that although the science is complex, her advice to parents is simple, summed up in what she calls "the grandmother principle," which is: "If you want a child to talk, you talk to them." And you read to them, too. When you do, your child experiences the structures, cadence, and eventually the content of language and reading in the context of relationship, imaginative storytelling, and playful interaction. "The little baby just loves the whole setup that reading gives," Wolf says. "You're listening to a sonorous, beloved one's voice reading, and they love to hear the sound of your voice. I can't tell you all that goes into that, but I can tell you it exists, it's powerful, and to neglect it is an absurd waste."

Wolf notes that studies show that in comparisons of children under two whose parents used an array of video and other auxiliary language learning aids and those who learned without the aids, language development was superior in the ones who learned without. "There is nothing better for the development of language than human language . . . people talking to them," she says.

The language development expert Lydia Soifer speaks of the intrinsic motivation that is lost, as well as the primary connection between language and human communication. It is in the textured dynamic or "charge" of interpersonal relationship that children learn language best.

"Online language programs are just tools, they are not dynamic enough," Soifer explains. "In fact, the foundations of literacy are in sounds, whether or not it's the speech sounds or the intonation patterns, or the pause and stress and juncture patterns. It's how we change meaning. Where there isn't a human dynamic, it isolates out the sounds from the language, from the intent, from the content. Yes, you need those skills to learn to read, but you want to do it with a certain level of motivation, and feedback that is affectively rich, not with bells and whistles that give you a reward. And I think part of what happens is that kids get addicted to the bells and whistles."

These early clues to the impact of tech on children's brains suggest changes in neurological processes that appear to alter children's intellectual, social, and emotional development. Some of those tech effects on the brain may be helpful for some children. We know, for example, that some children, typically a little older, who are medically diagnosed with nonverbal learning disabilities or other neurological deficits or differences may benefit from particular screen-based activities. However, for all other children, screen time may contribute to uneven brain development, as screen-based activities have been shown to stimulate visual processing more heavily than other parts of the sensorium. At the same time, by pulling their attention into screen-based play, tech distracts them from the most essential learning environment of all, the parent-child relationship and real-life experience in the family and the wider world. It also exposes them to potentially damaging content.

Extensive research since the 1970s has established that media violence contributes to anxiety, desensitization, and increased aggression; that among young children, violent media can trigger fear responses that are long lasting, that are linked to post-traumatic stress disorder (PTSD) symptoms, and that can occur after one exposure. Viewing frightening television, even programming deemed appropriate for preschoolers, raised children's heart rates and caused symptoms of PTSD. Sleep difficulties were one of the most common symptoms, and a recent study found that children who lose as little as half an hour of needed sleep per night can show behaviors typical of attention deficit and hyperactive disorder (ADHD). We don't yet know all the mechanisms for these links, but the outcomes are certainly cause for concern.

None of us wants to make our children anxious or aggressive or inattentive; we don't wish for them to be sleep deprived. But the power of the screen lures us into unintentionally exposing our infants to harm within the walls of our own homes.

All the research underscores our need for caution in how we introduce children to tech, especially in the first two years of life. Research as well as personal experience tell us that when grown-ups watch TV or any screen in the presence of children, they don't interact directly with the child with the same focus, pace, attention, or the visceral, immediate connection that stimulates a baby's healthy neurological and social development.

Experts have said most of this for decades, yet still parents somehow hold on to the belief that TV "mostly helps" rather than "mostly hurts" their children's learning. Perhaps an explanation lies in the conclusion of researchers who examined how many new words twelve-to-eighteen-month-old children learned from viewing a popular DVD several times a week for one month. The study found that the children viewing the DVD did not learn more words than the no-DVD control group. Further, the highest level of learning occurred in a no-DVD setting in which parents set out to teach their children the same target words during everyday activities. But the clincher was what the researchers discovered about the parents: those who liked the DVD tended to overestimate how much their children learned from it. The authors concluded that infants learn relatively little from the infant media, and parents sometimes overestimate what they do learn.

Bait and Switch: Things Are Not What They Seem

From the 1960s when TVs dominated American homes and before computers and cell phones became commonplace, experts warned about the dangers of too much screen time and how important it was to unplug kids and families. Even so, parents of that era came to depend on TV as a babysitter for dependable entertainment and the promise of educational benefits. The cautionary voices were duly noted and largely ignored. But at least a TV set didn't follow you out the door. When you went to the grocery store or the park or got stuck in traffic with your baby, you had to wing it. You sang and you played rhyming games. Your baby got lulled to sleep by the motion of the car or the stroller or your body as you rocked her or walked along. Who would have imagined that someday the teddy bear in the baby's crib would have a touch screen, digitalized voice, and apps that offered everything from virtual peek-a-boo games to bedtime lullabies? Or that tech for kids and adults would be so pervasive that we'd spend more time on screens or online than we would in the presence of each other.

The ease with which our screens travel with us today, and the ease with which we can hand them off to our babies anywhere anytime, adds a new urgency to the conversation. A father in the grocery store quiets his fussy toddler by handing her his iPhone with Angry Birds flashing across the screen. A mom asks me if it is all right that she straps her twenty-month-old son in his car seat and pops "his favorite" *Power Rangers* DVD on her iPad for him to watch while she catches up on work calls as they drive to day care. Another mom gives her two-year-old daughter a touch-screen storybook app to "get her interested in reading." All of them assume that tech is helping their child cope or play or learn in that moment.

We may think that their eagerness to engage with it or their quiet, focused attention means they are learning to calm themselves, learning to focus, learning to read, learning to draw, learning to do all the things on their developmental to-do list. But that is not what is happening. One reason TV and tech are especially risky from birth to two is that the stimulus-response mode—instant gratification—is so engaging that we may easily mistake our child's keen interest in it for "learning." Psychologically and educationally, a very different kind of learning is under way.

Core Lessons Come from Human Tone and Touch

When we hand our baby a touch screen to keep her occupied or entertained, she's missing the opportunity to engage herself—literally, to engage with her own inner self, her feelings, and processes for learning and adapting in the moment. A core aspect of infancy and childhood is the range of learning that comes from human touch and vocal interaction, from the rhythm and pacing of communication. It's as basic as the newborn learning the boundaries of physical self—the difference between his skin and his mom's or dad's. *You exist. I am here for you.* Forevermore the comfort of "the other" is a source of security, safety, and protection. When your child falls and runs into your arms, so many things happen in the physicality of that embrace. The voice, the word, the physical experience, the face, the gaze, the coo. The surround sound of human comfort and optimism. *You're having this feeling. And this is how you can manage that. You're going to be okay.* This is the beginning of how we teach our children to self-regulate—read their own emotional states and build their capacity for self-soothing and the foundation of emotional stability, optimism, and resilience.

We continue to teach our babies by calibrating our responses to fit the need. At first you pick them up all the time, then you differentiate between their signals; you encourage them to calm themselves. Guided by your growing intuitive sense about your baby, you discern when it's appropriate to wait a little longer to respond, teaching your child to trust that the response will come and to practice self-soothing in that pause. When your thirteen- or fourteen-month-old toddler tumbles, you don't look with your hurt look or your panicked look, you look with your calm and reassuring look and this helps your toddler learn that falling down happens and is no reason to panic. That we can feel angry, too, and work through that feeling rather than be defined by it. There's a big difference between *I feel angry* and *I am angry* as a definition of who we are. We are constantly giving them cues—verbally, expressively, emotionally in our tone, teaching them how to handle frustration, pain, and accidents, and how to self-regulate. These skills do more than curb temper tantrums. They are developmental stepping-stones that eventually will enable a child to control his bladder, use the toilet, tie his shoes, zip a jacket, learn to share. At each developmental stage the capacity to soothe yourself and calm yourself down, to deal with impulsivity, tolerate frustration, work through boredom to creativity, to make transitions from sleep to awake to play to eating develops through practice. It is in those human-to-human interactions day after day after day that our children learn these fundamental lessons that prepare them for all future learning.

Tech Goes Faster Than the Speed of Life

Tech creates an expectation of instant gratification. An app or screen game's instant response or a stimulating, fast-paced TV show delivers a happy hit to the baby brain that makes ordinary life sluggish by comparison. A popular YouTube clip shows a baby accustomed to playing with an iPad struggling to make a print magazine "work." The baby taps and swipes and thumps on the pages clearly attempting to make the pictures change. It doesn't work, of course, and the baby grows frustrated when nothing happens. When we train a baby or toddler to expect this kind of interaction and this tempo, we create an expectation that if something doesn't ping, whistle, light up, and move quickly on demand then it isn't fun or interesting or it's not working. Hands-on blocks, books, and puzzles lose their appeal. Coloring with a crayon takes a lot more effort and coordination than "touch and drag" coloring and it lacks the screen zing of responsiveness that makes interactive tech so fun.

Lost is the slow-paced hands-on practice that develops small motor skills, dexterity, and eye-hand coordination. The sensory experience that goes with that—the touch and smell and messy fun of play—is gone, too. When we give infants apps and entertaining games they are distracted but they aren't learning what we might think or hope, and when we routinely hand them a device—or a TV show—to occupy them, they don't learn how to entertain themselves. In those moments they miss the opportunity to think creatively, problem solve, or accomplish the important slower-paced baby tasks they need to: stacking, shaping, sorting, tearing, pushing, pulling, pouring, scooping. Your shared delight in their discovery is the best bells and whistles for their development.

Babies Need to Shake, Rattle, and Roll IRL

Tech is stimulating but sedentary. Children need to move to be happy. They learn and grow from being physically active. It builds muscle mass and has a positive effect on their appetite, their ability to sleep, their self-confidence, and their sense of competence in the world. Infants love physical activity: bouncing in your lap, scooching in their crib or across the floor, pulling themselves up, and reaching for and grabbing everything in sight. They like to shake, rattle, and roll. They like to flip and taste the floppy pages of soft books. They love to walk, crawl, or climb.

Even before they can walk, they push the limits of their physical development: they try, they succeed, they fail; you encourage and applaud and console; and they do it again and again. They learn about trying. In that process they not only develop their physical selves, they internalize your encouragement and your confidence in their ability to fall and get back up. These are the roots of the resiliency, grit, and optimism they need for life. The more we can share that with them and delight in each day's new adventure, the more our children will feel like explorers equipped to reach new heights in other aspects of their development.

On Screens Your Baby Is "in the Zone," but with Whom?

An infant may stare mesmerized at goofy cartoons or so-called educational programming, but not all children's programming is fit for children, especially infants and toddlers. When we put them in front of the

TV and their focus shifts into "the zone," it absolutely buys us time for a needed shower or uninterrupted work time. But we don't yet fully understand what the mesmerizing screen or tech zone means to the baby brain, and its relationship to future psychological health or unhealthy predispositions toward addiction, drugs, and the need for such stimulus later in life. We know that children who watch TV want to watch more TV, and that kids on screen games often have difficulty ending their play willingly.

The American Academy of Pediatrics, which has been firm with its recommendation of no screen time for infants under two, has come under fire for setting an unrealistic standard in this tech-driven era. The recommendation stands, meaning that if you're going to busy your baby with screen time or have your baby with you when you watch or use screens, be extremely picky about what they see and extremely watchful of the time they spend that way.

Choose shows that teach them about the world around them in a way that is kind, hopeful, and encouraging; not bratty, sarcastic, fast, or frightening. Make certain that content is not overstimulating (check the research-based media lists) and be sure the tone is always empathically tuned in to the developmental level of an infant or toddler. For the under-two crowd, look at the age two and up best bets like *Mr. Rogers*, *Sesame Street*, and *Blue's Clues* that share those qualities and remain favorites on the resource lists of safe media for toddlers. Pick any media exposure as carefully as you would pick a babysitter to leave alone with your baby. Remember, from your baby's natural point of view, Fisher-Price and Disney have nothing on you when it comes to dazzling educational toys and games. You are the gold standard.

Real Babies Need Real Read-Aloud Time

You wouldn't dream of substituting a touch-screen meal for real food, and no matter how sophisticated voice tech becomes, audio or digital books and reading games are no substitute for personally reading to your baby. Earlier in this chapter we looked at reading to your baby in the context of language development and establishing the neurological foundations for later reading—the nuts and bolts of brain development. Sharing a book together, reading aloud to your child brings you physically close in the happiest way. It also inspires conversation—even with children who aren't yet conversational in the strict sense. The timeless attraction of *The Runaway Bunny* is not just the story about love everlasting but the interaction it inspires between you and your baby. You read, you play, you listen and respond—you communicate. *Where, oh, where is that little bunny? Are you my runaway bunny?* Your baby points, grins, giggles, relishes your voice, your touch, your attention.

What about sharing story time on a Kindle or other e-reader or digital tablet? Does it really matter whether snuggle time and reading to your young child involves a screen or a printed page? Science simply doesn't have the answer to that question yet, so as parents we are left to reason for ourselves based on what we do know. On the plus side, any face-to-face snuggle time and read aloud time with your child is better than none. This includes the positive experience of being at the center of your attention, talking about a story, and interacting over it. For our infants and toddlers, though, other considerations include the unique window of opportunity for critical aspects of development that depend on the baby's physical interaction with the environment and the material world. For instance, an e-book eliminates the more textured tactile experience of a conventional book. Interaction with a three-dimensional book means that an infant sees and grasps different-size books and views different-size pages; feels the texture of paper and book covers and the weight and heft of a book; focuses the eyes on type and illustrations on a printed page rather than a lighted screen, developing eye-hand coordination by first watching you turn a page and eventually learning to do that herself. Research also suggests that the direct screen-viewing experience may narrow the neurological range of the young child's developing brain. That negative effect may show up differently for the older child. For the school-age child, we might want to consider how much time they spend on screens routinely, how easy it is to become trained for screen-reading, making physical books a more challenging medium for them to use for research or pleasure reading—a disincentive to engage the body of literature and information available only that way. These are just a few reasons you might choose to use three-dimensional books for that shared reading time, or at least consciously include traditional books in your child's everyday story time experience.

If parents absent themselves and outsource reading to tech—using audio books or other read-aloud devices—children lose opportunities to develop the focus, endurance, and the deeper sensibilities they need to move into more profound learning, says Maryanne Wolf. "We will have a generation of readers highly adept at handling multiple pieces of information streaming in at them every second, but they will lack the means—literally the very circuits in the brain—for deeper revelation." This may seem premature to mention in a conversation about infants and toddlers, but it is not. The growing array of digital devices that "talk" to us now—from touch-screen phones and self-check registers at the grocery store, to toys, games, books, and apps specifically target-ing the infant-toddler user—is only growing faster, more sophisticated, and more enticing. Your baby will join that culture soon enough. For now, take the TV out of the room, power down screens, pick up a book, and read with your child. Let your baby plug into you.

Turn Down the Volume on Commercial Pitches and Promises of Parenting Paradise

Advertising and marketing campaigns, upbeat blogs, and media buzz are designed to make you feel understood (*We know how exhausting/*

exciting parenting can be!), to win your trust (*You want what's best for your baby and so do we!*), to appeal to your parental aspirations and insecurities about their academic futures (*Build your baby's learning skills!*), and to seal the deal (*Here's a free game—click to buy the better version for just a buck!*). Our human predilection for instrumental parenting—the happy habit of giving our children material things—coupled with the desire to help our children prepare for success in school and among peers, sets us up. We are not only a target market, we are sitting ducks.

Science, medicine, and a world of caring experts advise caution about plugging your infant or toddler in to screens and tech, but advertisers and our tech-happy culture send a very different message—and loudly. It is hard enough to be a vulnerable, loving, sleep-deprived parent, but to be told by other parents and marketers that you're depriving your child by not exposing him to the latest and greatest Baby Einstein toys can feel overwhelming. There is a cautionary tale in Baby Einstein, the best-selling "educational" product Disney was eventually forced to recall for making false promises. Or the Federal Trade Commission's 2012 action against the makers of a product that promised, "Your baby can read." But that has not kept others from saying what they wish to promote products. While companies push products they claim will make babies smarter sooner and happier longer, most have no credible evidence to support their claims and often research specifically refutes them. Even among those which experts say show some learning potential, few are designed for children under two years old and none have been proven to make children smarter or more school ready, according to Common Sense Media.

The most insidious thing is that these voices are not just selling a product, they are selling a way of life, an unchallenged assumption that tech is fine for babies, great for parents, and the sooner you join the crowd, the happier you and your baby will be. An online blogger whose daughter is ten months old writes that, as parents, we've all wished at times that we had that "magic wand to make our babies happy" and apps for babies offer "simple, safe distractions for those times when you need it!"

He routinely uses his iPhone to entertain his baby girl as long as necessary, he says, reassuring readers that one particular game "offers up a fun, colorful educational delight." Fun, maybe, and colorful, no doubt. But educational? Safe? We know that from birth to two the brain is learning not just what to think but how to think, and we know that tech and TV pose risks for your baby's healthy development. With convenience and consumerism driving the conversation, empirical research can seem so remote, so out of step with what's popular.

Other reviewers and bloggers ramp up the conversation about "best apps" for babies and toddlers, hailing the smartphone as the smart parent's answer to the timeless need to entertain, educate, or pacify your child. Again, the claims ignore the science and the warnings from experts. When you filter out the flimsy advice, the flawed reasoning, and the manipulative language, what is left is an eerily honest pitch to make things easy on yourself and hook your child on screens as early as possible:

> If you have a small kid, the iPad bandwagon beckons . . .
> Touch screens were made for little kids, and the bigger the
> touch screen, the better.

> [My smartphone] gives me the ability to check in with
> e-mails, calendars, friends, and social networking sites without
> having to stop what I'm doing with my children.

> Many of these toys will do more than just suck those little
> brains in and get them glued to yet another screen. Choose
> right, and their favorite new tech toy could help teach them
> about math, science, physics, digital photography, computer

programming, or even motivate them to go outside and learn more about good ol' Mother Nature.

Not really. Your toddler isn't developmentally ready to understand content about math, science, and digital photography. And screen time—a mostly visual stimulus—fails to build the more robust neural configurations needed to do that later. As for "good ol' Mother Nature," the best way to learn more about her is to go outside and spend time with her.

Everywhere, the loose use of persuasive language covers sins of omission. Sure, a game may be "fun" or "entertaining" for a baby, but every minute they spend engaged that way is altering brain growth in ways that we know can be detrimental. "It's a culture of stimulation and that's the drug," says my colleague Ned Hallowell, a psychiatrist, noted authority, and author in the field of ADD and ADHD. "Stimulation has replaced connection, and I think that's what you need to watch out for."

An app is not a "safe distraction" like a stuffed animal or a musical mobile. An app is a stimulant, and overstimulating a baby's brain is not safe. Nor is an app educational in the way a parent would hope. I am hearing from parents of children two and a half and three years old—the age when children are old enough to benefit from high-quality educational toys—that having grown accustomed to touch-screen fun, they are less interested in those truly educational toys. They have already "learned"—to love audio, visual, and the fast-twitch stimulation of tech toys.

Few things could be more enticing than the prospect of flipping a switch to keep a baby happy. But when we give babies stimulants instead of calming attention and offer tech distractions from ordinary life instead of guidance through it, we teach them at a very young age to deal with life's up and downs by plugging into external sources to self-regulate rather than develop those skills within. It is so hard for ordinary, real-life academic research to compete with the magic-wand offerings of marketing wizards. Packaging on electronics for infants and toddlers should just say honestly: "This product may be hazardous to your baby's brain, to cognitive, social, and emotional development, and to early bonding, attachment, and attunement between parent and child." Until that day comes, you just have to say it to yourself—and spread the word. We owe it to our children to be as informed, sensible, and protective as possible.

Alice in Wonderland's Wake-Up Call

Alice came to see me when she realized that her careful choices to keep tech in check for her two young daughters had not kept her from tumbling down the rabbit hole into the "buy me" wonderland of the digital life.

In their peaceful rural Maine home Alice and her husband had diligently maintained a mostly screen-free existence for several years after their first daughter was born. Alice and little Meg spent happy unpressured hours outdoors. They folded laundry, pulled weeds, and pushed the vacuum cleaner around together. Reading, cooking, and arts and crafts were favorite indoor pastimes. When errands and playgroups took them into town, in the car they listened to music and sang or chatted. They had a computer, but Alice didn't use it much. She didn't have the time or interest.

Four years later, with the birth of her second daughter, Janie, Alice decided to get a smartphone. All her friends had one and e-mailed and texted one another throughout the day. Alice felt left out. She also imagined having the phone would help her organize and stay ahead of her to-do list, scheduling appointments and play-dates, shopping online, and keeping up with correspondence. Phone reception was spotty in their rural area and Alice soon found herself driving alternate routes on her way home, sometimes going twenty and thirty minutes out of her way so she'd never find herself without a signal. She went online more now at home, too, discovered parenting blogs and enjoyed the conversations there and on Facebook.

More open now to the fun and functionality of tech at home, Alice eased up on the whole screen-free thing for the girls, too. Meg, now five, often watched DVDs while her mom tended Janie. Depending on the chaos level of the day, Meg might watch one movie and listen to audio books or she might be given two movies and then play games on the Internet throughout the afternoon and evening. Propped on a blanket or buckled into her baby swing, Janie often watched with her while Alice did housework or went online. Sometimes when Alice played with Janie she also sneaked a peek at e-mail or Facebook, cruised online, or texted with her mom and sister who lived at a distance.

The days rolled along this way, the time with her girls and the time on screens merging into a seamless interaction between mother, daughters—and electronic devices. Then one day, at around eighteen months, Janie, in the midst of her toddler babbling, shouted what was clearly an effort at her first sentence: "My pone! My pone!" At first Alice couldn't decipher Janie's words, then her heart sank as she did. Janie screamed this every time Alice picked up her cell phone. Obviously she was trying to say "My phone! My phone!" Whether she wanted Alice to stop being distracted by the device or simply wanted to play with it wasn't clear and didn't matter. When Alice realized that this phrase would go down in their family history as Janie's first words, it was a turning point. She felt sad, guilty, and discouraged.

"I am an entirely different mother for Janie than I was for Meg as a baby," she tells me. "My attention now is divided not only between my two kids, but then it's cut in half again by my need-

ing to check my phone for e-mails or texts every three minutes."
She confesses that she'd come to think more routinely about when
she would be able to check her phone than to figure out what
game she might play that would be fun for both girls. "It's ridicu-
lous. I'm a stay-at-home mother," Alice exclaims, "and I barely
spend time with my kids."

"We'd come so far," she says of the earlier years of homespun
fun and outdoor play with Meg, "only for me to blow the whole
thing." Alice hadn't "blown the whole thing." The fact that she
is sitting in my office wanting to make a change was a clear sign
that this was not the case. The reality for us all is that we and our
children are together on the learning curve in the digital age, and
each developmental stage all along the way is going to bring new
challenges, different dilemmas, and opportunities for mistakes and
course corrections.

Alice was committed to making the shift and began by examining her own growing dependency on tech, how it was overpowering her and steering her choices for her girls. She could also see how
screen time had interrupted Meg's desire to do other things she had
begun to express an interest in earlier—read books on her own, play
with her sister, explore outside. Janie was plugged in far more than
her sister had ever been at that age and was losing out on just the
kind of discovery play and mommy-and-me time that Alice had once
guarded so fiercely for Meg. Among other things, Janie clearly had
developed a sense of competition with the phone for her mother's attention.

Janie's clarion call motivated Alice to refocus and reconnect
with what she intuitively knew: that the most important message
your infant or toddler needs to get is the message of "we" as strong
and powerful and what fundamentally matters. When they are
distressed they need to know they matter the most to you; when
they want to see themselves in your eyes and share the wonder

of discovery with you, they need to know you are their reliable partner.

New Pathways of the Heart and Mind

As hard as it is to get up night after night with your infant, I believe there is brilliant architecture in nature's design. As parents, our babies call us into a shared experience of solitude from the moment they arrive. They teach us how to find solace and comfort in dark hours as we comfort them. On a good night we do or, if that is not to be, we learn to stay the course and love them anyway. Whether in the nursery or the rocking chair, we create sanctuary with our babies. We make a sacred space. The moments and hours we spend in that close communion train the intuitive sense of mind and heart.

Those moments can also connect us to the "something larger," a sense of connection to mothers and fathers throughout time and to the spiritual dimension of parenting. This is an experience you cannot access online. Being with our babies has the enormous potential to help us as adults to rewire our busy brains and rediscover or develop new neural pathways that deepen our capacity for quiet reflection. Having a baby recalibrates your ability to commune with yourself in a new and satisfying way.

Nowadays when we gather in sanctuaries outside our home, instead of the officiant beginning with a spiritual welcome, so often they begin with the familiar invocation: "Would everybody please make sure their cell phones are off." We oblige because we understand that there are times and places that call for us to be fully present to the moment without the distraction of tech. We can let that be a cue for us as parents, too.

Let your baby's room be a screen-free room. Let the space between you be tech free. Read to your baby without interruption. Keep your eyes on your baby as she's crawling and climbing. When your left-brain multitasker says, "Just do it," let your right-brain relational self respond, "Just say no." Create rituals for you and your baby, and separate rituals for you and your screen. Preserve what your infant needs from you and give yourself the uninterrupted time to get your work done. Of course there will be times you will need to take a call, check an e-mail, or multitask. But the more we can do this mindfully and consistently for our children, the more likely we are to preserve the primacy of *we*.

Chapter 3

Mary Had a Little iPad

The Wisdom of Tradition, the Wonder of Tech: Ages Three to Five

*There is an inner life and an imaginal self that is develop-
ing in a child, and honoring that means not clouding it with
images that are prefab and pre-made. It means letting that
inner child's view evolve and be shaped through exploration
with real materials in the world that evokes the imaginative
mind of the child. It isn't that tech is necessarily a bad thing
for a child's mind, it's that you have a window of time in a
child's development where touch imagination, movement,
and language come together. There has to be time to
develop it.*

—JANICE TOBEN, AUTHOR AND SCHOOL CONSULTANT

Alissa is four and a half years old and delights in describing what she
likes to do when she gets home from school. "My favorite thing to do
is play dress-up," she tells me. "My favorite game is called, um, the
red carpet game."

When she plays it, which is most days after school, Alissa sits on
the couch or perhaps at the kitchen table, with the family iPad. She's
good at this game because she plays it a lot and her little index finger
darts expertly from space to space as she selects items and colors.

Tiara or barrette? *Tap!* Stiletto heels or peep-toe pumps? *Tap!* Pink or purple? Red belt or blue? Visor or sun hat? *Tap! Tap! Tap!* She also listens to the Tide or Claritin commercials that play before the game can begin. Like many of her preschool peers, she plays a variety of apps and screen games, often in the car driving home from school or while they wait to pick up a brother or sister from afterschool activities. Then some more at home for the afternoon.

Alissa says that playing dress-up is her favorite thing to do, but what she is doing is not playing dress-up. Playing dress-up means digging through the box for something you like; maybe you pick one shoe and then you have to find the other, but while you're looking you find a different one. All the while you're making up a story to go with it: you may start out as a princess, but then you find the wizard hat and suddenly you decide to become a sorcerer—if you can resist the pull of the shiny sheriff's badge and the fringed leather vest it's pinned onto. You might feel one fabric and then another, compare the texture and color, even smell them, to arrive at a conclusion about what you like and what you don't like; you might imagine yourself in one story or another and like or dislike the character or the story taking shape; you might not find that other matching shoe, and in that moment experience a pang of disappointment—and then, rather than give up, figure out an alternative to resolve your unhappiness. You might play dress-up with a friend and the two of you would trade shoes and tops and commentary, size up each other's costumes and ideas, pick up on one another's facial cues and body language expressing opinions and feelings about the play in progress, and eventually whether to keep playing or call it quits and move on to something else. You might argue and find out just how far you can boss a friend around before it spoils a good time. Maybe your mom or sibling would come in and ask what's up and you'd get to tell them the story unfolding from that box of dress-ups—or the falling out over it.

Handling things and fitting them together, changing in and out of clothing and imaginary scenarios, negotiating with a friend, parading around in your creation and imagining yourself as different than yourself, tattling to your mom and feeling heard, and having her talk with you about it: all those little movements and thought-filled moments developed your sense of self and your relationship with the things and people around you. All these iterations stretch creativity and develop imaginations. They deepen your capacity to learn and to keep learning, taking you on to new developmental challenges and growth.

"Children learn by touching and not just by passively clicking a mouse," says Nancy Schulman, director of the 92nd Street Y Nursery School, and a wise voice in the field of early childhood education. She describes the unique importance of the three-dimensional experience and how a child internalizes it very differently than the two-dimensional screen experience. "That's how children learn. They need to eat it, touch it, and move it, over and over again." The screen experience lacks the human interaction with other children and adults, the one-to-one response, eye contact, facial expression, voice, and tone. With screens and tech, Schulman's coauthor of *Practical Wisdom for Parents: Raising Self-Confident Children in the Preschool Years*, Ellen Birnbaum points out, "there's no room for language to expand or for children to deal with their feelings in that technology."

As a parent, if you can remember playing dress-up as a child, or perhaps having made a fort of couch cushions and blankets, then you know what a feeble substitute apps and screen games are for a child's imagination and the opportunity for genuine self-made play. Yet Alissa's language shows how easily and eerily the vocabulary of virtual experience and the virtual experiences themselves have become interchangeable for traditional play activities that involve so much more. The rich complexities of imagination and sensory, social, and emotional interactions of real dress-up go far beyond the simple hunt-and-tap experience of the digital environment. Alissa's favorite pastime is not dress-up; it is simply a screen game. She spends easily more than five hours a week on that click-and-play virtual reality.

The years from birth to five are often called the "magic years," reflecting the characteristic magical thinking of post-toddlerhood when children start to grapple with their irrational fears, fantasies, and fantastical interpretations of the world. Their growing language and logical reasoning powers are such new pathways in the brain that they exist alongside the magical thinking that makes an imagined monster under the bed as real as the family dog.

The progression from magical thinking to thinking that recognizes the more concrete "real world" marks a major developmental step of early childhood. It is through real-life experiences, specifically through loving interaction with parents and family, first experiences in friendship, and imaginative play with toys and each other that the mechanisms for development generate such dramatic growth during this time. These social and emotional experiences and the neurological pathways they cultivate create the engine for healthy development.

"The child's capacity for social interaction, higher-level thinking, problem-solving and logical thinking are largely set before they enter school," the child development researchers Stanley Greenspan and Stuart Shanker wrote in *The First Idea*, an exploration of how language and intelligence evolved to distinguish our species. They hold that "the foundation for this massive developmental process is the child's ability to be calmly focused and alert, or self-regulating" and that the years from birth to five are the first critical period for that learning. By age three, when children have the language to begin to express themselves and understand the world around them, this extraordinary period takes on even greater complexity and potential.

By age three today, children are immersed in the mainstream of media and tech available in their homes, their friends' homes, and nearly everywhere they go. They are physically able to manipulate it for play purposes; they can tap touch screens, push buttons on the TV remote, hit a "send" key, purchase apps, or happily experiment until they make something happen. As we consider the best way to introduce our preschoolers to TV, media, and tech, the questions we need to ask are not consumer questions—Mac or PC? Which brand or which game? Free app or upgrade? More important is the developmental question. These are the magic years for your child. Will you make it the magic of the playground or the magic of the iPad?

Subtle Changes Emerging in the Classic Picture of Preschoolers at Play

I walk toward the same kindergarten class that my children loved, sixteen and twenty years ago, and suddenly I'm in a time warp. It still looks like a friendly summer cottage, shingled, with budding bushes all around it, doors that open onto the large playground that is shared by the other cluster of classroom buildings. I'm sure it was updated in the past fifteen years, but it still has an old cozy feeling. It is the same room, the same light, warm, no computers, the same bins of tried-and-true toys and new ones I don't recognize. There is the fort structure, the dress-up area, the circle time space, and books all around the room at five-year-old reach, with the children's fanciful artwork hanging above. A room rich with simplicity. As I quietly observe this year's students interact with Ms. Jacobson I am once again filled with an overwhelming sense of gratitude that my children got to learn "everything they needed to know about life" from this sublimely wise teacher. At first glance Ms. Jacobson's classroom looks almost unchanged from the way it was when my children were here two decades ago.

There is the marble run set in a box in the corner. The assorted wood pieces can be assembled any way a child wishes, to make a stand-alone tower of chutes, ramps, and funnels through which they send a marble rolling down and out the makeshift maze. Surprisingly, no one has had it out all morning. It was such a coveted classroom toy in my children's era that the kids had a waitlist for who would get it next. Later Ms. Jacobson explains that the game has lost its fan base. "It takes a lot of precision and practice to make it work really well," she says. It requires a "stick-to-it kind of attention," a level of patience and greater tolerance for frustration than many newer toys. "We introduce it to every class in the same way, but I've noticed in the past few years it's not being used," she says.

Magna Tiles are the new thing. The brightly colored magnetized tiles make it easy for young hands to craft three-dimensional structures. The children love Magna Tiles and most have them at home, Ms. Jacobson says. Unlike the traditional wooden or plastic building blocks, when the magnetic tile structures fall down they are also very easy to rebuild. All a child need do is place one tile close to another and the pieces practically jump together, eliminating the need for more skilled placement and the frustration that comes with trial and error.

The room feels bright and cheery as ever, but the children's social activity seems less intensely purposeful than I remember. I see a cluster of girls talking to each other, hanging out in a way that makes me think of middle school girls clustering around their lockers. Ms. Jacobson says she sees less of the elaborate play scenarios that once dominated child's play for this age and more "just chillin'."

She nods toward a cluster nearby. "There are kids who sit and watch and talk and fiddle like this, and then there are those who are so eager to play and engage with learning," she says. She wonders about the fiddlers and what it means when children this age seem uninterested in initiating active, creative play. "Kids are always social, and they will always be curious about each other and that isn't going to change. But it takes practice to be able to sustain play, and I wonder if they're playing at home or after school in the same way that kids used to play, practicing at just playing," she says.

All these subtle changes over time present a troubling pattern. Children are showing up for kindergarten more savvy about Spider-Man and Strawberry Shortcake and mom's smartphone than versed in the ABCs of free play and social interaction that used to be assumed for the age. Meanwhile, preschool teachers report a similar flat spot in the social-emotional bell curve, with fewer three-, four-, and five-year-olds showing the traits of persistent learners and original thinkers eager to engage. More wait for direction from teachers or they use TV or screen game ideas or scripts to structure their play with other children. The level of creativity and originality in their play has diminished, too, reflecting less sophisticated thought processing and imagination.

"We see kids playing animals crashing into each other over and over, like video game and cartoon action, making simple structures rather than building more complex structures, which takes practice," Ms. Jacobson says.

Teachers will work to help a hesitant child engage in unstructured play and draw from within to find the confidence and creativity to do so. But they take care to avoid doing too much for the child, since the problem is already a symptom of adults and adult-organized activities and entertainment options having kept children from learning to do things for themselves. "As an adult you want to support kids, and don't want them to be uncomfortable in a harmful way, but you want to guide them through discomfort when it comes to problem solving and not solve the problem for them," Ms. Jacobson says.

Many young children do not have as deep reservoirs for managing the discomfort of learning, the teachers say. And learning is full of uncomfortable moments. In fact those are some of the most valuable teachable moments for us and for children themselves, when they persist through the discomfort to a resolution of it by their own action. They manage to tie the shoelaces. They zip the zipper. They find the puzzle piece. They figure out what they need to do next. However clumsy, this moment of discovery and the realization that they can manage discomfort and move through it is a big one in children's minds. This is a critical step in psychological development, and when they take charge of it, they own it. They need to experience the challenge, the struggle, the mistake, the discomfort, and the gain, again and again.

Compared to peers of a decade ago, preschool-age children today are more impulsive, less practiced at waiting their turn in play or discussion, and slower to pick up on the need for it in the group setting, teachers tell me. They have more difficulty making transitions from one activity to another; they make less eye contact with teachers and one another, and, perhaps related, they have more difficulty identifying emotions—their own or other children's. To manage our emotions, we need to be able to read them; to understand our impact on someone else, we need to be able to read theirs. This basic emotional literacy is essential for learning how to get along—why "plays well with others" is more than just a nicety on the preschool report card. A child who cannot do this by age five or six will have a steep learning curve upon starting elementary school. That learning curve is getting longer for children whose experience of face-to-face interactions and conversation with adults about feelings is shrinking.

New research on school readiness at five years old and on how to evaluate a preschool points heavily to the quality of relationships a child has in preschool as more important than what he or she learns there academically. Similarly, studies show that children who come into kindergarten with good social-emotional awareness are better liked by teachers and peers, more academically engaged, and more successful overall than children who come in skill-drilled for reading readiness but lacking the social and emotional awareness.

Some research also suggests that preschool children who are overly skill-drilled are more stressed out, and more passive and anxious learners.

Robin Shapiro, director of the LEAP (Language Enrichment Arts Program) Schools outside Boston, tells me that her school used to have computers in the classrooms for children to use with teacher supervision but it removed the screens—videos, handhelds, and TVs—several years ago. The school scrutinized the cost-benefit ratio in terms of preschool children's developmental needs and decided that screen time took away more than it contributed. Kids love the rewards of advancing levels in screen games, but the games fail to provide the kind of challenge that children need to advance levels in their own critical and creative thinking. "I think the foundation of all young children is developing strong positive attachments to the adults in their world—parents, teachers, caregivers, and siblings. Feeling safe, well cared for, and secure allows children to take risks in their learning at both home and school," she told me. Screen time doesn't do it.

Nor do the skills acquired—finger taps on two-dimensional visual images—contribute to the real-life integrated skills that enable a child to advance to more sophisticated physical movement and cognitive learning. As Greenspan and Shanker wrote, "Thinking means not just having a verbal or visual image, but being able to take that image, manipulate it, combine it with other images, and then organize these at different levels." Angry Birds, Baby Einstein, and screen games don't do that. The social and emotional toolkit your five-year-old has as she comes into kindergarten is much more important than whether she knows her way around a keypad or can recite her numbers or letters. Most important, that little toolkit comes from you, delivered through your everyday conversations, play, routine, and rituals. What media and tech do so much in the emotional lives of young children is to come between parent and child with those constant mini-moments of disconnect or distraction—on both sides. When that happens, our parenting signal weakens and media moves in with its own messages.

Media Mixes Signals for Learning Emotional Literacy

Barely out of diapers and able to pump and swing solo at the playground, our nursery and preschool-age children are already seasoned travelers in the media universe. By age three more than one-third of children have a TV in their bedroom and watch on average two hours a day. Including TV, they are on some screen for approximately two hours a day. Twenty-nine percent of children ages two to four use (presumably play with) a touch-screen device. Screens and electronic devices are routine accessories in home and carpool life, and "quiet time"—even with a favorite book—is as likely to include screen time as not for many children. Favorite TV characters and children's books have read-aloud apps with digital voice features, complete with occasional voice prompts to encourage your child to respond to the game or story—or to click "buy now" for an upgrade or accessory.

What does it mean to be three, four, or five years old in this stimulating milieu of family, media, and technology? Given what a child needs to learn about life at this young age, how do media and tech fit into that picture—or change it in ways that should be of concern to us? In these few short years, phenomenal growth is under way that will take your child from a wobbly toddler to a not-still-for-a-moment child, from first words of *no* to defiant declarations—*You're not the boss of me!* In everyday experiences your child is constantly taking in information—experiential data—and struggling to break it down into bite-size portions that she or he can digest developmentally.

In these years from three to five, with loving guidance your child progresses remarkably from being dominated by the glorious intensity of a robust little self—*I'm mad, I hate you!* and the biting, kicking, screaming, running away, chasing game—to a child who can modulate his impulses so he doesn't run away and can use his words instead of having a tantrum. These cues are the ABCs of emotional literacy, developing the emotional self-regulation that every child needs in order to grow into a well-adjusted individual more likely to enjoy success in school and life.

Neurologically during this period, the brain continues to build the vital scaffolding of sensorium experiences. It does so most richly through outdoor play, building, dancing, skipping, singing, coloring, conversing, imagining—all the hands-on real-life experiences that come so naturally to children and that allow them to learn from direct interaction with their environment and the people around them. This multisensory engagement stimulates the brain centers for language development, for cognitive processing and for deep thinking, and for social and emotional intelligence. Passively watching TV or using this precious time for early skill drilling or pressured academic preparation for preschool is the antithesis of what preschool-age children need and how they actually learn best. Most of all, it is the connection you create with your child through everyday caring and closeness that gives rise to all the other learning.

These are the years when you begin to teach your child what it means to be more fully human: that we have a wide range of emotions—a wonderful thing—but that if we let our emotions run wild then our behavior can hurt us or other people. Emotional intelligence begins when we learn to identify emotions—recognize when we are sad or angry or happy or disappointed—and develops as we learn to make sense of emotions and express them in healthy, productive ways.

A young child's emotional reactions to everything are immediate, strong, and often unfiltered, much like the cartoons and the screen characters they watch who act on raw emotions. In real life, your response filters experience for your child and invests it with meaning. Your child counts on you to bring order to emotionality. Your responses teach him how to relate to his own inner experience. Do you take mishaps in stride or treat them as disasters? It is in this reassuring give-and-take, the way you give meaning to moments and calibrate the emotional tone, that you teach your child the nuances of emotional self-regulation.

Managing emotions is about teaching children how to contain feelings and process them, calm themselves, and self-regulate when emotions run high. Sadness, anger, frustration, disappointment, and excitement: you want your child to feel able to experience the full range of human emotions and not feel confused or threatened by them. *Of course, you're sad to leave but it'll be okay. We'll see Grandma again.* They develop social awareness when you help them recognize how other people feel: *That's good sharing—see how happy your brother is when you share?* They learn relationship skills when you help them understand the connection between behavior and consequences: *Look what happened when you grabbed the shovel back.* And personal decision making: *How can you both play in the sandbox together? Let's think about good choices so everybody feels okay.*

Day by day, for your child that cueing becomes the voice of the inner ally, the voice of encouragement and resilience, the sense that says, *I'm cared for as I struggle to learn.* In those millions of moments in the life of a toddler and preschool child, whether the cues come from mom, dad, grandparent, day-care provider, or schoolteacher, a child experiences the sensations of safety, of being known and cared for and protected by the grown-ups. This becomes part of his inner well of emotional reserve and emotional intelligence.

The natural curiosity and imagination of this age, children's developing capacity for language, their desire to play and make friends, and their interest in objects that interest us also make them eager to engage with any media or tech within reach. The infant who delighted in exploring your smartphone by gumming it, by age three has developed more sophisticated new skills for exploration and wants to use them. To be able to tap a touch screen and make funny things happen is powerful stuff! To "play computer" like your big sister or dad is fun and makes you feel big.

Four-year-old Karl enjoys frequent visits with his grandmother in Finland—via Skype—and this New York boy is growing up bilingual, speaking Finnish. He is ecstatic when Malle comes to visit in person once a year, and their screen time in between is imbued with the loving relationship they share. When his family visits Finland, he plays readily with his cousins, all the easier because they are so familiar to him from Skyping.

At its best, tech can reinforce what it means to be family, watching shows or going online together at home, as well as in the sense, as with Karl, that we are family and we stay in touch wherever we are. There is nothing like doting grandparents or an aunt or older cousin showing up on the screen and just lovin' you up. I know many parents who, when they travel and leave the kids at home, use Skype or other video chat platforms to read to them at bedtime or make a screen appearance at the breakfast table to touch base before the school day begins. Tech is fantastic for this.

Developmentally appropriate educational programs and content are good, too, but more often than not they are accompanied by junk values from advertisers who aggressively target preschool-age children with cross-marketing schemes to boost sales. According to a CBS report, in 2009 companies spent nearly $17 billion marketing to kids, increasingly targeting tweens and younger children. Advertisers target the very young because research shows that, beginning at age two, a child can recognize a brand. And of course your child is the most direct conduit to you and your deep pockets—more than $150 billion in parent spending by one recent estimate. Because many educational toys and web sites—schools, too, increasingly—have commercial backing, the branding and messages are pervasive. Research shows that children this age can't differentiate between the motives behind commercial and noncommercial content. So Mr. Rogers's neighbors and SpongeBob's sponsors have the same credibility in your four-year-old's mind.

Along with that commercial world and much of the program content for young viewers come early gender messages about what it means to be a little boy or a little girl. At three, four, and five, children are just learning what it means to be human in the world—how animal babies are so different from us, or how people celebrate different holidays around the world—and they are learning what our culture expects of boys or girls. Their notions of gender are not fixed at this age, but what they are learning about gender is powerful. It might seem extreme to describe Rugrats or Power Rangers as proxies for the harsh, cynical, misogynistic adult culture that is so harmful to boys and girls. But those characters and plotlines make up the starter set of stereotypes that begin to train children away from the full range of human potential by steering them toward limited, and often exploitive, gender roles.

It's so easy to think entertainment is educational—or that it's "just kid stuff." But that kid stuff is teaching kids lots of stuff. So whatever enlightened values your family may have about men's and women's roles, media and screen games intervene with their own version of role modeling. There, males dominate through power and violence and females tend to be powerless and are portrayed as sexualized objects.

Common media fare and screen games portray a kiddified model of masculinity that places a premium on aggressive, impulsive behavior and a no-tears policy. It teaches our sons that it's not okay to feel sad and if somebody hits, you hit back harder; that when you want something it's okay to use force to take it; and that when things don't go your way, you can bully people to get your way. Research suggests that as few as nine minutes of SpongeBob SquarePants can lead to aggressive behavior after you turn the TV off.

The mainstream media messages and screen games for preschool girls so often come packaged in pink—figuratively, if not actually—and though it's not a bad color, the standard-issue stereotypes teach our daughters that looks are a girl's best friend and dressing for cute is a girl's top priority. Strong, smart, assertive girl characters are out there, but while Dora the Explorer may go adventuring with trail gear, in her special birthday pack she's toting nail polish and lip gloss. *Rugrats'* Angelica, the quintessential mean girl, who is just a three-year-old herself, offers that cynical alternative; she may come off as satire to older viewers, but a child of three, four, or five doesn't understand satire. Angelica and Mr. Rogers are equally credible characters in a preschool-age viewer's mind.

Like the storybook Hungry Caterpillar who consumes every edible thing it encounters, children of three, four, or five are voracious learners. Hungry for experience and novelty, they are naturally attracted to media and small-screen gadgets. Once they get a pleasurable taste, they tend to want more. Repetition and positive reinforcement are, after all, the way young children learn anything, from coloring with crayons to potty training. The neural map is always open to something new and exciting. This is how, neurologically, play turns into pathways and pathways into preferences for more of the same. The brain patterns itself after the "environmental input" it receives, be it cuddling or computer games. Tech can quickly establish itself as preferred territory in the young developing brain and come to dominate it at the expense of other essential but slower-growing connections that involve the complexities of thought, emotional signaling, and the distinctly human rhythmic back-and-forth of communication.

Protecting Our Zones of Interaction
with Our Young Children

The afternoon pickup scene outside almost any preschool is so familiar that it hardly seems remarkable, unless you compare it to the same scene from a decade ago. Today as the car-pool conga line forms, many parents are so busy texting or talking on their cell phones and tablets that they don't even see their child come out of the building or across the lawn to the waiting car.

Teachers running the car-pool line have to shout out parents' names to signal that it is time to scoot the car forward. Or a teacher will have to tap hard on the window to alert the tech-focused parent to unlock the door. I watched a three-year-old boy bang on the passenger door of his mother's van as his mom appeared completely lost in thought with a call on her phone. Her gaze was fixed toward the distant horizon, not greeting and beaming at her child three feet away. A teacher later told me that the scene is "the new norm" to the point that tech-distracted parents have become oblivious to children's safety at clearly marked crosswalks. Once children are buckled in, they often plug in as well, to handheld screens and digital games or any available smartphone. Because tech is so present and integral in our lives, our relationship with it can become invisible to us. But the effects of our distractedness, its role in our parenting, and its presence in our parent-child zones of interaction have become increasingly visible.

In my interviews and focus groups with children as young as four years old, kids have told me how disheartening it is to have to vie for their parents' attention and often come in second. They describe feelings of isolation, loneliness, anger, and sadness around the waiting game. Or a parent's routine multitasking through bedtimes, mealtimes, and playtimes that were promised as "mommy and me" or "daddy and me." Now tech makes three, and most of the children couldn't remember a time when it wasn't that way at home.

Alex, the father of a three-year-old and two older children, quit his executive-level job for less pay and less pressure when he realized that the continuous multitasking demands of his job made uninterrupted time with his children impossible.

> I could look them in the eye and have a conversation, but I realized that I was not having that conversation with my whole head. It was like 2 percent of it because I was thinking about the next e-mail coming in or the other things I needed to do, and was so addicted . . . I was 0 percent present, really. I could see it in their eyes. They knew I wasn't really there with them. It was awful, but I got so used to it. I'd say, "Oh, I'm sorry" and I'd take a call or get up midgame to check my e-mail.

He thought hard about it and concluded that being emotionally present for his children was "a parental imperative." Once the aha moment hit, he couldn't pretend that it hadn't and he reprioritized.

These are hard parenting years because the demands are high and so are the stakes. Young children can be frustrating. Potty training takes time; dressing kids in snowsuits and boots or strapping them into car seats takes patience, especially when dealing with kids who have sensory or attentional issues. If you are rushing around tending to all things all the time, you may be a multitasking maven, but eventually something's got to give. And it is almost always the kids. As a parent, it is hard enough to be patient and it is very hard to remember who you are in the life of this little person as they interrupt you in the middle of what you're doing. So they tug on our sleeve, whine a little, cry, or storm around, and at some point, if we don't think to calm ourselves and resist the impulse, we just think, *Fix it. What is the quickest thing I can do to make it stop so I can get back to what I was doing?* We put them in front of the TV or a touch screen or some other electronic diversion, so we can return to more demanding or entertaining things, often on screens ourselves. In the moment, our mutual distraction seems to solve the problem. Instead, the pattern is setting us up for more. Are there times when this is the inevitable solution? Of course. If you are taking care of a sick child or a special commitment requires your focused attention temporarily, then all bets are off. The danger is not the once in a while but creating a pattern where the quick fix is the default position.

LUKE: ATTENTION PROBLEMS AT SCHOOL
REQUIRE ATTENTION CLOSER TO HOME

Susan, forty-three and the mother of two young children, comes to me in a panic after her five-year-old son Luke's prekindergarten teacher suggests that he is showing early signs of attention deficit and hyperactivity disorder (ADHD) in school and might need to be medicated. Susan is polished, poised, and articulate, thorough and well organized as she outlines the details. She explains that her husband works long hours in a very demanding job and they had agreed to somewhat traditional family roles—that she would be the primary parent. In addition to her part-time duties as a departmental administrator at a bank and her responsibilities as a wife and mother, she was the primary support for her aging mother and her father, who had Parkinson's disease.

She had been struggling with Luke's impulsive, annoying behavior a bit at home, but the call from his teacher and the suggestion that he might need to be medicated prompted her quick follow-up. In our first visit she lists for me the examples of her son's behavior that has led the teacher to make this recommendation: Luke has trouble listening to and following instructions. He is frequently unable to sit still and often interruptive. He has great difficulty letting his mother leave the classroom when she drops him off in the mornings and even more difficulty transitioning after she leaves. As Susan describes her son with affection, I hear her longing to understand her son and her pain at not knowing how to be a more effective mother. I ask her to describe a typical day for her and her family.

A routine day for Susan is a nonstop juggling act—taking Luke to and from school, getting Anna to day care and herself to work, and researching health issues for her father. All this while managing the many logistical details of her own life, her father's decline, and her mother's struggle to care for him at home. The medical issues are nonstop and she is the person in charge of everything. As a result, she often finds herself multitasking while she is with her kids. She spends time online while they get dressed in the morning. She texts her husband or the night nurse while walking Luke to school. She catches up on e-mails while running errands with the kids in the late afternoon.

Additionally, Susan has come to rely on TV cartoons and video/screen entertainment or handheld tech games to occupy Luke and Anna when she needs to cook or clean up at home. She frequently bribes them, bypassing conversation in favor of moving them quickly from one moment to the next: "If you're quiet in the car while I'm on the phone," she'll tell them, "I'll let you play a video game when we get home." Or watch cartoons. Or play Club Penguin. Everything is a negotiation and most of them are about watching TV or movies or playing on the computer. There is a never-ending stream of distractions on offer but almost no time for a simple conversation.

It wasn't until she took the time to describe her daily routine that she realized the only time she regularly spent with her children untethered from tech was when she gave them a bath in the evening. Afterward, her husband, who had been at the office all day, would read to the kids and help put them to bed. Susan's life had become an overwhelming blur of obligations and technological reliance.

Given these family circumstances and some other factors, my hunch is that Luke might not have ADHD; he might simply need more focused time with his mother, more creative play, and much less time in front of a screen. The fact that he is having trouble letting her go in the mornings at school makes me wonder if he is expressing a sense of frustration at not being able to get her attention long before she says good-bye.

Susan has always been "the organizer" in her family and at work and she tends to bring any conversation about her mothering style back to practical aspects of organizing life to "keep everything together." She feels effective dealing with the tasks required to care for her father—they are fact based and results oriented—and it is understandable why she prefers to talk about things that make her feel competent. Luke isn't one of them.

She feels out of her league with Luke's problems. She has lost confidence in her own maternal instincts, and the hurried, fragmented quality of time she spends with Luke do little to cultivate the deeper emotional connection and intuitive sense. It tends to be stressful; she rarely enjoys the kind of relaxed, pleasurable playtime together, rich in the back-and-forth nuances of emotional dialogue that helps a parent develop an intuitive sense about her child. Susan's own parents had been strict with her as a child, with a more emotionally distant and formal approach to raising children. She knows that she wants to be closer and more involved in her children's lives than her mother was in hers but she doesn't know how to create that closeness. And she was unprepared for the hard early years of parenting when children are so uneven and unpredictable, the tasks are so repetitive, and each day requires endless patience and attentive guidance of the children.

We speak first about changes she might make in her schedule. As tough as it might be, this appeals to her sense of practicality, especially if it would help calm some of Luke's stormy emotional behavior. She decides to wake up twenty to thirty minutes earlier to start her day. She initially cringed at the idea, understandably, but she is willing to try using that time while the children are still sleeping to take care of her online morning tasks—it might be research, e-mails, arranging a grocery delivery, anything that requires she be on the computer. If she needs more time, then she gets up even earlier. At 7 a.m., she wakes the kids and helps them get ready for their day, giving them her undivided attention and affection while doing so. She makes a point of focusing this way with them at other times, too. Not always, but often.

Luke and Anna reveled in Susan's new attentiveness; they felt the embrace of her undivided attention and of her more frequent and affectionate hugs. Susan also felt herself blossoming in her interactions with Luke and Anna. When they were together she practiced a more attentive kind of listening and responding to them.

How did you build that very tall tower? You used so many blue blocks this time. I can tell you worked really hard on that!

She learned how to understand and accept her children in the moment as little works in progress who needed her help:

You're angry and upset that Anna ran into your tower. I know that's hard but it's not okay to hit her. When you hit her, she is sad and gets scared of you.

She worked on new ways of being patient with Luke's desires to be independent, his frustrations, and to help him find solutions:

I see you're helping Anna now—you're being such a good big brother!

She was able to maintain a more accepting presence with him, knowing that she had checked in with her other commitments online earlier and would reconnect with them after drop-off. She fought back her desire for sneak peeks at e-mail or the urge to text herself with reminders. She now saw this compulsive checking as unhealthy, a red flag of addictive behavior.

Before long, Luke was slipping his hand into his mother's as they walked to school and was chattering away about the things they saw as they strolled or what he wanted to do when he got home that afternoon. As these positive developments occurred, Susan became more creative in pursuing ways to help Luke feel more settled at school and at home. She hated that she had become so dependent on bribing him when she needed him to cooperate. She saw that she would interrupt his meltdowns and rush in with an offer of screen time to quiet his upset, rather than try to understand, teach, and redirect him and help him problem-solve. We worked on helping her learn how to talk with Luke about his feelings, learn how to calm himself down, and express himself, especially his strong feelings, in appropriate ways.

Susan still felt that Luke was spending too much time on screens and that she was enabling it by giving in to his whining and letting him plug in to TV and video games. She learned to refrain from offering them so consistently as a diversion. We made a reminder list of four or five different nonscreen activities, such as drawing, building with Legos, playing in their small yard, or playing a single-player memory game with Susan nearby. At first, Susan kept this list on an index card that she carried with her everywhere she went, as a reminder of how to keep from giving in to whining. Eventually she didn't need it in hand anymore. But she kept it posted on the refrigerator door at eye level as a reminder to herself that healthier options exist. Sometimes parents and children alike need a nudge to make these choices in our relationship with tech.

At the same time, Susan committed to scaling back her own tech habit throughout the day. She set aside a small pocket of time in her schedule to make calls and respond to e-mails just before picking Luke up at school. Then she would stay off-line and stay present to Luke from the moment she greeted him—and for at least forty-five minutes afterward. This allowed them an undistracted transition when Luke could have a snack, talk to her about his day, unpack his backpack, and ease into family and home time. On some days, Susan was able to play longer with both Luke and Anna.

Sometimes the needs of the day intervened and occasionally she would allow them to watch a movie or play a game on a web site. But she was more thoughtful about what they could watch and when, because Luke took them as a cue to beg for junk cereal and advertised toys. She eliminated *Power Rangers* and *SpongeBob SquarePants* and fast-food-sponsored cartoons and limited their TV diet to *Sesame Street, Blue's Clues, Mr. Rogers,* and other research-vetted recommended programs that help children build a vocabulary for their emotions. She learned to be more discriminating in selecting other programs, attuned now to the type of social-emotional learning (SEL) they promoted. She chose programs with story lines and characters that model the kind of social and emotional skills a child needs to develop, especially around anger and negative feelings. Luke wasn't as agitated after watching them, the most welcome evidence that the changes were making a difference. Now she recognized how many of the entertaining games are not developmentally oriented and do not address these essential life skills and she avoided them.

When Luke would get angry, Susan worked at engaging empathetically rather than disengaging or getting angry back. She tried harder to treat his upsets like a passing storm, not a permanent character flaw. Those stormy moments are our opportunity to help children safely experience anger and learn from it. As tempting as it is for us to want to dismiss our children's anger, to get angry at them for being angry, or to try to distract them, children learn how to handle anger and disappointment and frustration best in relationships with grown-ups who are clearly not angry at them for feeling angry. They need to know that you love them when they are angry, that you will stay with them when they are frustrated, and that you will not get frustrated with them. They need to see that you can hold their anger and help them hold it and resolve it by soothing them and empathizing with them. Good educational TV and other media can teach and reinforce those messages, but our children need to learn these first and foremost from the people closest to them.

Finally, Susan took steps to shift away from the urgent pull of tech for herself and her family. In addition to structuring her computer time, she reconfigured her relationship with her smartphone, recognizing when it was a necessary tool and when it was a distraction. She instilled good rituals around bedtime so Luke and Anna got enough sleep and she created more opportunities for socializing as a family. The climate at home shifted. It was easier, happier, more connected. Not perfect; it never is. But Susan had a newfound confidence in herself as a mother and now had strategies in place to balance life's demands while staying connected to her children in the ways she hoped she could be.

Luke continued to make significant improvements. He was no longer as clingy when she dropped him off at school in the mornings; his teachers noticed that he shifted into class time more smoothly. His teacher was stunned by his progress and backtracked from her earlier suggestion that he might be struggling with ADHD.

Some children do have attention disorders that are not fully resolved by the types of changes that Susan and Luke made. A proper diagnosis of ADD/ADHD meets clear clinical criteria, and my own bias is that a comprehensive psychological evaluation is enormously helpful and should be part of that workup. I have worked with children with ADD/ADHD and for many, proper medication makes all the difference in their ability to live up to their potential and thrive in school. But given the dramatic rise in the diagnosis of ADD/ADHD in recent years and growing concerns about the overmedication of children, we need to be more discerning. Given what we already know from research about the negative effects of media and screen play on children's self-regulation, attention, aggressive behaviors, sleep, and play patterns, we cannot ignore the possible link.

Luke's "attention problem" improved once Susan made changes at home and thought carefully about the quality of their time together. Whether a child is already exhibiting signs of attentional problems or you want to do everything possible to eliminate risk factors for him, there are things you can do. The first step is to slow down and really study your child to more fully understand his or her individual makeup—the unique rhythms, receptivity, and sensitivities to surroundings, personal interactions—to understand the child's internal experience of emotions or physical sensations. Every child is different, and you will find some ways to tailor a child's day or home space or your interaction style to accommodate those needs, while family life will require compromise in other ways. These are the opportunities to help your child learn how to do that—learn how to self-regulate and do the inner work needed to adjust to circumstances in healthy ways, practicing making choices that support his or her well-being. Whatever individual needs your child may have, a supportive home environment generally is one in which we limit screen time, provide ample physical activity (from casual solo play to team sports), offer plenty of time for social activities (hobbies and personal interests), repeatedly reinforce self-management, especially around boring everyday routines (make your bed, brush your teeth, clear your dish), and coach the child on social skills (maintain eye contact, no interrupting). Without a conscious effort to create family rituals, personal and practical routines, creative outlets, and opportunity for healthy play, tech can easily displace family relationships with experiences that overstimulate and undermine healthy development for a young child. When we know from stories like Luke and Susan's that shifts at home can make such a profound difference, we owe it to our children to try.

When Our Tech Habits Say
I Need You to Stop Needing Me

In order for us to be the best parent we want to be for our child—engaged and fully present for back-and-forth emotional communication and focused and playful interactions and explorations—our tech habits are just as important as our child's. If we are constantly disappearing into our screens and absenting ourselves from the flow of dialogue with our child, as parents, we are playing hooky.

In my clinical practice and in conversations with parents of young children, this is a clear and growing problem. Tech makes us get sloppy about our child's needs at this age for good routines around play, rest, meals, and bedtime. So many parents explain their child's bad day or problem behavior as part of a larger picture that—when they take a moment to reconstruct it objectively—includes putting the child's basic needs aside to accommodate a parent's evening e-mailing or TV time. The daunting task for parents, then, includes not only learning to regulate their own emotional responses to a child of this age, but to stay organized and in control of their attachment to tech and screens and their real-world work requirements. Since there are fewer boundaries between work and family time, those competing demands now penetrate bath and dinner time, even bedtime. The child's competition for a parent's attention and the resulting tension for both are damaging because at this age children magnetize to parental frustration and anxiety, anger, and inattention in a way that resonates powerfully in a child's psyche.

At three, four, or five years old a child cannot say to herself or understand that *oh, Mommy has to finish her report because her boss is in a different time zone.* They just get irritable and cranky when you put them on hold with your explanations. All they hear is (to borrow from my favorite literary canine Martha, from the children's book *Martha Speaks*) *blah-blah-blah.* And that is at best. Perhaps you say *two more minutes* in a calm voice first, then several minutes later, *I told you—two more minutes* in a tense voice. Sometimes it is *can you just leave me alone?* Or *I really need you to stop crying.* Or *I really need you two to get along* now. Or *You will lose TV if you don't stop interrupting me!*

We all say these things from time to time. But as a consistent feedback, the child of three or four or five hears: *I need you to remove yourself. I need you to stop having the feelings you're having. I need you to stop needing me. I need you to disappear yourself. I don't have room in my heart or mind for you right now; you are an annoyance. Please go away.*

As parents, we must find the social and emotional head space to calm ourselves and react to all of these minimoments in nuanced ways. In our everyday interactions with our children, from doing errands to playtime to bedtime, we are teaching the same thing repeatedly—self-regulation and social and emotional skills—but we have to be able to focus on these tasks with genuine interest and presence.

Protecting the Magic of "the Magic Years"

We know that children ages three to five integrate everything they are learning from life through their imaginative play. They play doctor, they play house, they play heroes and villains. They also begin to understand that life isn't always easy; it can be sad and scary and hard. There's illness, there's shots. There's death. All sorts of stuff happens, but when left to their own imaginations to understand these concepts, children are very good at pacing themselves and taking in as much information as they can handle, gradually approaching a deepening understanding in a way that doesn't crush, stifle, or harm them. When we expose children at too young an age to overwhelming visual stimulation and adult definitions of good and evil, they are outpaced psychologically and have difficulty processing the experience.

Selma Fraiberg in *The Magic Years* pointed out that the imaginary crocodile under the bed is a different beast for the five-year-old child than it is for the eight-year-old child. She was referring to the way a child of any age naturally has of imagining precisely what is manageable for him in the moment, given his individual and unique developmental readiness. We don't always know when something our child sees on the screen is overwhelming. This is true as well for background television grown-ups may be watching. Young children in particular often do not have the language to explain their upset or fears to us. We see them expressed in problems around sleep, attention, irritability, aggression, or anxiety. At times, actions speak louder than words.

JACK: A BOY, A VILLAIN, AND A SUPERHERO

My friend Cassie, a teacher and mother of two, is furious and near tears when she calls. "I can't believe it!" she rants about her four-year-old son, Jack. Jack had just thrown her iPhone into their brand-new, major-splurge, $3,000 flat-screen TV. Money was tight and she was undone.

"Why would he do this?!" she wails. "He knows better than that."

I know Jack. He has a vivid imagination, is expressive and wonderfully affectionate, loves to be read to, and has a robust sense of his little self in the world. He is not normally a destructive fellow. Cassie explains that Jack had been watching a Disney video, *101 Dalmatians*, for the first time. As he watched his movie on the giant screen, he was holding his mother's new iPhone, as she sometimes lets him do. When the evil Cruella de Vil made her grand entrance, Jack flung the phone into the TV, breaking both in a nanosecond.

Understandably, Cassie's upset was over the shattered TV and iPhone. But Jack's upset had nothing to do with the phone; had he been holding a teddy bear or a juice box, he would have thrown that. Overwhelmed by what he saw and developmentally unprepared to manage those feelings, Jack's behavior had all the hallmarks of a young child's magical thinking. Many young children find Cruella de Vil frightening, especially on a TV screen where, unlike in a book, she seems to come suddenly and terrifyingly alive.

The page and the screen are two very different experiences, especially for very young children. Michael Rich at the Center on Media and Child Health explains that we can't compare fairy tales on the page to screen action where violence or threatening characters come to life. "A kid only imagines what his or her life experience allows, [and] watching violence is different from imagining violence," he says. Just so with wicked witches and the very young child.

Cruella on the page is static. She doesn't have a voice; it is Mommy's voice in that leading role and Mommy knows just how to modulate the right amount of scariness. Moms and dads and favorite babysitters are always intuitively editing the scary stuff or tempering it with the tone of their voice or silly asides. If, as research tells us, infants are capable of intuiting the intention of the adults around them, then imagine the vibes a three- or four-year-old might pick up seeing Cruella bigger than life—*right there in your own living room.* Or so it may have seemed to Jack.

And that is the point. I imagine that the image of Cruella on the screen for Jack was much bigger, scarier, and more threatening than his own imagination had gotten to in its capacity to envision, and thus confront evil. In that instant of sensory overload, the sound and the sight of her triggered the full-body reaction of a terrified four-year-old. Not a seven-year-old reaction: *That's scary so I'm gonna turn the TV off* or *That's scary but I know it's not real.* Not even a five-year-old's, *That's scary, I'm gonna hide my face in my hands till this part is over.* At just four, a frightened Jack was old enough to understand that Cruella was very bad, but too young to have the understanding and self-control not to throw the phone.

We cannot know exactly what was going through Jack's mind when he reacted so wildly, but his was not the thoughtless act of a misbehaving child. Research suggests that we should be concerned about the impact of tech in areas of our children's development that we don't yet understand, that we should be cautious about potential negative effects between tech and the young developing brain. As we begin to explore the wonders of media and tech with our three- and four-year-olds, we need to remind ourselves to let a game of dress-up *be* a real game of dress-up and let our children make their *own* magic.

Chapter 4

Fast-Forward Childhood

When to Push Pause, Delete,
and Play: Ages Six to Ten

Kids are so obsessed with sitting inside and playing with their iPod Touch and it's so useless. I watch my cousin, who's six, and she sits on the couch and plays Scooby Doo with her friends on the iPad, and I'm like, jeez, when I was six years old I was figuring out how to tie teddy bears to gate posts or flinging them over the banisters. I just think of all the fun things my sisters and I would do, all these fun memories that I have and my cousins won't because they are sitting on the couch and video-chatting. My cousins aren't having any childhood.

—SUZANNAH, AGE THIRTEEN

Trevor was on spring break, enjoying all the things a ten-year-old boy does, when he checked his e-mail and found an odd one waiting for him. It was from nurfmadnessXYZ.

"I just thought it was random—you know, spam—or a joke," he says. "I didn't tell anybody. I just deleted it right away."

About a month went by and he had forgotten all about it, when two more e-mails popped up one day from the same address. This time nurfmadnessXYZ made sexually explicit taunts about his genitalia along with calling him "a fucking asshole."

"That's when I started to believe that someone was trying to make me feel as bad as they could," Trevor says. "I got really freaked out. I didn't know what to do. I mean, the first time I didn't respond because I thought it was random and that was what everybody said to do. But now it was persistent. I started pacing my room trying to think of what to do. For about five minutes I was just trying to pull myself together, and then I thought I should really tell my parents."

His parents shared the e-mails with the school principal. They tapped the student grapevine and within a week found nurfmadnessXYZ, a ten-year-old girl who had been nursing a grudge against Trevor for more than a year since he had openly ridiculed her loose grammar on the bus a few times. She had talked to friends about "horrible things to do to him," her friends reported later, but none of them ever thought she would do anything. What she did do eventually was sign on to a friend's e-mail account to send the sexually harassing e-mails.

"When I found out I sort of broke down, like in a movie when the main character just breaks down and everything is in slow motion, that's what it felt like to me," Trevor tells me the day he comes with his mother for our first session. He is struggling with anger and depression.

It has been a year since that last e-mail and the whodunit disclosure that followed, which became the subject of hushed conversations among students and the school community. Secrets are hard to keep at school and in an online universe. The girl was suspended for several days and when she returned she was told to steer clear of Trevor. But her clique of friends rallied around her and school for her is fundamentally unaltered. Trevor, on the other hand, continues to suffer. He has been alternately angry, uneasy, and unfocused at school since it happened. His good grades have dropped; he used to love going to school, now he hates it. Every time he checks his e-mail, he looks first for the name he doesn't want to see there with a haunting sense of *ick*.

For Trevor, that the girl keeps her distance at school doesn't change the fact that for all he knows, the nasty conversation about him may continue forever in cyberspace. Sexual harassment via e-mail would be disturbing to most adults. It is nothing Trevor or his parents were prepared for in the life of a ten-year-old child.

In an elementary school a few thousand miles away, the bright side of tech and its interface with young schoolchildren offers a more reassuring view of schoolchildren and screens. Two dozen second-graders are gathered for story time. They've scooted their chairs into strategic position with all eyes on the wall-mounted whiteboard as their teacher, Ms. Davis, prepares to read a story aloud. Her calm, welcoming demeanor and thoughtful responses to each child are very much in the Mr. Rogers tradition.

The students are always offered a choice of the old-fashioned cozy corner or the whiteboard setting for story time; they often choose the screen. They clearly love both, however, and at free time they swarm the shelves for their favorite books just as generations before them did. Unlike most of their parents at this age, they each have a personal online library account on which they keep track of the books they have checked out and the due dates; they also use it to access online resources. Later in the day, fourth-graders take these seats for a lesson in finding and citing research sources. After a brief discussion of online how-to, including the meaning of copyright and how to avoid plagiarism, they move to the nearby bank of computers and log on to continue their work for an assignment in progress.

As at most schools that have a strong media and tech curriculum, the program promotes reading and literature, research skills, and smart, responsible use of tech by students in kindergarten through fifth grade. Over the course of their elementary school education, they will learn how to use computers, the Internet, and aspects of social media (blogging, for instance) for schoolwork and everyday purposes. They'll be taught what it means to be good digital citizens and they'll review standards of behavior to make sure that their tech world is a safe and civil place for them and their friends. Many of them are already familiar with tech from recreational use at home. Few have discussed digital citizenship and netiquette with their parents.

In the calm, structured space of the media center and the measured pace of its developmentally based curriculum, children's interaction with tech provides an optimal orientation at every grade level. But as Trevor's experience and those of so many children show, beyond the carefully managed environment that schools take pains to provide, the old information highway has become a fast lane into broadening communities of peers and, inevitably, the vast online world. At times, they find themselves in difficult situations that weren't previously part of life for children in grade school—and many that shouldn't be. Inexperience is a problem. So is childish thinking and the ordinary recklessness that goes with it. Parents feel hard-pressed to get up to speed in new ways as gatekeepers, screen monitors, tech support, and cyberlife referees, in addition to the just plain human side of parenting.

Traditionally and by nature, so much of child development at this age is driven by children's desire to connect with each other and the exciting world beyond; to fit in, stand out, look older, smarter, cooler. If the computer has become the new playground for our children, then we must ask what they are playing, who they are meeting there, and what they are learning. It is certainly a faster crowd online than ever graced a neighborhood playground. For all the good they can find there, other influences, from screen games and commercial pop-ups to YouTube, social media, and online erotica, introduce them to images and information they are not developmentally equipped to understand. The combination of their innate eagerness to mimic what's cool, and the R- to X-rated quality of the cool they see, has collapsed childhood to the point that we see second-graders mimicking sexy teens and fourth-graders hanging out with online "friends" and gamers far older and more worldly. Life for six- to ten-year-olds has taken on a pseudosophisticated zeitgeist far beyond the normal developmental readiness of the age. Outwardly, that compression may seem superficial, but inwardly, many children experience a suffocating squeeze on developmental growth that is essential for these early school years.

There's Nothing Elementary about It

When parents, teachers, and school administrators call to talk with me about circumstances that have gotten out of hand or that threaten to, they often begin by explaining that these are good kids, good schools, good and caring parents. "But just last week . . . " they say before launching into stories of troubling new behavior and attitudes among these young students that have everyone struggling to understand how it could have happened and what to do now that it has.

"We had to expel a seven-year-old boy for threatening to kill a girl's father," a principal says. "This boy has used violent language and imagery that clearly comes from overexposure to TV or movies and had a psychological intensity that is beyond what our school culture is comfortable with."

"The buses have always been tricky, but now they seem out of control," an administrator at another school tells me. "A fifth-grader showed a second-grader an obscene YouTube video on a phone and thought it was funny. Neither child should have seen it."

"My fourth-grade daughter got an e-mail from a boy in her grade who said, 'I think you're hot,'" a mother tells me. "She told him 'I don't like you saying this.' I helped her write an e-mail to him trying to explain why it made her so uncomfortable. But now she feels like she doesn't want to be his friend. She doesn't feel safe with him. He

never would have said that to her in person, and reading it felt so jar-
ring to her."

Once upon a time, the canon for childhood was fairly simple, particularly at this age: do your schoolwork, play fair, obey your parents, and honor your family. This was the age of friendship songs, a
budding sense of justice, and knowing right from wrong. Of taking
pride in practicing values like respect, kindness, and sharing because
they represented maturity. Childhood has never been as simple as
nostalgia portrays it, but the years from five to ten do represent a
significant developmental transition that ushers a child from a protective home base into the larger world of school and society. Erik
Erikson observed that at this age a child's "inner stage seems all set
for 'entrance into life,'" as off they go to school. Every child enacts his
own version of this passage. From the brave anticipation of five- or
six-year-olds on the first day of school, we see our children grow in
confidence and character day by day, sometimes in a wondrous moment right before our eyes. One morning the good-bye wave comes
without a second look back at you. Your shy child tells you about his
new friend. Your me-first child stops to lend a sympathetic hand to
a struggling classmate. And from that, some five or six years later
emerges a preadolescent with the seeds of moral conscience, identity,
empathy, and agency firmly planted.

Healthy cognitive, social, and emotional growth through this period continues to be grounded in hands-on play, curiosity, imagination, and the layering of those experiences in the widening realms
of family, school, and friendship. Children are eager to explore new
ideas. They want to develop mastery, often in line with an interest,
be it ballet, bugs, or baseball. They are open to new relationships
in a widening social circle and with teachers. Technology is naturally enticing. Children tend to be fearless tinkerers and navigators

in a realm that stumps their parents. However, even as they broaden their base of operations, they are still forming critical attachments to you, your family, friends, and community. This is the time to build a strong foundation for family primacy and values, for self-discovery, and for children's natural curiosity about the people and world that surround them. Children need to develop their capacity to engage in life. That's how they develop resilience and self-motivation.

This is a critical time for moral development. In these formative years from five to ten, kids are developmentally ready to grapple with issues of right and wrong, being accountable for their words and actions as they affect other people, and learning to reason on their own what it means to be a person of character. They have the mental muscle to wrestle with tough questions of conscience. With the infusion of computers, cell phones, and online activities at younger ages, elementary school also now has become the training ground for people relating to each other through tech. At a developmental time when children need to be learning how to effectively interact directly, the tech-mediated environment is not an adequate substitute for the human one.

The Inner Critic: New Pressure to Measure Up Bigger, Better, Faster

No matter how fierce play may look on the playground or in the social scrimmage of the school day, the more grueling competition is the one your child faces each day to measure up in her peer group. At around age eight, children start to compare themselves to each other in more competitive ways. They develop the voice of the inner critic. *I don't have a best friend . . . I'm not fast enough. He's a better reader than me.* There is nothing new about this, but media and much of life online introduce an adult context for a child's self-assessment.

The behaviors they see there that set the bar for cool, cute, bold, and daring come from the wrong age and life stage. The mix suddenly includes adolescents and adults, media coverage of fame-addled celebrities and jaded politicians, teen magazines, and Victoria's Secret at the mall and in the mail. The inner critic's presence becomes larger and louder. And it isn't only inside their heads anymore. It is everywhere—in online chats, texts, e-mails, commercials, TV programs and screen games, and the vast 24/7 media milieu that crowds a child's inner stage.

As the inner critic grows, parents become indispensable as the voice of the inner ally, the voice that helps balance a child's innermost sense of himself. For the child, a parent's optimistic steady encouragement becomes the child's internalized voice that says *I can handle this, I'm a good friend, I'm a good person, I know right from wrong, I take good care of myself, I'm a good helper.*

This sturdy sense of self and self-esteem, the start of a core identity, takes time to develop. It's pretty amazing when you consider the difference between a first-grader and a fifth-grader. Day by day, kids need time to process their experiences intellectually and emotionally, to integrate new information with their existing body of knowledge and experience. They need time to consolidate it all so that it has meaning and relevance for them. Ideally, they do that with their parents and in the context of family and community. It happens in debriefing time after school over a snack with mom or dad, in extended-day with a teacher, a caregiver or with someone else at home. Dinnertimes when thoughts about the day are processed through family conversations that are supportive and nonjudgmental. Bedtime reading and rituals that offer a quieter space for reflection and an opportunity to bring the events of the day to a peaceful close.

Time for all this was more readily available in the predigital age. With unstructured play on the wane and immersed as they are now in media and tech through these formative years, our children have lost that protected time for reflection and conversation, especially with parents and family. Instead, they often plug in for the ride home from school, watching handheld screens, circulating pictures, texting, or e-mailing friends en route. By the time they arrive home, their social network has moved on to the next thing and they with it, still plugged in to the trending conversation with peers. Kids don't get home from school anymore; they bring school—and an even larger online community—home with them.

"I tell you, it feels overwhelming," says a mother who used to look forward to the drive time as talk time but has seen it devolve into a futile exercise in screen censorship. "All of a sudden I'm driving and hear them in the backseat—they're looking at YouTube and I think, *What the heck?* I mean, how much screening and censoring can a parent do?"

Nancy is concerned about her daughter Alex, a second-grader, feeling the pressure from peers to leave childhood behind before she's ready. Alex still writes notes to the tooth fairy, but the other girls are already talking about makeup and teen-oriented online sites and TV shows they watch with their older siblings. "They're talking about things she isn't ready to know about, and of course she wants to be friends with these girls," Nancy says. No wonder as a parent she feels trapped.

In Over Their Heads:
Too Much, Too Soon, Too Fast

Long-established insights into children's learning and their inner lives tell us that in the ways that matter most, speed derails the natural pace of development. Pressure to grow up faster or exposing children to content or influences beyond their developmental ken does not make them smarter or savvier sooner. Instead, it fast-forwards them past critical steps in the developmental process. Job No. 1 in elementary school is learning the rules of social engagement: how to make your way in the larger social group, how to make friends and be a friend, compete fairly, read social cues, and find your niche in the boy and girl cultures at your school. Doing so much of that through tech-mediated correspondence changes the playing field, significantly complicating the task.

So often when I am called in to help a child or a school with concerns, this developmental fast-forward tech effect has played a role. Sometimes a child has missed certain relational experiences— learning how to share, how to disagree without getting mad, make eye contact, and not put others down or engage in inappropriate touch- ing. Or perhaps the child hasn't heard important messages about what's okay and what's not when it comes to expressing our feelings or acting on them. Sometimes children are clearly copying behaviors they have seen on popular TV shows or YouTube pranks. Kids are still being kids, in that they act and react in fundamental ways as they have forever. These are not "bad kids" or emotionally disturbed children, necessarily. The third-grade boy who invited a girl he likes into a closet to lift their shirts and kiss each other's nipples, which he had seen the night before on the family computer; a fourth-grade boy who sent sexually explicit rap lyrics to a girl he wanted to impress; the ten-year-old girl who sent Trevor the e-mails to get back at him for embarrassing her—they didn't dream up those ideas or lyrics; they picked up disturbing content from a media and online environment that is saturated with it. They are in over their heads and they need adult help.

Developmentally, this is the time children need parents and teachers to help them learn to tame impulsivity—learning to wait their turn, not cut in line, not call out in a class discussion—and for developing the capacity to feel happy and alone, connected to oneself and empathetic toward others. Some things in life you just have to do in order to learn, and do a lot of to grow adept at it. Like learning to ride a bike, developing these inner qualities of character and contemplation calls for real-life practice. In the absence of that immersion-style learning, time on screens can undermine a child's development of these important social skills and the capacity to feel empathy. Studies already suggest that media and social networking play a role in loneliness, depression, attention problems, and tech addiction among adolescents. Other findings also show media exposure contributing to impulsivity and aggression among younger children. With nature pressing for human interaction and a child's world of possibility expanding in the new school environment, to trade it all for screen time is a terrible waste of a child's early school years.

Emotional and social development, like cognitive development, can benefit from "judicious use" of tech, as described by Jordan Grafman, chief of cognitive neuroscience at the National Institute of Neurological Disorders and Stroke and member of the Dana Foundation Alliance for Brain Initiatives. "But if it is used in a nonjudicious fashion, it will shape the brain in what I think will actually be a negative way," he wrote.

In a Dana Foundation report, he and his coauthors concluded that "the problem is that judicious thinking is among the frontal-lobe skills that are still developing way past the teenage years. In the meantime, the pull of technology is capturing kids at an ever earlier age, when they are not generally able to step back and decide what's appropriate or necessary, or how much is too much."

Michael Friedlander, the head of neuroscience at Baylor College of Medicine and another member of the Dana think tank, said that given the dearth of hard data as yet on long-term effects of media and tech on children, "the best we can do at this point is look at a lot of the science that has been done in much more controlled settings and try to extrapolate that to the real world of kids interacting with these technologies."

I work deeply in that real world of kids interacting with tech and see the fast-forward phenomenon affecting grade-school children in four fundamental ways that set them up for bigger trouble in their tweens and teens:

- An increasingly destructive gender code starts younger than ever.
- Social cruelty has become in vogue and more intense via social media.
- Popular culture deletes childhood by normalizing violence, sexual exploitation, and pornography, once adults-only domains.
- The tech accessible to children at this age far exceeds their capacity to manage their use of it or anticipate the consequences of their misuse.

The Gender Code Starts Younger, Sexier, and More Aggressive Than Ever

Tara, the mother of three children in elementary school, describes her seven-year-old daughter's tale from recess one day: The girls had sung "I'm Sexy and I Know It" in a self-styled *American Idol* competition between "girl bands" and "boy bands." Classmates cheered boys and girls bumping and grinding—truly grinding together—or booed performances. They humiliated one boy to near tears with their jeers, intimating even though they didn't fully understand it, that he couldn't get an erection—couldn't "get it up." The girls didn't even know what the language meant. They just knew the words from lyrics they had heard.

Long before they can read a chapter book, boys and girls are already well versed in the gender code for popularity, sexuality, and the everyday ways in which boys and girls are supposed to differentiate and define themselves. Just moments ago, they were rocking to Raffi and the constant chorus of parental coaching at home; suddenly, it's R-rated music videos and popular songs with lyrics unprintable in family newspapers.

But in school, they take their cues from the crowd-sourced conversations they hear among friends and on social media. For girls, even seven-year-olds on the school playground, sexy is the new cute. Thin is still in, but for ever younger girls. In a study of the effects of media images on gender perceptions, one study reported that by age three, children view fatness negatively, and free online computer games for girls trend toward fashion, beauty, and dress-up games, reinforcing messages that your body is your most important asset.

Fran calls me, upset and frightened about her ten-year-old daughter, Channa, who has begun stopping in front of every mirror and pinching her tummy. She asks too often, "Mommy, am I fat?" She has stopped eating some of her favorite foods and brings home a half-eaten sandwich from lunch. She's obsessed with *America's Next Top Model*, sneaks to "how can I lose weight quickly" sites, and surfs fashion web sites. There is a history of eating disorders in their family. I meet with Channa one afternoon to assess what is going on. She is clearly in the early stages of the disordered thinking that paves the way to disordered eating. She has "good" and "bad" foods. She is afraid of being fat. She compares her body in parts—hair, legs, waist—to a popular girl in her class with that (frequently computer edited) cover girl look. Beneath these behaviors are early signs of social anxiety, emerging perfectionism, and insecurities that in therapy we will work to uncover and deal with directly.

Unfortunately, there is nothing new about girls being taught that the primary source of their power is their body and looks. But prior to Britney Spears, most girls had ten years of running around, rid-ing their bikes, and experiencing their bodies as a source of energy, movement, confidence, and skills. That was before children's fashions included thong panties for kindergarten girls, stylish bras for girls not much older, lipstick or lip gloss as a top accessory for nearly half of six- to nine-year-old girls, and "Future Pimp" T-shirts for schoolboys.

Boys, too, are under pressure. They must measure up to the supermasculine ideal of the day, portrayed and defined by more graphic, sadistic, and sexual violence than the superheroes of yesterday. Homophobia and the slurs used to express it remain a common part of boy culture, but now at an earlier age, as does a derogatory view of all things female and an increasingly sexualized attitude toward girls.

Developmentally, these are formative years for learning and practicing "what it means to be me"—in this body, with this brain and learning style, with these likes and dislikes, and these fears and dreams. Children do best when they are free and flexible to try on and cross over the gender codes—girls who skateboard and play ice hockey, boys who draw or dance, boys and girls who enjoy each other without "dating" overtones.

However, that's not the message they get from media images. When University of Indiana researchers Nicole Martins and Kristen Harrison studied the short-term impact of TV viewing on children's self-esteem, comparing results by gender, they found that TV viewing helped white boys feel better about themselves, and left white girls, black girls, and black boys feeling worse. White boys saw male media comparisons as having it good: "positions of power, prestigious jobs, high education, glamorous houses, a beautiful wife" all easily attained, as if prepackaged. Girls and women saw female media comparisons in more simplistic and limited roles, "focused on the success they have because of how they look, not what they do, what they think or how they got there." Black boys also saw their media comparisons in the negative, limited roles of "criminals, hoodlums and buffoons, with no other future options." Video games, the researchers said, "are the worst offenders when it comes to representation of gender and ethnicity."

Research shows that a majority of TV content reinforces traditional gender stereotypes, although new media are ramping up some gender traits. Males are ever more aggressive. Females are sometimes strong but always highly sexualized (think Warrior Woman fighting in Victoria's Secret underwear). Researchers Karen Dill and Kathryn Thill analyzed computer gaming magazines and images of males and females taken from popular computer games, and with other research concluded there is "a clear link between media violence exposure and aggression" as well as to other damaging consequences including eating disorders, poor body image, and unhealthy practices in an effort to achieve idealized appearances. "Failure to live up to the specific media stereotypes for one's sex is a blow to a person's sense of social desirability," the authors said.

A GRADE-SCHOOL WORKSHOP DECODES
THE GENDER CODE

I was asked to work with faculty, parents, and fourth-, fifth-, and sixth-graders at a small elementary school where, just two months into the new school year, several incidents had the faculty concerned. Girls were competitively not eating their lunches, some boys had picked up on gay-bashing language, and others were teasing two boys for being "brainiac nerds." There had been some sexual threesome jokes on the playground and talks of Internet porn that were confusing and uncomfortable to some students and the faculty.

Much of my work as a school consultant focuses on assessing school culture, curriculum, and social-emotional learning opportunities, and helping schools and parents develop new approaches to this essential piece of children's development. The faculty and parents at this school wanted to be proactive and address the concerns early rather than wait and hope that nothing more serious would occur. The school curriculum for kindergarten through fifth grade already had some social-emotional learning components woven throughout, but not as consistently as needed in the context of today's cultural influences. Recent developments suggested it was time to renew that conversation as a group, and strengthen the school's curriculum.

The children didn't know I had been called to address these concerns and that I would be designing a more comprehensive kindergarten-through-eighth-grade SEL curriculum for them. My class was billed as a "media literacy workshop" to empower children in their relationship with media and technology. Every child is an expert on his or her own experience of media and tech, and every child likes to talk about it.

We gather in a classroom, forty students around six tables, and for the next ninety minutes talk about their screen experiences. All say they routinely watch some TV each week, most at least an hour each day. Some already have cell phones and texting privileges, though most did not. All of them are Internet users and many use online sites daily for social networking. We talk about how many things fall under the rubric of media and how we are always receiving messages through advertising and other content. I bring out my Ken doll collection and we laugh at how the original Ken debuted in 1961 with a smooth, bland torso—no suggestion of muscles at all—and now he is all abs and looks jacked up on steroids. They can see parallels to the way media and life online has jacked up other aspects of gender roles for them and their friends.

They are eager to talk about the gender code and what everyone feels the media ideals are for masculine and feminine. It's a vibrant conversation, everyone participating and listening to one another. They are at their best, a group of students talking about something that matters to them. They write these words on index cards and read them aloud. Cool for boys: sports, action, baggy clothes, dominant, strong, tough, aggressive, video games, mean voice, funny, violence, muscles, fashion/Abercrombie, anti-intellectual/not cool to be smart, protective, guns, don't value education, alcohol, "always hanging with guys," disrespectful to women, and Axe deodorant. Cool for girls: skinny, attractive, embarrassed about their bodies, weak, self-centered, need boys, mean to each other, tan, attitude, popular, pretty, chest size, cute, hot, sexy, pink colors, clothes, dumb, snobs, don't like who they are. We talk about how hard it is to like your own body. Boys are as vocal on this as girls, as the image of the hypersexualized six-pack jock is as daunting to them as the Victoria's Secret image is for girls. The girls raise the subject of eating disorders and how scary that is. I talk about how the pressure can be so intense that a person might do really unhealthy things to make herself feel better.

Then I ask, "What's missing from the mix—what qualities about you are missing there?" For boys, what do they see in their fathers, uncles, and male teachers that is missing from the list? Similarly for girls, what do they see in their mothers and other women in their lives, and in themselves, that is missing from the media stereotype? Again, they are eager to share. They see themselves as respectful of people; they place a high value on education, being nice, creative, a good person, well mannered, healthy, smart; they value their education, have manners, are generous and loving, and creative.

As often happens, when you address children's sense of social justice, it pulls them into a critical perspective and empowers them to take control rather than be passive. When they see how they are being manipulated to be untrue to their selves, they get mad—and honest about their own behaviors. They volunteer ways they put each other down when they use sarcasm and trash-talk, weight-related comments and other insults about looks, smartness, and other kids. We talk about how the gender code becomes the basis for a lot of social cruelty and for unhealthy attitudes, behaviors, and relationships. This casual cruelty didn't originate with today's media, but the media's pervasive presence has normalized an ever more aggressive form of it.

We talk about how they—not the media—are in charge of their school community. And how if they acted in the ways the media wanted them to act it wouldn't be a safe place. They brainstorm about ways they could reclaim their culture at school and make it a counterculture to the "bad ways" the dominant culture tells them it's cool to act. I am impressed by their being able to have this conversation and tell them so. This was another way they are making sure that being part of their school feels good and safe. We close the session with their suggestions for how to act on the insights they had generated together. What if they saw a classmate struggling with these issues or had concerns about someone, or perhaps saw one classmate treating another poorly?

"Maybe I can speak up when I see someone saying something mean to someone else," a fifth-grade boy says.

"Or when someone's being a jerk," says another boy.

"If we all do it, we can make a difference," a fourth-grade girl says.

And about the messages they tell themselves: "This is what I look like in my natural beauty, not what TV tells me," a fifth-grade girl says. "If I'm curvy when I grow up, that's got to be okay."

Here was a school pushing the pause button for conversation at the early signs of potential trouble rather than waiting for a crisis to require it. Our talk time provided students with the opportunity for meaningful discussion—not a lecture but a genuine dialogue and call to action—about how to filter the gender messages and revise the gender code for a healthier social environment. These kinds of workshops and comprehensive SEL curricular experiences teach children to deconstruct complex issues and think morally, think of themselves as prosocial activists. These are the conversations at home and at school when our children get to think of themselves as a force for good; where they experience being recognized and rewarded for doing the right thing; where they can take pride in their knowing that they are capable of doing the right thing and acting; and when they learn that personal power and efficacy in the name of doing the right thing feel good and are recognized. This is where they experience and learn about accountability. This is how children build character.

When given the chance, kids can learn to outsmart cultural stereotypes, but they need constant reinforcement in these prosocial, upbeat ways. Short of meaningful conversations about what it means to be a man or a woman in our culture and how to define yourself and others by a more respectful standard than the media provides, we leave kids to learn from those same skewed sources. This is moral education in action.

Cool-to-Be-Cruel Rules from Playground to Blog Posts

A third-grade girl tells me how she lies in bed at night unable to sleep as her mind replays a message she received on a social networking site from a girl in her class at school: "u are not popular. no one likes you. we all have secret names and u cant sit at our [lunch] table. we r not friends anymore. i m deleting u from my contacts."

"Even though I erased all those really mean things the girl said to me, how do I know they erased them?" she asks me. "How do I know they're not still laughing and adding to them and having their private mini mean fest? I can't stop it. I wish it were like a chalkboard and you could erase it, but every night I go to bed thinking about their mean words."

Before *The Simpson, South Park, Family Guy*, reality TV, and YouTube redefined family and children's screen viewing, TV shows generally reflected societal norms that cast childhood as a protected time before "coming of age." YouTube, Tumblr, MySpace, Formspring, and ask.fm weren't an option, nor were the Kardashians. You couldn't find crass behavior, swearing, sexualized humor or drama, or humiliating entertainment for kids. Now it is just a click away. The result is that a more heightened kind of meanness—humiliating, sarcastic, crude, and often prejudice-based—has transformed the intensity of teasing and power plays among kids.

It is hard enough in second grade to have someone be mean to you directly to your face, but to read something online—when you're having to work to sound the words out—has a larger-than-life quality to it because it doesn't go away. It becomes etched in a child's young psyche in a very different way: *I am deleting you.* The cool-to-be-cruel ethos undermines the development of empathy, an important developmental facet in these early school years. *How do you think she feels?* is one of the most common refrains we repeat to children. When you act from a distance or anonymously, you remove that critical piece of social and emotional learning—how to read the impact of your behavior on someone else. This undermines not only the development of empathy but also of understanding and accountability. Further, when someone delivers a cruel comment and then disappears off-line, the victim is left alone holding the injury with no way to shake it or respond back. So much of what we learn from fighting with each other involves the ebb and flow of a fight, how you escalate and deescalate, what makes it better or worse. The shared experience is the meaningful learning experience. In the push-send-sign-off world, kids don't have to do that and they miss out on something very important.

In the context of this age and children's trusting vulnerability, this hit-and-run cruelty can wound a child more deeply than the words alone might suggest. Research using rats to study chronic fear showed that "unpredictable shock produced more fear than predictable shock because the rat never knew when it was safe." Forwarding mean gossip to the child being maligned or a party planning post to a child who wasn't invited, posting texts or pics that make fun of a child's looks or abilities: this is the stuff that keeps kids awake past bedtime, and can make school—or home—feel unsafe. When children experience offensive ambushes through tech, the impact is more intense because they can't psychologically protect themselves in the same way they could if they saw an obnoxious kid coming at them in the lunchroom or on the playground. The startle factor is much bigger and the impact can be, too. It doesn't even have to be a malevolent act—it could be your good friend showing you something she thought was hilarious but which is deeply disturbing to you. The problem is the shock of seeing something bad from someone you trust and your inability to step away or filter it out.

Then there is the exposure to the actual content, which is often sadistic or pornographic. Sometimes they can't "just forget it." Think about how sad your child has been at the loss of a grandparent, a pet, or a friendship, and how she has needed time to move through the emotions of that experience step by step. Especially at this age, kids process things over and over and over to integrate them. One conversation doesn't do it. So when something is shocking and traumatic, the brain's ability to process it repeatedly and integrate it is even more challenged. For a child, that can mean replaying the offensive image in the mind's eye, reexperiencing it over and over again.

As a parent, you may never know when your child has seen something that is traumatizing for him, especially if you are no longer monitoring his screen activity. I worked with a ten-year-old boy whose normal curiosity delivered more than he could handle. He had looked up the word "pornography" online because he'd heard boys talking about it at school and was curious. But what he saw there was so disturbing to him that for the next several months he had nightmares and couldn't sleep alone in his room.

The boy told his mother, but often children don't. They may see something disturbing yet hesitate to tell us, fearful of our reaction if we have never broached these subjects before. It's a raw world out there and we need to let them know we know that. If they run into it on their own, they may feel shame or worry if we haven't prepared them to share with us. This is one instance in which the developmental gap between readiness and reality, no matter how wide or at what age, can be bridged with a single message: *You can always tell me. I will never be mad at you. Even if you have gone someplace online where you weren't supposed to go, it's more important that you tell me. There is nothing you can tell me that I can't handle.* From kindergarten on, kids need to hear this message again and again, because at different ages a child will hear it differently. But it will always be reassuring and will make your child more likely to come to you for help.

Social cruelty is nothing new. It has always hurt to be called ugly, and it has always hurt to know it has become public—even just a note passed in class that lands in the wrong hands. But there is a haunting, potentially traumatizing quality to knowing that the hurtful material is forever in existence on the Internet. And kids use the threat of eternity against each other: "I haven't deleted it and you can't make me."

In one sense, Trevor's story of the sexually harassing e-mail could have come straight from the script of an R-rated thriller: the anonymity, the hostility, the explicit sexual language chosen to shock and humiliate. But she is ten and he is ten and this was, in fact, the story of a ten-year-old girl offended by a ten-year-old boy correcting her grammar in front of everyone, and she was out for revenge. Her choice of crude, sexual language was disturbing, as was her choice to send the anonymous e-mail instead of confronting him directly over his earlier insulting behavior. But the incident says more about the culture we have created for children than about any one child's misguided use of it.

Children Have Too Much Access to Ugly Stuff

Mimi Ito, a cultural anthropologist whose cutting-edge research has focused on new media use among young people in the United States and Japan, sees today's media accelerating the pace at which children are connected to "harsh, ugly and frightening aspects of the adult world" at younger ages and more widely than was possible in the predigital world. As children are introduced so young to a darker side of life, they are at once drawn into that world of images and ideas and at times scared or overwhelmed by what they see.

Graphic screen violence and pornography have become a normalized part of our cultural landscape. We know that children—more boys than girls—are viewing screen violence at very young ages. Many kids report feeling addicted and no wonder. Playing video games triggers and doubles the amount of dopamine in the brain, roughly akin to a dose of speed. In many video or online games, kids "win" by killing, raping, and bullying. We also know that many boys by fourth grade have viewed pornography and their involvement with it tends to grow as they enter preadolescence and adolescence.

However, any thought that these younger children are oblivious to imagery or dialogue that is developmentally over their heads is mistaken. What we do know is that they don't experience it as we do. That doesn't mean they don't experience it and attach their own meaning to it.

So, for instance, a YouTube clip that shows four naked men in a shower in behavior suggestive of anal sex was making the rounds of schools last year. I got calls about it from multiple schools and parents in different parts of the country. In that instance, boys were sending the video to girls and to other boys. It wasn't a sexual come-on; it was a shock come-on, sharing the newest, most outrageous thing they had seen.

Andrea and her husband, as part of letting their fifth-grade daughter Olivia use e-mail, instructed her to never open a link unless it came from someone she knew and trusted. Olivia was good about following the rules. So she thought nothing of clicking on a link sent to her one day by a boy she knew. "Check this out for homework," he had written in the subject line. He was a pesky boy with a bit of crush on her, but she had no reason not to open his e-mail. When she clicked on the link it opened to "the four guys in a shower" video clip. It was extremely upsetting for her. Like other children I've worked with who have been exposed this way to graphic sexual content, she was frightened by it and remained generally anxious and fearful for months after.

A second-grade girl described how she dreaded bedtime because her father routinely watched violent action movies in the family room after she went upstairs to bed and she could hear the sounds of gunfire, screams, swearing, and angry and terrified voices. She was afraid to ask him to turn the TV off because she thought he would get mad at her.

In a focus group one day with fourth- and fifth-graders, two fourth-grade girls describe their problem with violent screen games and some online content they'd seen. Katarina and her friend Cora say they'd seen things they wished they hadn't—mostly violent games the boys play, which they see either at school or at friends' houses where there are older brothers. "Whenever I see something bad I feel nauseous," says Cora.

Marten, a fourth-grader, says that he often plays Gangstar, a game based on L.A. gang life, even though he has mixed feelings about it after he saw two teenagers in a fistfight "for real" on a street corner one afternoon the previous year. Now whenever he watches screen violence or plays those games, he says, "I feel nauseous and I feel a bit scared."

As the other children listen patiently, he describes the street fight he had witnessed from the car as his parents drove by. It disturbed him to see people on the street walk past instead of intervening. He felt guilty as he and his family drove away. And now, months later, the screen violence and the real violence he saw that day remain linked. By bringing it up in our focus group, he is continuing to work through his feelings.

Even when children this age know that something isn't "real," their experience of virtual reality feels real and, for some, disturbing. The connection between virtual and real life experience isn't fully understood, but the notion that pornography and screen violence have no effect on children is "wildly naïve in the twenty-first century," says researcher Jackson Katz. "To think otherwise is to live in a fantasy world. Media structures and shapes our psyches and our fantasy life."

Katz believes we must be concerned about repeated exposure to screen violence and pornography because children seem to have become increasingly desensitized to it. He describes a typical scenario:

> You can have these . . . nine-year-old kids at a movie
> theater watching incredible brutality and not even blinking,
> and bragging about the fact that they can just sit there and see
> somebody disemboweled . . . right in front of their face, and
> computer-aided graphics are so sophisticated that it actually
> looks like it's really happening.

Katz, like most experts in this field, is confronting the limits of what is known about the effects of violence as the digital revolution advances through our children's childhoods. "There is no precedent in human history for any of this," he says on the subject of violent video games, "so we're just trying to figure out, okay, what effect does it have that these young boys are doing this for hour after hour. They're vicariously committing these brutal acts and the only consequences are, do you get more points? Do you get to the next level in the game? But in terms of the real-life consequences of violence, and brutality, and misogyny . . . you just turn the game off."

The question is whether the experience creates a greater tolerance for real violence, even an acceptance of it. Research tells us that when a child is engaged in violent-action video games, the part of the brain that experiences empathy disengages.

There are different ways of looking at this. Sociologists on the broad scale say we're not seeing an increase in gun violence on the street, but that's not the only issue. First of all, we are indeed seeing more horrific acts of gun violence in public gathering places—at schools, at shopping centers, at movie theaters, even on the roads in road rage incidents and assaults on police. And when gun violence of this kind occurs, the news—and sometimes video footage—is now instantly broadcast in every taxi, waiting room, and over and over online and in public places where children see it. Furthermore, teachers and psychologists and parents are concerned about whether exposure to gross violence desensitizes kids to meanness or ordinary forms of socially provocative, mean-spirited behavior beneath the radar—more aggressive behavior in ordinary circumstances, and a lack of empathy and understanding of the impact of antisocial behavior. Just looking at statistics of increased gun violence is not sufficient to justify the idea that viewing screen violence as entertainment has no impact.

Researcher and pediatrician Dimitri Christakis suggests that the vivid sensory features of what he calls "gorenography" (the ultrarealistic screen hybrid of violence and pornography) increase the risk that real-world violence will mimic screen violence. He describes a recent promotional game trailer as an example of this kind of "glamorized violence, staged with beautiful classical music playing in the background, and extraordinarily realistic." And while research is under way to more clearly identify the effects on children of hyperrealistic screen violence and pornography, Christakis sees no reason for parents to wait and wonder whether to take action. Some parents "already let their children engage in many of these things and they are just weary of the battle it would take to disengage them. Or they feel like it's impossible to fight that fight with their child so they choose to believe that it will not be harmful." However, as a dad himself, he suggests the question that any parent should ask: "Is this really the best you want your kids to be getting from playing a game?"

Some years ago it was reasonable to say "trust your gut" about what to allow your child to see, especially anything involving violence or sex. I still say that, but would add that the way the media has normalized this content has put us all at a disadvantage in remembering what we deeply, truly wish for our children to grow up seeing and hearing as young children or preadolescents or teens. When was the last time you looked and listened objectively to contemporary family TV—including programs you may enjoy watching—for the role that violence plays in story lines and screen action, or the casual use of sexual innuendo or sometimes explicit language or scenes, or physical or relational abuse between characters? Not to mention stereotypes about gender, race, and sexual identity. If we're comfortable with it because "it's just TV," then we may shrug off any hesitation about letting our children see it, too, not because we've thought through what they'll make of it but because we assume it doesn't matter.

Tech and media standards reflecting popular culture will always precede the research. But already, every day more and more research and our own experiences are saying this is damaging stuff. So when in doubt, use caution. Only you know your child's sensitivity or the extent to which he or she is mimicking cultural content that doesn't feel age appropriate to you. Pay attention to that, think clearly about where your child is developmentally, and in that context, choose what feels appropriate for exposure.

This is a situation where "getting there first" with early exposure to this content can mean getting hurt, not winning. Overexposure or premature exposure is not the same as an advantage. This is particularly hard for families with widely spaced children. I'm extremely sympathetic to the difficulties of protecting childhood this way in families where the children are at different stages of development and maturity. That said, you are the boss and you can make older children's access to screens dependent on their responsible use, which includes not showing younger siblings material that you don't want the younger people in your family seeing. Part of being a good older brother or sister is also protecting the childhood of the younger siblings and members of your family. That's something we can tell our children and need to tell our children.

Kids Can't Pull the Plug on a Good Time— It's Our Job to Show Them How

Amanda tells our focus group the Christmas story of her seven-year-old brother Clay and the Wii. They had opened the gift from their parents very early that morning, but the rest of the day Amanda had turned to other fun with family and friends. Her parents were busy, too. Clay eagerly unpacked the Wii, figured out how to set it up, and began to play. She wasn't surprised to see him playing it throughout the day when she'd pass through the family room. "That's normal," she tells us, "it's like a new, exciting video game, and he's so little and it's so exciting. He gets like obsessed."

The family stayed up unusually late enjoying rounds of company and conversations well past midnight. At some point his mom noticed Clay was still playing and she made him go to bed. Then out of curiosity, she checked the Wii's play history: he had played for eighteen hours.

Our children are quick to learn how to use tech, but how to set limits and use it wisely requires more self-discipline and emotional maturity than most grade-school children (and many adults) have developed. They don't think about self-monitoring for time or inappropriate content, exposure to violence, or addictive potential the same way we do. If we listen closely to our children, however, we can hear how their media and screen activities affect them, even when they can't see it themselves.

"It's real hard to stop," fourth-grader Katarina tells the focus group, "because it's like you always want to beat these levels and . . . it's so fun, and it's hard to stop because the game doesn't stop. It just goes on and on and on until you lose."

Ryan, also in fourth grade, boasts that "pretty much every video game I play is violent." He has heard that "people say it rots your brain in the sense that you're just going to copy them. But I don't do that. Like when I see someone shoot lightning and run up a wall I'm not like, oh, I can do that, too. I'm not crazy."

This makes literal sense but it doesn't mean that the uptick in aggression and attention problems that research suggests comes with repeated exposure or the risk of desensitization that Jackson Katz and Dimitri Christakis describe are not concerns. All it means is that we cannot expect children Ryan's age to see the relevance of those concerns and dial down their screen time by themselves.

Craig Anderson, an Iowa State University professor, director of the department of psychology, and author, has been involved in cutting-edge research on the effects of screen violence and video games on children. In addition to its desensitizing effect, Anderson says, addiction among children is a growing concern because of gaming's immediate rewards.

"They can see themselves improve and they get all the media reinforcement for playing the game correctly, whether it's shooting the bad guys or whatever, building a better roller coaster—whatever kind of game it is, there is that element of the immediate rewards that starts to get into things like addiction," he says. Tech addiction, often called Internet addiction in other parts of the world, is emerging as a serious concern for children and adults. South Korea and many other countries now offer treatment programs for children as young as five years old who are addicted to technology.

In a focus group with parents, the mother of two boys, ages six and eight, says she and her husband struggle with their sons over the gaming even though they have limited the time allowed for screen play. "When they aren't actually playing it, they're begging to play it," she says. "And it is completely addictive—the kids can't stop once they're plugged in and we don't know how to stop them either. Trying to take the Nintendo away is like taking a bottle away from an alcoholic." New research suggests that one out of eight children who play video games shows signs of addiction. And it shows how patterns of use affect the developing young brain.

"Neuroimaging suggests that when kids play violent video games, the medial prefrontal cortex—the part of the brain that allows us to balance our emotions, be empathetic, and make thoughtful decisions—becomes *less* active," says Tina Payne Bryson, director of parenting education for the Mindsight Institute and coauthor of *The Whole-Brain Child*. "And at the same time, the amygdala—the part of the brain that causes us to act before we think, be territorial, and reactive—becomes *more* active. Keep in mind that repeated activation in the brain impacts how the brain is wired. The brain develops what it gets practice doing. The content matters, not just in terms of what kids' eyes and minds are exposed to, but in terms of how the circuitry of their brains gets activated and wired."

Nor do the emerging child tech addicts match the familiar stereotype of the glaze-eyed gamer. As more and more children are given phones, laptops, and largely unsupervised and unlimited time on them, they are becoming as tech-dependent as the rest of us. They take their phones everywhere—to school, to bed, to the bathroom—and if they can't, or if the battery is low, or they discover they're in a location with poor reception they get a little anxious. This is a kind of "separation anxiety" unique to the digital age, and the problem is, without thoughtful attention to tech time management, it can take over.

Elementary school kids are impulsive, have little sense of what is age appropriate, and can easily make innocent mistakes using the not-at-all-innocent content they pick up. The only way they can learn right from wrong is from parents and teachers who care about them. There is no way that children of this age can adequately protect themselves. This is a time when kids need real limits and parental supervision. They can't self-monitor. They need us to monitor and set limits in a calm and thoughtful way—not in a rush of upset over ordinary missteps.

Slowing It Down: Strudel Theory and the Alchemy of Time

My colleague JoAnn Deak, a developmental psychologist, describes in her so-called Strudel Theory how the layering of nature, nurture, life experience, and the time to integrate and consolidate it all is unique and critical to every child. "We can and must layer experience over time to help children grow strong and resilient," she says.

We know from our own life experience that time is, literally, of the essence. Rush a cake and it falls. Rush childhood and the time for layered learning at an individual pace is lost forever. There is no going back. The way we parent our children at this age is so important because if we cannot slow ourselves down to do the essential work of connecting to our kids and teaching them in these years, tech will raise them instead. More than ever we are all under pressure to pick up the pace of work and life, using tech as both a tool and a taskmaster. It allows more people to ask more of us, and we do the same to ourselves, often to a degree that we know is not good for us.

Speed works against relationships for all of us, but for children at this particular stage of development, research shows that the core social and emotional skills they develop now and throughout their education will define their success in school and life more powerfully than almost any other factors, including GPAs, SATs, and where they go to college. With stakes that high, we need to push pause, play, and reset to give our children the time they need for the natural pace of development. Our challenge is this: to intentionally and comprehensively teach social and emotional skills, and create opportunities at home, at school, and in the community for our children to develop the character traits that generate the psychological strength and resilience necessary for success.

You will never again have greater control over your home environment, the tech your child uses, and the content he or she sees than when they are five to ten years old. You will never again have greater access to your child's social and emotional life, or closer connections with their friends, their friends' parents, their teachers, their counselors, and other resources to support your effort. Faster than you might imagine, as your child moves on to the tween and teen years, your chance to stack the deck in their favor will be gone.

You want your kids at this age to experience you as being available and present, open to knowing them deeply, and committed to helping them navigate the troubling new world of tech with boundaries and conviction. This is the best way to make sure they will turn to you when they have a problem in sixth or seventh or eighth grade.

When your children are still in single-digit ages and they are little people having little problems, it's hard to anticipate the kinds of difficulties they may encounter when they are teenagers. In our rush to prepare our kids as early as possible for academic and future professional success, it's easy to overlook the hazards ahead for which emotional sturdiness and resilience are the best protection. The picture of troubled "outcomes"—a euphemistic way to refer to deeply troubled adolescent children and college-age kids—could not be more clearly drawn than in the incidence of alcohol and substance abuse, anxiety and depression, eating disorders, high-risk sex and sexual violence, academic cheating, and other deeply concerning behaviors. Yet we fail to connect the dots between a child's deep learning in these early school years, the powerful impact of media and tech on them, and how it all affects a child's inner resources for creating an authentic, healthy, loving, responsible, satisfying life.

Chapter 5

Going, Going, Gone

Tweens, Screens, and the Perils of
Independence: Ages Eleven to Thirteen

*Don't let your kids have computers in their room before age
thirteen—you lose them if you let them have them earlier. I
haven't seen my brother the entire year since he has a computer in his room. It's really sad. He's in sixth grade.*

—DAVE, FIFTEEN

Not long ago, a mother told me a story about her daughter, Danielle,
and her daughter's friend, Katie—two responsible sixth-grade girls
who made a plan to go to a local pizza parlor for lunch. This was a
privilege—to go to the restaurant on their own—that the girls had
been granted in honor of entering middle school. Their parents knew
they were going and were happy for them to do so. But what their
parents didn't know was that the girls had also sent a text to Allie, a
friend from summer camp, asking her to join them. On the way, Allie
ran into three boys who were her friends and invited them to come
along. While at lunch, one of these boys, Eric, took photos of the girls
with his phone. No one thought anything of it. After all, kids take
photos of themselves and each other all the time these days.

Later, however, Eric cropped the heads from the photo of Danielle and Katie and Allie—eleven-year-old girls—and attached them

to photos he found online of naked women; he then sent this image to six other boys from his school. Someone posted the photo on another boy's Facebook page, and in no time it was reaching everyone who knew any of the participants, including a boy who knew these girls from camp. This young man told his mother, who called Katie's mother. The girls were devastated at what had happened, and mortified.

Meanwhile, the girls' parents fought bitterly about what they should do and who was to blame. Allie and her parents, who were friends of Eric's family, insisted that Eric was a nice kid from a good family—he'd just done a stupid thing. Katie's father, on the other hand, was in a rage. He wanted to go straight over to Eric's house and "scare the living daylights out of that kid, call the cops and his school principal." The debate went on for days, and it became a much-gossiped-about incident among the families in their community. This hubbub arose from an innocent lunch at a pizza parlor, a once-benign childhood ritual for preteens.

The idea of middle school as a transition zone between elementary and high school was created to punctuate the notoriously complicated three-year period in which children by eleven or twelve are no longer thinking or acting like they're eight or nine but they are also not yet the relatively more sophisticated abstract thinkers they'll be at fourteen and fifteen. It is the age of awkwardness and uncertainty. At home it's a time of snuggling one minute and hating mom or dad the next. At school it is a world of hormonally charged changelings as puberty transforms the body and bathes the brain in hormones that affect how they think and how their emotions color the moment.

Preadolescents are painfully self-conscious and their capacity for cruelty is legendary. The mix of vulnerability, confusion, and emotional volatility is the reason some teachers fondly refer to middle school as "a holding tank," a contained educational space designed to see kids through puberty when academic endeavor can hardly compete with the roiling physical and emotional action of the age. Others have called it a "bridge" between the developmental territories of elementary and high school. One superintendent called it "the Bermuda Triangle of education. . . . Hormones are flying all over the place."

The age was challenging enough in the days when school dances and pizza parlor dates were the most exciting new thing on a preteen's social agenda, the social getaways from parents where girls and boys braved the awkward encounters of curiosity, crushes, and social cruising. Summer camps introduced the first coed dance. Youth groups brought them together at church or temple. These were tame, well-chaperoned, structured opportunities for boys and girls to hang out. The kids were "pre"—not yet teens—and their social interaction reflected that. All of this was relatively manageable—for them and for parents.

The Internet blew the gates off this holding tank and swept us all out to cyber sea, into a new Bermuda Triangle of tweens, screens, and no limits. Now with laptops and smartphones, texting, sexting, and online social networking, we've lost all control, transforming what traditionally had been a de facto if unheralded rite of passage through a physical, social, emotional, and developmental metamorphosis into an online spectator sport.

In the new tweens-without-borders environment, when we lost control, tweens lost the essential preadolescent experience of hitting up against the limits set by the adults in their lives. We tell them, *Don't e-mail while you're doing your homework,* or *Don't look at the J. Crew site in class,* or *Don't write anything online you wouldn't say in person,* but many do and they get away with it because we cannot stop them. They aren't internalizing the early formative lessons in morality that come from thinking *it's wrong, don't do it, I'll get caught,* or *my parents will kill me.* Now it's *my parents are clueless and this is so much fun.* Parental warnings sail past to no effect; the lived experience is what they internalize and their lived experience is of getting away with it.

We call them "tweens" to designate their developmental locus between two worlds, childhood and adolescence. But in truth that notion of two worlds is outdated; access to the online world and social media has jettisoned preteens unprotected into the adult world. Our kids are gamely making their way as kids have always done. But psychologically, they face a far more complex and often confusing challenge in the digital culture where those defining elements of selfhood—friendships, interests, and sexuality—are so malleable. The casual yet hurtful hallway insult is magnified exponentially when you read about yourself on Facebook or online chats. Online information about sex, including content that would be X-rated on the street, is confusing, disturbing, and, for some, addictive. At a time when developmentally they need to be advancing their early relationship skills, tweens begin the drift away from face-to-face interactions that build social and emotional awareness and instead opt for texting and scrolling superficially in relationships online.

Nor are we as likely to hear about their adventures until they become misadventures. We are also often clueless about their adversities if they are venting on channels we aren't tuned in to. It used to be in middle school that if your child was having a meltdown you knew it because she would wait to decompress with you at home. Now that they have unprecedented total access to each other 24/7, they don't wait to debrief with you anymore. They are growing up in a culture that tells them to have those conversations with a friend, or a friend of a friend, or a multitude of "friends" online. The Internet is the primary source that millions of tweens go to for their education about sexuality, relationships, drug and substance abuse, what's cool, and how to deal with all of life's ups and downs. Facebook alone is the after-school study group for 7.5 million American kids under age thirteen—at a time when the legal age to establish a page *is* thirteen. Unfortunately the lessons being learned there and elsewhere online are not what we would sign our kids up for. And yet many parents do just that, as one lawyer mom said, "Putting aside the moral torpor of knowingly breaking the law with my child and getting her, at twelve, a Facebook account. All her friends had one."

Children tell me they are also less likely to talk with their parents because some parents' workplaces are stricter now in response to cell phones; it's easier to text, harder to call. Other tweens say their parents are routinely absorbed in their own work, interests, or online activities, or they are out and about. Some parents tell me that even though they are home and available to their children, they are intimidated by their children's problems and complaints, unnerved by their exploits. They are silenced by their tweens' insistence that "it's their world online," while worried about their children's obliviousness to what seem obvious risks. The parents are embarrassed to show their ignorance or helplessness, as well as afraid to ignite a storm with questions. At times, even when they know and take heroic measures to protect their children, they are helpless to stop the viral mob that can descend on a child with stunning ferocity.

Many schools and religions have a rite of passage to punctuate the developmental passage: bar mitzvah, confirmation year, or a big youth-group project. In school, the literature for kids this age is all about the hero quest, meeting the limits, and having character-defining experiences that clarify and fortify core values. Girl Scout and Boy Scout badges were once emblematic of the age. In order for a rite of passage to happen, the ritual has to contain a clear beginning, a middle passage, and an end. Without the ritual learning and preparation that a rite of passage implies, the tween world now bleeds directly into the teen world because they all roam the same media and online world together. The epic quest has been rewritten; online the odyssey of self-discovery is now a daily competition to see who can find the edgiest new YouTube clip or bring down an online foe, whether that's a dragon or a kid who dissed you in the hall. Meanwhile, as researchers like Dimitri Christakis and Jackson Katz point out, the preadolescent's character is being informed not only by virtual experiences but by all too real life experiences.

Tweens Meet World: Coming of Age in the Age of Butt-Dialing

If you're not on top of this, and even if you are, the reality is that while you might have put good privacy settings on your phone and computer and you have not given a smartphone to your child by sixth or seventh grade, chances are your child has access to what has become the standard fare of games and online life. The 2012 Pew Internet and American Life Project survey found that young people (ages twelve to eighteen) are "routinely able to get their hands on games" rated M for mature content and AO for adults only. Three-quarters of parents reported that they "always" or "sometimes" check the ratings on the games their kids play, yet the study found that half the boys questioned rated an M or AO game as their favorite. This compared with 14 percent of girls.

Our children are never far from a friend or schoolmate or kid on the corner who's got the newest phone or tablet or game or app. Somewhere, at home or in someone else's home, there is a computer or tablet unguarded, and those who want to can go online and go anywhere they wish almost any time they wish. Once online they may seek out risky business or stumble onto it purely by accident. And there is little we can do to protect them in the old-fashioned sense. As schools are stepping up their efforts to educate children about responsible use of technology, the parenting mission now, as well, is education and prevention: talking about family expectations and netiquette; clear guiding principles, family contracts, and rules; monitoring children's access to media and their use of tech whenever possible; and teaching them how to be safe and responsible. When that isn't enough—and inevitably it isn't—we move on to damage control, that is, teaching our children to take responsibility and learn from their mistakes. This becomes all the more difficult at an age when children need more independence, not less, to practice being responsible for themselves as they take the first small steps toward adulthood.

My friend says she used to believe that tech offered "relatively benign tools and exciting new ways to experiment," for kids like her three daughters. "It really wasn't all that different, I thought, than the notes I exchanged all day, every day, at school with my girlfriends, only to go home and spend the entire night on the phone with the same girls," she says. Or so she thought until the day her daughter Alexa, thirteen, unknowingly sat on her phone and "butt-dialed" a boy she had met once several years before at a fourth-grade dance. When her phone buzzed, there was a text from a number she didn't recognize and the blunt question: "Who is this?" Unaware that she'd started the exchange when she pocket-dialed the number, she shot back, "Well who is this?" After a testy texting exchange in which the unknown texter was "getting rude" and threatened to post her number to Craigslist, she told the texter off (via texting, of course) and turned off her phone.

The next several days exploded in a texting frenzy and the discovery that her phone number had indeed been posted on Craigslist with a sexually suggestive photo of a woman and the message that she was only in town for two days. Alexa says:

> I went to sleep, thinking it was over and I woke up the next morning to thirty-six text messages and twelve missed calls, and there were all these pictures of men with their shirts pulled up saying things like "Hey cutie, I hope we can meet up" . . . so then the panic started to set in, I'm feeling a little out of my range.

She called her parents but they were en route home on a flight, so she called her two older sisters, first Stella, eighteen, who was scared and knew it was serious. Together, they called Rose, twenty-one. Alexa tracked the number in her contacts and identified it as belonging to the boy she'd met in fourth grade. Stepping up, Rose called the number to speak to him about what he had done. It turned out that the number no longer belonged to the boy at all. Rose met with an angry response from the new person on the other end of the line. It turned out that Alexa had pocket-dialed a transsexual prostitute, who was offended by her comments and decided that Alexa needed "to learn a lesson." In what became an X-rated comedy of errors, before the day was over, Alexa's phone number was posted on an online site for a transsexual escort service in Los Angeles. Her parents arrived home to the drama unfolding and brought the curtain down quickly. They changed Alexa's number and got her a phone that was password protected so she couldn't accidentally dial again. It was indeed a learning experience for all, says Alexa.

> I didn't do anything mean or start off trying to annoy anyone. I didn't send naked pictures to anyone. It was just something that spun out of control. It showed me how fast you can lose control in a situation online and how much the Internet and social networking can conceal people's identities. I used to laugh at girls who got into situations talking to fifty-year-old men online, like how could you not know who you were talking to? But I really had no idea who I was talking to.

"Mea culpa," her mom said later. "I was completely wrong. It's not the same thing at all as writing notes or talking on the Princess phone." Alexa's is a story in which there was no real damage done—it was more funny than scary. But this is not always the case. For a year and a half, Cindy in Connecticut had been "best friends" with a young man she thought was her age, who lived in Wisconsin. Practically every afternoon and evening they chatted online and she shared her day's events with him and she had come to trust him and think of "Rusty" as one of her closest bestest friends. Then came summer and a special out-of-town trip for the most senior campers, which included Cindy and a group of her longtime summer campmates. It was a huge deal to make this trip, and the destination was always kept top secret to add to the suspense. When the destination was announced—a city in British Columbia—she shared this exciting news with Rusty and, true to form, he wanted to know all the details—what day they'd arrive, where, and their trip itinerary. On the day of the trip, when the girls and their counselors all arrived at their destination, there was Rusty—not a boy her age at all, but older and "creepy" and accompanied by a couple of his unsavory friends.

Cindy's fellow campers and counselors were freaked out by the sleazy visitors and furious at Cindy. How could she have been so stupid to tell a stranger where they were going? What was she thinking, having an online relationship with a boy she'd never met, telling him all the details of her life? The boys were sent away and the trip went on, but Cindy grew anxious, depressed, and—with no history of disordered eating—went into an anorexic slide. After the trip, she still felt upended by all that happened and it was scary to her: her friends' anger and lack of support for her, the adults' difficulty understanding how this could have happened, and the way it had added to the counselors' worries about keeping the girls safe. It altered the rest of her summer and shook up her sense of herself, rocked her world in a very bad way.

We talk about "stranger danger" to our kids, but more often than not a predator is someone known to the child. In this way, tech becomes the network where creepy people can get to know your kid in just the way "Rusty" zeroed in on Cindy. Further, the morass of privacy issues around Facebook, social networking, and online marketing makes it hard to know who has what kinds of information about your child. It is a natural and wonderful inclination of tweens to want to reach out and connect to the world at large. While so many wonderful, healthy online social groups are available to kids, the questions we used to ask our teens before we handed them the keys to the car—who, what, where, when—we now have to ask at an earlier age when our kids have access to the keyboard.

"A large part of this generation's social and emotional development is occurring while on the Internet and on cellphones," notes the 2011 report of the American Academy of Pediatrics Council on Communications and Media. It describes enhanced social connections and communication as potential benefits of social media. But the report goes on to paint a familiar troubling picture of the gap between what preadolescent children need for healthy development and the impact of extensive everyday use of media and tech.

Many teens have easygoing, positive, relatively uncomplicated experiences online. They don't get caught up in heavy usage or heavy drama. And research confirms that when you do comprehensive social and emotional education you reduce the drama significantly. But even the simplest moments hold the potential for trouble, as Lyla found out while doing homework with friends one evening.

Lyla, in eighth grade, was doing her homework online with her two best friends since kindergarten, Paula and Melanie. As they often did, the girls were simultaneously Skyping, IMing, and doing homework. In the flow of their screen chatter, Lyla mentioned that her parents were going away for the weekend. The rapid-fire IM chat went like this:

Paula: let's have a party at your house, haha
Melanie: great idea!
Paula: yay I'm going to tell everyone now
Lyla: don't you dare!

Paula: but it would be so cool!

Melanie: so much fun!

Before Lyla could get out another "don't you dare" Paula had hit send and the e-vite went out:

Party at Lyla's Saturday at 8. RENTS free (no parents)

What Paula and Melanie had forgotten was that they also Facebooked, IM'd, and Google-chatted with Lyla's mom, so she received the invitation along with the more than fifty kids in the eighth-grade class. In the middle of her own evening screen time, Lyla's mom saw it instantly and strode upstairs to find Lyla crying, still struggling online with the girls. Mom cut in on the screen chat and told the girls in no uncertain terms: "You have to clean this up right away. How could you do this to Lyla? I'm really disappointed in you. Take care of it NOW and I won't tell the school."

They did. But instead of apologizing to Lyla, the girls were punitive toward her: "Way to go running to your mother!" When she explained, "No you idiots sent it to my mother cause your just that smart," again, instead of apologizing, they got even huffier. For the next three weeks Lyla cried every day before going to school. Every day Paula and Melanie iced her out in the most obvious, hurtful ways. Her attempts to get the girls to "just deal and be friends again" were futile. And so it went, with verbal jabs—along with that in-your-face IM text lines—delivering low blows and sucker punches, an IM version of the "girls fighting" meme videos on YouTube. It was an experience of friendship, trust, and unexpected betrayal that was searing for Lyla for a couple of weeks. It was a lesson for all of them in how fast and furious online connectivity can amplify social dynamics.

A 2011 study of teens, kindness, and cruelty on social network sites found that while teens across all demographic groups generally had positive experiences watching how their peers treated each other on social network sites, younger teenage girls (ages twelve to thirteen) stood out as considerably more likely to say their experience was that people are mostly unkind. One in three younger teen girls who used social media said that people her age are mostly unkind to one another on social network sites, compared with 9 percent of social-media-using boys twelve to thirteen and 18 percent of boys fourteen to seventeen. The unkindness they witness ranges from snarky comments about someone's hair to brutal verbal assaults or humiliating disclosures.

"Hanging Out, Messing Around, and Geeking Out"

I like Mimi Ito and her colleagues' description of kids' use of tech for three basic purposes: "hanging out, messing around, and geeking out." In my work with children, I see those three patterns in healthy socializing, gaming, entertainment, and pursuit of special interests.

Ito describes hanging out as that "ongoing, lightweight social contact" that allows kids to connect with peers and experience some independence from parents, teachers, and the world of adults. They hang out and create their own social world and that has always included the language of their culture—the music, movies, and cultural reference points—as well as the ongoing sagas of their own development, their struggles and victories with relationships, responsibilities and the emotional intensity of their age.

Hanging out online makes it easier to be part of the ongoing conversation with friends. For some tweens, especially those who feel socially isolated at school, finding friends online provides a healthy sense of connection and belonging that is otherwise painfully missing for them and can be lifesaving. The online hangout is how kids socialize today and they learn the "rules" and values from their peers in that space. Social networking provides creative venues for learning as kids join social network sites, follow blogs, and post on YouTube.

Especially at this age, kids with special interests often "geek out" online, endlessly absorbed as they discover sources of content, experts, or peers with similar interests. My daughter Lily and her friends—a wonderful posse of boys and girls—would make hilarious movies and entertain themselves for hours writing and filming dramas. Kids in middle school often use technology extremely creatively, introducing arts, videos, the wonderful way of saving a narrative, documenting the world as they see it. If you are the parent of a geeky tween, you have no doubt been grateful for such opportunities. There are intricate strategic ARGs (alternative reality games) that kids— and many adults—are playing, some designed to create solutions to real-world challenges such as urban planning, or limited water and oil. The intellectual content is high, the ethic is prosocial, and the experience of collaborating is empowering and certainly better than endless hours of World of Warcraft.

As for messing around online, we all know what that is and we do plenty of it. As one middle school boy explained to me, his complaint about his parents' "obsession" with their iPads and the home computer had nothing to do with his desire to spend more time with his mom or dad; he wanted more time with the computer. "They're kind of hypocritical," he said of his parents, for rationing his time so closely while spending unlimited time online themselves.

Hanging out, geeking out, and messing around all have a place in a healthy media diet for children this age. But as a therapist, I also see the dark side as preteens struggle with issues of body image and identity and flex their social power and capacity for cruelty more boldly, and often anonymously, online. A few that my patients have shared:

Quit trying to be everyone's friend. Can't you see how much
    your new "friends" hate you? If you had eyes in the back of

your head you'd see all the faces they make every time you
talk.
You're such a slore (slut-whore)

Hanging out is how they keep tabs on their friends, yes, but it
is not always positive—even between supposedly good, supportive
friends. The notion that others are always watching, always waiting
for a reply, always quick and sometimes sharp with their own, car-
ries an edge of expectation. The constant need to be in touch is itself
stressful and works against the sturdy sense of self and the capacity
for reflection and inner dialogue that we want our young adolescents
to develop.

Tweens struggle with the new vicissitudes of relationships. On the
one hand, despite some possible awkwardness at times, face-to-face
friendships are richer in the three-dimensional way that life is. On
the other, as one tween girl explained to me, you can talk more openly
in the two-dimensional world online. She had been friends with two
boys online who had upsetting things going on in their lives. They
talked about it all openly in their online exchanges, and, she said,
"We got really close." She soon discovered that created awkwardness
in person:

> I mean, I know so much about you and you know so much
> about me, but we've never had this conversation face-to-face.
> But you actually exist and you know all my secrets and I feel
> really vulnerable, and so I turn on you . . . it's like realizing
> that the person you've been talking to is actually a—*a person*—
> and you realize the person you were talking to online is more
> "perfect" than the person in real life. Being online allows you to
> create a persona, an identity and a person that don't necessarily
> match up. You know how to deal with the identity but not actu-
> ally the actual person.

Children sometimes find themselves trapped by the persona they've created and they struggle with the consequences of needing to live it out in real life in school. A wallflower of a boy who was marginalized in the social hierarchy at school created a brash comic persona online—rude, crude, sarcastic, and willing to say anything for a laugh or just to shock—to say what others wouldn't. An observant classmate of his who had known him since elementary school told me,

> The Internet is sort of trapping him into being this person, everyone expects him to say the most horrifying thing or the rudest thing to make them laugh, and any time he does something different everyone is like, *Whoooa what are you doing? We have you in this place, you can't change!* So he is trapped because no one wants him to be a nicer person. In sixth grade, he was nicer and I bet if no one was around he would say "sorry," but the more people are watching this whole show, the harder that gets.

When Hanging Out Turns to Zoning Out

Imagine an opportunity to work your way up from an entry-level job to senior management, even CEO, at a place where you love to go to work every day, where the mission is clear, the objectives straightforward, and success is yours for the taking—all this by age twelve or thirteen. That's the attraction of screen games for many kids who play them. They may play a lot, especially by the standards of worried parents, but if their gaming is part of a larger balanced life that includes nonscreen activities and interests, a core of good friends, reasonable attention to school and grades, and a lifestyle that supports good physical and psychological health, it's probably fine. I've talked with so many young adults—like my son Daniel and others in that first generation of gamers—now immersed in careers or graduate school, who say that screen games were simply an enjoyable social activity and a place where they could be, more so than at home or school, masters of their own fate. Facebook and social networking have a game quality, too: kids post, they play, they collect social capital, they lead the equivalent of raids to rescue or punish others. They fire off quips and "likes" and gossip and innuendo. Words become the currency of the realm and can be used to make allies or punish enemies.

Beyond a reasonable, healthy social and recreational experience, screen and social media time also often serve as coping techniques for kids under pressure in an increasingly high-pressured early adolescence. Healthy coping strategies help us deal more effectively with the large and small stresses of life. But when a tween's coping mechanism leads to excessive or compulsive TV or screen time, texting, or social networking, what began as a fun helpful pastime can suddenly hijack a life. Until recently, we associated problems of excessive use with older teens. We are now seeing them at younger ages, and we are learning from clinical experience that the seeds of addiction in the teen years often are sown in the middle school years, as more tweens have access to tech and are developing habits that can easily tip toward dependence.

JACOB: WHEN GEEKING OUT IS FOR A GOOD REASON

Jacob's mother was so worried about her thirteen-year-old son's fixation with his online interests—science and news from publications around the world—that she got materials from Al Anon, the support organization for families of alcoholics, to learn more about how to deal with him more effectively. Then she called me. She listed some of the warning signs she observed in his screen use:

It's hard for him to go long periods without it. He knows
when he turns it on that it regulates his mood and it does
something to his body chemistry. "Trust me," he'll say—which
is what addicts say—it's always the language of trust. And I
understand that it is in some ways about trust, but that could
also be a sign that he is hiding something.

At the same time, she understood that certain other truths about
Jacob put those behaviors in a different, possibly more reasonable
light. "He is so smart," she said, "and he understands things about
the world that I will never understand." It was true. Jacob was an
extremely bright boy by any academic measure and he had a passion
for world news and for science. He had always had those passions,
but his extreme online pursuit of them had begun during an extended period of serious illness when he couldn't attend school and was
homebound. During that time he had been somewhat depressed at
the turn of fate that had changed his life so dramatically, his mother
told me. Now he was back in school and doing well enough, but he
would return home each day and stay glued to the screen for hours
and hours—more than seemed good for him, she said.

"I worry about the multitasking, having so many windows open,"
she said. "I remember how I learned to learn, and he was very on
track to discover that—that you put in a sustained amount of work
and have that *aha* moment that comes from developing the power
of concentration. I want him to have the skill of analyzing content
deeply," she said. She was worried that his screen time was eroding
his capacity to do that.

I met with Jacob a short time later to discuss his mother's concerns and how he himself viewed his relationship with tech. He could
see his mother's point of view, but he did not feel he had developed
an addiction. He explained:

I have to use my computer a lot for school. It doesn't really
pop into my head that I'm distracted on my computer 'cuz I'm
on it for homework and I just lose track of the time, but I don't
think "has it been too long" . . . I don't really monitor it, but
my eyes get tired and that's usually a sign, then I'll take a break
from it.

I asked him to describe a normal day, including his online and
any other tech time. Like many students, he typically got out of
school at 4 p.m., would hang out with friends for about forty-five
minutes, then come home by 5 p.m. He'd say hi to his dog and
go on the computer, break for dinner with his parents "and other
stuff," then be back on the computer from about 8 p.m. to 10:30
p.m., when he went to bed. What did he do on the computer during
those blocks of time? Homework usually amounted to about three
hours of work, he said, though it was interspersed with other diversions. He would visit an average of thirty news sites—he is a news
junkie—and then watch Truth or Fail, a trivia game and "a few"
YouTube game shows. And TV. "I consume the majority of my TV
on my computer and I'll watch my three shows or five on Hulu."
He also had gotten an iPhone, which he used as a phone and would
call friends, as an online connection for playing games (often with a
friend), for texting, and for photography.

As Jacob shared more about his health issues and the impact they
had on him physically, it was also clear that his condition had been
emotionally challenging and that plugging in had been a helpful coping mechanism. Jacob's passion for world affairs had deepened, although he also acknowledged that he had become "a skimmer," not
reading as deeply as he once had. Nonetheless, in Jacob I heard a
very mature capacity for reflective critical thinking: a cool, creative,
intelligent kid.

As we talked about tech use versus tech dependence or addiction, it seemed that while he did devote significant time to those activities, which can be a sign of dependence, he was also able to be off tech comfortably when circumstances called for it. He wasn't restless when unplugged, didn't feel a craving for it, and easily left it behind for tech-free time with his parents. Once when he'd developed symptoms of repetitive stress injury in his wrists he dropped gaming for a month until the symptoms went away. He had friends—face-to-face ones, good ones, and not only online friends—and he didn't neglect them. He didn't lie about the time he spent on tech nor did his activity interfere with his schoolwork. When his mother would ask him to get off-line or come to dinner, he did as she asked.

I have worked with several young boys who, like Jacob, developed a useful, temporary dependency on tech during a time when they were socially isolated at school or circumstances forced them to change their familiar routines—an illness or injury, a move, or a best friend's move away. Jacob was eager and ready to begin high school, and signs were good that he would be fully engaged there. His passion for news and politics may ultimately distinguish him as a serious student of debate, perhaps of international relations or politics. Or maybe some night I'll see him on CNN. As it was for Jacob, for many middle school boys life online is a lifeline, a world where they can connect with interests and with others who share them.

KAREN: WHEN COMPUTER ADDICTION IS A COPING MECHANISM

Karen had always been the family firebrand. She had been "the challenging one" from the time she was born, her mother, Rona, explains. Hard to feed, hard to settle for the night, hard to keep occupied and happy through the day. Rona struggled as an at-home mom through Karen's early childhood. Through elementary school,

Rona found it easier to park Karen in front of the TV because it was one of the few things that kept her happily occupied. "I feel so bad about it now," Rona tells me. "The function of it, if you got to the very bottom of it, was that it kept her out of my hair. When she was in front of a screen, she wasn't fighting with her sister, teasing her brother, causing a disturbance in our house. It's not a great reason." But so understandable.

She goes on to explain how she didn't get help early enough with her second child, and while struggling with her, lost footing with regard to tech in the home. By the time Karen reached middle school, the family had four TVs in the house. TV and computer games had become the salve, exposing Karen to everything Rona hated in the culture—"junk programs, junk values, junk language, and junk attitudes." Her daughter's most recent obsession: the Kardashians.

"My thirteen-year-old knows every last fact about the Kardashians, and I want to throw up every time I see her watching the Kardashians, and the things they talk about, and the morals that they exhibit on that TV show."

Things got worse when Karen began seventh grade and the school issued laptops to students for school use. Instantly, she became even more screen-obsessed, Rona says. She was glued to her laptop after school, claiming to do homework but clearly multitasking on noneducational web sites, chat rooms, and social media. She would contact her mother online from school and beg her to play games with her on her iPad. The school had no control and did nothing to monitor students' computer use during the school day. Rona felt overwhelmed and now very concerned about her daughter.

Karen arrives for her session a bit defensive about her behavior; she thinks her mother was overreacting *as usual*. She wiggles and fidgets through our session, yet seems unaware that her body is in constant motion. Her eyes are huge and expressive and she can go from sunshine to stormy in a nanosecond. Unlike Jacob, Karen does show early signs of addictive behavior: she describes how her mood will change when she is plugged in; she feels happier. When her parents pressure her to get off screens, she responds angrily; she sees it as an intrusion, after all. In diagnostic terms—and practical ones any parent would recognize—she is unable to transition away from screens in a reasonable way. She also prefers tech to any nontech activity with family and friends. She will hedge about the time when Rona asks her how long she has been watching TV or playing online. She also admits that she sneaks her laptop into bed and watches more there, often late into the night, costing her sleep and making the next day at school all the more difficult.

Shortly after we met, I recommended neuropsychological testing and Karen was diagnosed with ADHD, which her parents now see as having been at the heart of much of the tension at home from the start and which most likely played a role in her screen addiction. For children with attention and some other learning disorders, screens can become quickly addictive as they soothe the anxious or agitated mind. The plan of treatment for Karen's ADHD would include constructive attention to her media and screen time.

For any child using screen time to de-stress, the risk is that he or she isn't getting the kind of conversation and interaction with parents, friends, or family that help develop the self-regulation skills and social and emotional insights they need. The more they depend on their computers to cope with underlying, often unidentified problems, the greater the chances their dependency can turn into an addiction. Connected as they may be to so many online, the experience of this kind of immersion is often one of isolation. We know good, strong friendships and support from parents are critical at this age;

when computer time disrupts that personal connection, it's time to reboot.

Sexual Development, Drives, and Diversions Color the Tween Scene

A teacher told me the story of how proud she was when her thirteen-year-old son took his laptop into the bathroom during homework time. "I thought, 'Wow, he's really into his work!'" Then she discovered he was masturbating to online porn. "Okay, so I guess it's not under the covers anymore," she told me.

Our children are entering puberty at an increasingly earlier age, some at ten or younger, whether due to hormones in the food chain or other environmental factors, and many are arriving at changing bodies earlier than their parents did. They are doing so in an era that overexposes them to the adult world of sexuality. Puppy love in the age of sexting, vying for social capital on Instagram and Facebook, grinding and freaking at your first middle school dance—all of it fast-forwards tweens into a world of confusing messages.

As wonderful as the Internet is as a primary resource for academic and a lot of everyday information, clearly it has also become the primary source of sex education. This is especially so for children in families uncomfortable with having thorough, ongoing, positive conversations about all aspects of love, intimacy, good sex, bad sex. And that is a great many families. Ironically, we live in a culture where sex is all around us, yet parents are reluctant to talk about sex with their tweens, as if talking about it will plant prurient ideas in their minds. Meanwhile middle-schoolers are fluent in the language of sexting, booty call, and FWB (friends with benefits—that is, sexual ones). They sit with us and watch a Superbowl commercial in which a beautiful woman in a bubble bath texts a picture of herself and men fall off ladders and, what?—we all laugh together. Popular culture exposes kids to a virtually unlimited range of sexual behavior, a great deal of it unreal, often unhealthy and trending toward emotionally damaging and abusive and violent behavior. Parents feel utterly ill equipped to deal with it. Very few of us had the kinds of conversations with our parents that we should be having with our children today.

Kids today know far more about sex at an earlier age than generations past and are often more sexually aware than their parents because of their exposure to TV—including so-called family TV—YouTube, and the sites their friends show them online. When we hit the mute button on candid conversation about sex and what they see online or hear from their peers, they lose the opportunity to talk thoughtfully about their own sexual development with adults who care about them, and we lose the chance to talk with them about how to protect their own introduction to healthy sexuality. "Safe sex" means more than condoms. Without conversation we leave them psychologically unprotected for exposure to sexual content in the media and adolescent culture.

When the psychologist Michael Thompson interviewed a group of high school boys for his *Raising Cain* PBS documentary several years ago, the boys spoke openly about viewing porn pretty much daily and about their masturbatory lives. When parents see the film, he told me, "What is unnerving for them are the stories the boys told about seeing porn in third or fourth grade. The average American child sees pornography now at eleven, and their parents are still not able to talk to them about it when they're thirteen or fourteen. So they're getting information unmediated by any adult guidance. And what that requires is for parents to stop hoping that their boys don't see porn."

With Internet porn so accessible, Thompson notes, "boys are seeing everything and they're seeing everything early. And they get that it is exciting, and they get that it is forbidden. And when you are ten years old, how can you go to your parents and say, *I saw these explicit pictures, and they were very exciting to me and baffling and overwhelming?* To whom do you turn? You can only turn to your friends or to older boys."

Now that peer culture accepts slut-chic and views sexting as cool, marketers have stepped up offerings that express and encourage that kind of ambient sexuality. Halloween costumes for tweens now include the "sexy nurse" and "sexy warrior woman," or outfits clearly designed to look like hookers or strippers.

From his long work with boys, Michael Thompson sees the power of the family culture and parents'—especially fathers'—influence for good. "Pornography stokes a boy's fantasies. Does it tap into every boy's future possibilities as a rapist? I think not," he says. "I think loved and well-raised boys who haven't seen women exploited and who have seen their fathers and the other men in their life respect women—I think they have the fantasies, but they understand that it's a zinger, it's just a fantasy." I've had conversations with thoughtful fathers who upon discovery of their sons' use of online sadistic porn have given them more respectful nonviolent erotic images of women for their viewing enjoyment.

Research suggests that for many boys it is harder to separate out fantasy and reality, entertainment and exploitation, in the era of reality TV shows where teens are entertained watching adults behaving horribly to each other. A study of 1,430 seventh-grade students found that many are dating and experiencing physical, psychological, and electronic dating violence. Teen-dating violence and abuse is a major public health problem nationally and prevention needs to be a public health priority, the report said.

While the hookup culture may seem hip, it isn't benign. FWB usually means boys convincing girls to bring them to orgasm without any commitment or relationship. When asked if it is reciprocal, girls often say *eww gross, no way*, revealing their real level of psychological, social, and sexual development. The disconnect between healthy sexuality and a caring connection, between who they are and what they are doing, is striking and worrisome. Between adults, casual, recreational sex is a choice. But it is not a value we want to promote for preadolescent children. It's also not legal, and a boy who engages in sexual activity with an underage girl—whether it is consensual or not—runs the risk of criminal charges and potentially a permanent record as a sex offender. Who benefits from FWB? It might seem that boys do. But let's step back and ask if boys really benefit from a culture that teaches them to treat women as sexual objects, to disconnect their own capacity for love and romance from sexual intimacy, and at worse to be frankly demeaning and abusive to girls. Nobody benefits from FWB. It has always been the province of preadolescents to figure themselves out sexually as they approach adolescence, but this is the first generation to grow up in a culture where being sexually intimate is understood to be disconnected from the context of a relationship and caring interpersonal intimacy. Using the word "friend" in this context can be confusing, especially to girls because most often they are the "friend" providing the benefits. In middle school, would a true friend really ask a friend to sexually gratify them this way? We need to talk with our kids about what it means to be friends in the context of FWB.

Jackson Katz says media has become "the great . . . teaching force of our time," and that the porn culture "has shaped a whole generation of boys' understanding of their sexuality and the way they interact with girls sexually." In the absence of thoughtful discussions about sexuality and in the absence of thoughtful, information-focused education, Katz says, the pornography industry rushes in and introduces "an incredibly brutal form of men's sexuality."

Parents and most adults in teens' lives tiptoe around the issue,

embarrassed by it or in denial. We hide behind language that pretti-
fies and minimizes the disturbing reality, Katz says.

> Boys are not "looking at porn" or "watching porn videos."
> They're masturbating to porn. And that's much more active
> engagement with it than the words "look at" or "watching"
> suggest. They're actually having orgasms to it. So again, think
> about the profound experience of a boy, say a twelve-year-old
> boy who's just through puberty and really charged and sexually
> vibrant, and they can easily get to the Internet, and on the In-
> ternet this is what they get, right? They get scenes of men gang-
> raping women, scenes of guys calling women "F-ing bitch"
> and all this stuff, while they're having sex with them, grabbing
> their head and putting it down to their penis, the stereotypical
> "you're going to give me a blow job" kind of thing. And these
> young guys who are just charged up sexually and heterosexu-
> ally attracted to women and girls—this is what they're seeing as
> normative and they're obviously getting turned on by it.

This is damaging to boys and girls alike. The public conversation among adults so often becomes polarized and unproductive in
that boys get cast as aggressors and girls as victims. The argument
is also made that erotic content has always been out there and there
is nothing to be done about it. This is not the place to take up that
debate. But what gets overlooked in all that is what is happening to
our children. And that is this: children are learning a model of sexual
intimacy and romance from pornography. All of them are ill served.
It is a shame when we let the conversation stop there and ignore the
fuller emotional life of boys and girls, their need to find love and be
loved, to be responsible and appreciated, to make a difference and
matter in the best way.

There is a world of difference between normative sexual arousal and gorenography. Looking at a model—whether it's in a swim-suit ad or the *Sports Illustrated* swimsuit edition or a *Playboy*—can serve as a fantasy that stimulates the erotic arousal process, which is the inner fantasy world of a twelve- or twenty-year-old. In a boy or young man's own imagination, they take an image and make a movie. That's been going on forever. That's a world apart from triangulating those basic sexual responses to sadistic and violent movies that you're actually watching, participating in a script that's being acted and likely introduces a level of sexual content that is developmentally and psychologically inappropriate for a young "viewer." When boys link their sexual response cycle to gorenography, it ruins their capacity for normative sexual arousal. Boys are confused: a thirteen-year-old boy at a school one day asked me, "I don't get it—why would a woman get turned on being choked?" In focus groups, in schools where I've facilitated other kinds of group discussions, and in the privacy of the therapy sessions they have asked these questions and shared these pornography-based assumptions about girls and expectations of girls. They see these images that imprint on them the arousal and they don't realize this is not the kind of sexual intimacy that most girls long for or would ever want to get involved in. If they see this over and over they think: this is real. I've worked with boys who have been surprised, disappointed, and confused by girls' unwillingness in real life to role-play sexually violent scenarios or otherwise satisfy the fantasies the boys have internalized from porn.

When my friend Nina's daughter and her girlfriends posted Facebook photos of themselves posing model-like in bikinis at the beach on spring break, they liked that they looked cool and sexy like movie stars. Nina could see the obvious pin-up-girl potential and suggested to her daughter that they might want to reconsider posting them online. Her daughter insisted it was fine because "only our Facebook friends can see them." Nina decided she needed to clue her in.

"I just said it. 'I hate to tell you this, but eighth-grade boys do a lot of masturbating. They are gonna be looking at you in that bikini and they're gonna be touching their penises.' She said, 'I'll be taking that down.' I told her, 'I know they're your friends, but this is what they're doing right now.' And she was just like, 'Oh my God.' Part of her was like, 'Please don't tell me that.' "

"Throwing Them into This World . . . Kids Make Terrible Mistakes Every Day"

On a crisp fall morning I sit with Robert Warren, a seasoned head of a middle school at an independent school in the Midwest. A dad of four young children, he's a thoughtful man, always open to rethinking assumptions and always approachable. He is firm and fair with students and laughs easily with them.

His school, in a progressive initiative just before he came aboard, had embraced laptops for all students. In the decade since, he has seen a paradigm shift in the tween world, as the gathering ground for students' social life has shifted from school, sports, after-school playdates, and the neighborhood mall to online and social media. The revolution in the living room has sent educators like Warren rushing to broaden efforts to help children, parents, and communities navigate the challenges when kids and the media culture collide.

When they first implemented their laptop program, the biggest issues had to do with kids wanting to play video games in class or use the school e-mail system to send messages to each other, which was not allowed, Warren says. "The worst thing somebody would do was send somebody else a nasty e-mail and call them a, you know, 'fucking bitch' or whatever. And then we'd have tears the next day and we'd talk about it."

As it has been at so many schools, this was just the beginning of degrees of difficulty in school dynamics that popped up with laptops. The situation has gone from name calling online to far more complicated problems. Now that the honeymoon with laptops is over, many educators are becoming increasingly concerned about the complexities of giving middle school students unlimited access to a computer when they are not developmentally ready for that kind of responsibility. More and more schools are developing policies, contracts, and expectations of what can and cannot be done on school laptops. And as is always the case, some kids will follow the rules and some kids won't.

The preadolescent brain is not ready for the responsibility that comes with open access to online media that can take an image or message viral in seconds and that have such serious consequences for sender and subject alike, Warren says. Given unfettered access to the Internet, "kids make terrible decisions every day . . . and why not? It's more fun, it's more exciting, it's more set up, it's more—*everything*. And I'm not just talking about watching a lot of pornography. I'm talking about the social networking and how they interact with one another in a way that's not nice."

The idea that online anonymity provides a safe cover for meanness is true but not the whole truth, says Warren. Contrary to the idea that kids find it easier to send nasty notes by text or e-mail because they don't see the impact on the recipient and aren't aware of how hurtful it is, Warren thinks part of the thrill for them is that they are well aware of the pain they're inflicting. "I really think that kids know what the impact is, and they're doing it specifically to generate that impact. I don't think it takes away the capacity for empathy, I think it just makes it easier to be worse and to be more powerful the more removed you are."

In one of the worst situations to arise at his school, students set up an ask.fm page with a child's name on it, a socially awkward new girl living with her grandparents who was struggling to make friends. On the ask.fm page the students described her as "a miserable, soul-sucking, stupid, ugly fucking loser" and they warned other kids to stay away from her. After consulting with the family, the school leadership decided to share the most offensive portion of this hateful post with the entire grade in a strongly worded assembly that drove home these points: the post was egregious, there would be consequences, and everyone was responsible for any behavior that hurts someone and undermines the community. When reckless behavior becomes routine, children are emboldened. This head of school was unwilling to let these kids think that their habit of using online posts to gossip and post grievances about each other was benign.

Warren urges parents to establish house rules and clear guidelines for media use. Programmable or other installed parental controls that limit a child's use of devices are essential, he says, "because what kids can do is not what kids should be doing. It's not okay."

Parents often tell me they feel lost defining limits on computer use because their kids are on screens for so many different reasons and seem to assume they're entitled to do as they wish. The basic message you want to send then is this: *This is not your computer—I know it has your name on it, but this is my computer (or your school's computer). I'm your parent and I reserve the right to see everything that's going on there. You need to be on the computer in an open place. I have the right to know what your homework assignment is. You can't be in your room with the door closed. You can't take it to bed with you. You can't collapse a screen when I walk by. We have a code of conduct and we expect you to stick with it: don't be mean, don't lie, don't embarrass other people, don't pretend to be someone you're not, don't go places you're not allowed to go. Don't post pictures that Grandma wouldn't love. Don't do anything I wouldn't approve of.*

Fear, anxiety, and the thrill of competition amplify the online schadenfreude into a spectator sport. In the digital world, where children this age need boundaries, there are none. Where they need more experience in face-to-face communication and relationships, they are getting less. And where they need the connection to family to ground them as they move into the wider world, many find a growing isolation. So they turn back to their screens.

Given the reality of our culture and the digital environment, our best shot at protecting our children comes in educating them early and continually, as they become increasingly independent and active participants in the digital culture. Caught between the cozy world of elementary school and the propulsion to the world of YouTube, they need our time, attention, and courage to discuss subjects that make us uncomfortable—whatever that may be for you. They need the kind of "hanging out" family time and conversation in which you and your tweens can talk about values, love, flirting and hurting, civility and cruelty, and what constitutes crossing the line. Your tweens really do want to understand how you see the world, what matters to you, your values. And they want to feel that you want to understand what matters to them. They want to be able to bring all the identities they are trying on that day, without you overreacting or getting too preachy. They need you to be curious and clear, to set limits, and to be flexible. It's an in-between time for everyone.

Chapter 6

Teens, Tech, Temptation, and Trouble

Acting Out on the Big (and Little) Screen

*My generation is so comfortable at communicating electroni-
cally, but we are terrible at actual relationships.*

—CHARLOTTE, EIGHTEEN

For two years through the thick of high school, the three girlfriends were there for each other day in and day out, around the clock. Texting, instant messaging, and posting on Facebook, they kept each other emotionally afloat through freshman year and beyond, through tough tests and the inevitable social drama that comes with teen life.

"We were all very close, and communicated a lot by text," says Mattie. One of the girls, Jill, almost always communicated via text. All three attended the same high school, but since the girls texted more than talked, it was a sign of devotion—and a sign of the times—that these best friends hardly spoke of important things when they saw one another socially.

"Jill just couldn't communicate in person and so we would have all these serious conversations texting and, you know, whenever she was upset—not even with me, but just about life," Mattie says. "She just couldn't do it in person."

The plot thickened, as did the intensity of the messaging, when Jill began sharing the ups and downs of life with her new boyfriend Alex, a camp counselor she'd met a previous summer. The girls hadn't met him yet, but he had professed his love to Jill the summer after freshman year and over time their relationship had taken on a kind of epic quality in the running texts and posts among the three girls.

"She had this boyfriend for two years and she would tell us about all these dramatic interactions they would have. You know, he would do this or that, they would break up, they would fight, they would get back together," Mattie says. Later, there was talk of some abusive behavior.

Mattie cared about Jill and her troubles, worried about her well-being as a friend does, and took special care to "be there" for her, texting whenever Jill needed to vent or needed support. "I came from a sort of rocky middle school background so I was just happy to have like a really constant friend," she says.

Then one day Jill finally posted photos of the counselor she was dating. At last the mystery boyfriend had a face.

"But then somebody else in my grade said, 'No, the kid in that picture is my friend and I know he went to that camp, but he was just a camper and I know him and that's not her boyfriend.'" Jill had also said that her boyfriend attended a particular school, so Mattie did some social sleuthing and discovered he didn't exist there, either. Eventually the girls confronted Jill—by text of course—and after much dissembling she confessed that she had lied. That wasn't a picture of her boyfriend; she had lifted the picture from a camp yearbook. And her boyfriend wasn't unattentive or abusive. He wasn't anything. He was just made up.

Mattie was dumbfounded. She felt foolish for having believed her friend for so long, even when Jill had been evasive and her boyfriend hadn't materialized beyond texts for two years.

"Mostly I felt betrayed and sort of in shock because I had really spent two years kind of living through her. You know, she had this exciting dramatic life and I thought it was cool that she was my friend . . . but then it was all basically a lie."

It wasn't the lie itself but its long life and dramatic presence in the friendship that made the gullible girlfriends feel so manipulated. As Mattie explained, "I think that certainly people could make up lies like this [before texting], but I don't think it could go on for as long because eventually, you know, you'd have to start having these conversations face-to-face and I think it's harder to keep up the lie when you're having more spontaneous interactions."

Literature is rife with classic tales of miscues and lies about love and the story of the phantom boyfriend or girlfriend is one I hear often and certainly is not the most disturbing scenario from the digital teen world, as we'll see shortly. But in its quiet way, this minidrama of deception exemplifies the challenge and vulnerability of adolescence as teens struggle to define themselves and explore relationships. In their search for meaning, connection, and the relationship skills to take them into young adult life, tech turns it into a shell game.

At a time when developmentally they are wired for exploration, experimentation, and self-discovery in the context of human relationship, teens' relationship with tech has itself become the context for all else, the experiential platform on which life plays out. What nature intended to be an embodied experience of one's self in relation to others—face-to-face, sensorily rich, and physically, emotionally, sexually, even spiritually textured—is now a mediated experience. Tech, for all the wonderful ways it can enhance life, at this particular developmental juncture of adolescence sometimes does more to obscure and confuse than contribute helpfully to the human connectivity a teen needs. Instead of opening a way into deeper experience and understanding and deeper social and emotional learning, tech and social media can become a digital detour that takes them far-ther from, not closer to, a meaningful relationship with themselves or others. In extreme cases, detached from meaningful connections, some teens create a chilling world of their own in which self-injury or addiction precludes human connection, where porn replaces inti-macy, and where disturbed thinking can turn tech connections into a weapon for social or emotional cruelty.

Avatars, Angst, Ambition, and Adventure

Adolescence has always been characterized by emotional intensity, impulsivity and risk-taking, the bid for autonomy, the quest for iden-tity, and the desire for intimacy, sexual and otherwise. To a teen, tech is the perfect accessory. It's the do-it-all tool, the Swiss Army knife for modern adolescence. Its key selling features:

Tech plays to the powerful adolescent drive for independence, providing opportunities through texting and online where teens can create a world of their own largely inaccessible to adults. Without stepping out of the house, without so much as glancing down as they text, they chat amongst themselves and can roam the world and hang out with anyone they please, from a secret flame to flaming radicals trolling for recruits.

As they rush to live large, tech plays to the superficiality of teen relationships and teens' chimerical tendencies. Everyone's an avatar. Teens' efforts to grow up, fit in, find and define themselves make them mercurial, at times seemingly unknowable—even to them-selves. Facebook and other online platforms allow for endlessly mal-leable identities.

Tech is the perfect accomplice to their classic risk-taking behav-ior, putting them a mere mouse click away from disaster. They can video-chat themselves into a dangerous multiplayer game of "truth or dare," only to discover that it has been recorded. They can arrange hookups with strangers, or steal your identity and pretend to be you online, or provide text or phone cover for friends who want out from under a parent's watchful eye. It's all just that easy.

Tech offers a round-the-clock stage for teen drama. Like an app for emotionality, tech connects and quickens, heightens and amplifies all things. Even silence. Texting, video chat, sexting, and social media have created a streaming soap opera for teens, a scintillating reality-show subculture in which everyone gets to play both celebrity and paparazzi. The story lines reflect the miscommunications and misunderstandings, disturbing sexual content online and in text exchanges, and adolescents' angsty mix of self-critical, self-doubting, and mean-spirited inclinations.

Tech offers a quick fix, whether for entertainment or escape. Overwhelming intense and persistent emotions of feeling inadequate and hopeless about ever measuring up, feeling lonely, empty, like a reject, powerless, and/or anxious about it, make a teen ripe for addiction. As do a lack of self-control, impulsivity, and vulnerabilities for anxiety, depression and mood disorders, and ADD/ADHD.

Finally, tech plays to the natural sexual drives and desires of adolescence but deletes the pause between impulse and action. The fast-twitch ethos creates a realm where reckless sex, pornography, and loveless liaisons are the accepted norm. This, perhaps more than any other single factor, has transformed teen life and teen risks so profoundly. There is no app for emotional intimacy, no digital shortcut to the deep, rich knowing of another human being—or of ourselves in that context. To the extent that texting and social media dominate a teen's time and attention, he or she is missing the kinds of conversation and face-to-face interactions that develop the relational skills for friendship and emotional intimacy. Many teens describe a palpable longing for meaningful relationships, for healthy romantic relationships where sexuality and intimacy go together. Others don't speak of longing, but are confused or disappointed, some already bitter or cynical, about love and sex with a loving partner.

Evolutionarily speaking, it has always been adolescents' job to join the dominant culture, so their rush for the border should come as no surprise. What has changed is that there is no longer any border, not even speed bumps slowing their entry into a dominant culture with destructive elements that can so swiftly overwhelm them. Teens like to project an image of sophistication in their ramped-up realm, but as a therapist and educator in schools I spend a lot of time helping them make sense of the disconnect between what they felt compelled or pressured to do, what they are emotionally and developmentally ready to do, and their desire to stay true to their self. I hear the stories of video chats, where two or three girls are in one room, chatting away with another boy or two, full of that intoxicating false courage, and maybe a shot or two of vodka as well, and suddenly they are flashing each other, and then some. This, with no clue whether anyone is recording it, or who else is in the room, or who will remember, record, or reveal what took place the next day or the next year.

The tech may be easy to use, but that doesn't mean it makes the real work of adolescence any easier. Online you can use Photoshop for identity formation. In real life, it takes discovering who you are, whom you love, who loves you, how to love, what you think, and who you want to become. Then there is figuring out who you are in relation to others and to yourself, nurturing self-acceptance and dealing with body image. Photoshop isn't up to that task.

When teens come to see me in my office, I watch them attempt to ease into their unplugged self as they fall onto my sofa. Some hold on to their phones as they put their feet up on the huge ottoman we share, unable to be fully present without their technological touchstone near at hand. Like the faint electrical buzz that hums around high-voltage lines, there is a subtle, suspenseful charge to the energy around teens and their tech. You never know whether, in the next moment, that power will illuminate the darkness or light the fuse on a meltdown. Neither do they.

Texting Pushes the Mute Button on Emotional Nuance

Nora, at fifteen an engaged, thoughtful, and accomplished girl, has a problem at school. She shows me a text conversation from earlier in the day with Mike, a boy in her class whom she doesn't know well:

> Mike: so, are you good at hooking up?
> Nora: Um idk. I don't really think about that.
> Mike: well, I want my dick in your mouth? Will you at least be
>   my girlfriend
> Nora: I don't really think of relationships like that. I want to get
>   to know you first
> Mike: Don't be such a prude. I really like you
> Nora: I think we should be friends first. I'm not gonna just do
>   anything. I'm not like that so make up your mind.

Nora is frustrated, to say the least. There she was, she says, writing her English paper and up crops this "stupid, disgusting exchange that is so not like me, like any of my close girlfriends, and certainly not anything that I would respond to. Is this what romance is for my generation? I *hate* this, I hate that a relatively 'nice' boy who sits two seats away from me thinks he can just e-mail me like this and it's 'fine.' That I should be flattered that he likes me? I don't know what to think about him, this is such a disgusting come-on—but, honestly, it's typical for the boys at our school."

She wants to talk about how she can find out what Mike is really like. Apparently his text has not been a deal breaker, but an icebreaker. Is there more to him than this? What does he mean when he says that he "really likes" her? She plans to tell him that she is offended by his calling her a prude, when *his* behavior is disgusting and immature. How will he take that? She wonders, tearfully, if it is worth the effort. Like so many girls I listen to, she tells me how trust and friendship are her lifeblood, and how she longs for a boyfriend and to be in a real relationship. As much as she relies on texting and enjoys social media, she sees tech seriously complicating and compromising relationships, she says. "I think it's taken away, like, this generation's ability to socialize like normal human beings."

Texting, especially for teens, has become a substitute for direct, live conversation in a way unlike any other medium in history. Nothing matches its capacity for rapid, continuous interpersonal communication, multitasking that allows for unlimited conversations to be in progress at one time, or the way it makes distance disappear. Yet no one would consider staying on a single phone call all day, every day. Everyone agrees that texting is great for making plans, meeting up, last-minute changes, quick kudos, or cybersmiles. But for any meaningful conversation, a lot gets lost when the translation consists of only a few lines of text on a screen. The traditional Zen koan asks, "What is the sound of one hand clapping?" Texting is the digital age answer: a conversation without tone or emotional complexity, content without substance. It is a shallow language of inference, not insight, and certainly not intimacy.

Psychologically, texting often promotes a pseudointimacy that easily becomes a stand-in for the real thing. Teenagers are so afraid of intimacy, the vulnerability of someone knowing and seeing you and the risk of rejection. The more children begin to use texting at an earlier age rather than speaking or reading, the more a printed word replaces listening to the human voice and absorbing and understanding nonverbal social cues such as facial expression and body language. The more they text, the less opportunity they have to develop basic relationship skills in face time conversation and hanging out. The less practice they get at face-to-face interaction over everyday things, communicating ideas and feelings in person, the less ready they are for relationships of greater emotional complexity.

"We can text for hours on end, or video-chat while doing home-work so we're never alone," Isabel says. "Sometimes I've fallen asleep with my Skype camera on and my boyfriend can see photos of me asleep in a video chat! But still, when it comes to being really honest and open when we are together, it's so hard. It's hard to say really important, really personal things. We're so used to texting 'cool!' and just being witty and sarcastic."

I sit after dinner one evening with several teens I know well. I ask them to share their experiences and thoughts about adolescence à la tech, especially texting and its impact on relationships.

Nichole, nineteen, tells me she had exchanged *twelve hundred texts* with a friend who is a boy, not her boyfriend, when she had to travel by train for a day and "he kept me company." When I express surprise at the number of texts they'd sent, she responds pragmatically without missing a beat, "That's why I have unlimited text messaging."

The steady patter of texts can be a reassuring connection to the teen's inner circle and social world, they said. But the steady patter of reassuring texts can also become a drumbeat of obligation many teens find exhausting. For any message with emotional content, text-ing is as likely to confuse as clarify. Texts that read like meaning-less banter can seem laden with emotional subtext. Yet many teens will choose to have conversations about emotional situations—fights, for instance—by text because they feel "it's easier," Shayna explains. She is not one of them, preferring instead to deal with emotionally charged conversations by phone, video chat, or in person because "you can't get points across when you text." However, she says, most of her friends feel differently:

It seems like a foreign concept to a lot of people to have a
fight or something in person. Girls are like, "I don't want to—
it's too awkward in person." People will say "Yeah, do it over
text if you're going to confront her." And if you ask to talk to
someone face-to-face or talk to them on the phone, they won't
do it. I was just in a fight with someone and I was texting them,
and I asked, "Can I call you, or can we video-chat?" and they
were like, "No." And I knew they were doing nothing—it's just
too awkward for them to do it face-to-face.

"Text is easier because you can think your thought," Laine in-
terjects, "you can think it through more and you can plan out what
you want to say, and you don't have to deal with their face or see their
reaction; you just hear what their reaction is."

Learning how to communicate is one of life's greatest challenges
and gifts. The capacity to know and communicate what you are
thinking and feeling, when someone else has their different thoughts
and feelings and you both are upset, is a core life skill. Many teens
tell me that not having to see the other person's response makes it
easier to stay connected to their own reactions without feeling silenced or activated by the other person's visceral and verbal reactions.
This is all good, but what is missing is the ability to see the impact of what you are saying on the other person, to listen to what the
other person is feeling without losing touch with yourself, and to be
able to figure out together how to move forward. At best, all written
communication provides a personal pause, time to think and reflect,
to literally and metaphorically think before you speak. When and if
texting leads to better real-life direct communication, then that is a
good thing. Ultimately, being in a relationship means just that, being
present, to yourself and the other. But the nature of texting is so often
the antithesis of slow reflective thinking about one's self and the other
who will receive the message.

The very things that texting eliminates, making it easier, are the lessons teens need to learn: how to calm yourself, express yourself clearly and respectfully, understand your impact on the other with empathy and not just with anger, read their physical and emotional cues as you listen to their side of it, and together figure out how to go forward, each accountable for your own behavior and its effect on the other. Even if a teen wants to communicate emotional nuance, the space constraints and rapid response cycles of texting as a medium militate against it. Text excuses you from dealing with the human complexity of communication.

Tech often amplifies gender differences in communication styles. Girls search for a signal of personal interest (*do you think he likes me?*) and boys play their cards with their best poker face. Girls often come into therapy appointments clearly upset and show me a text from a boy that says, *hey sup* (what's up?) or *sorry, not tonight.* They ask me what I think the boy means. *What does he mean? What does he want? Is it just tonight or is it something more?* It is such torture to not hear a tone, to not have a chance to say *how come?* without looking needy or desperate or too into him. It is so hard to know how to play your cards when communication is so indirect, so easily misunderstood— and sometimes intentionally manipulative.

"It's so confusing to know what a text means," says Serena. "When a guy texts you at a party or on the way to one, to invite you to a party, they don't feel an obligation to keep texting you—or even come up to you at the party! Sometimes you know it's really a booty call and if you text back you're saying, 'Yeah, I'll hook up with you . . . FWB.'"

A sixteen-year-old boy tells me that boys aren't confusing, they simply have a different agenda in the conversation: "It's kinda true that we like to entertain ourselves by seeing how far we can push the limits with girls online . . . it's a guy thing, we do it to impress each other."

Leigh, seventeen and a senior in high school, comes in for therapy and tells me, "Today when I walked out of English class I felt dead inside. I wanted to curl up on the couch in my hoodie and disappear." She has had depressive episodes and says now, "I just feel so disconnected from everybody around me and I have no energy to do anything about it." She is scared to be honest with people, and when I suggest she talk to her boyfriend or one of her girlfriends she says she can't. "Kids don't talk like that anymore." She just texts them that she is "fine" and she tells me that she'll continue to be "kinda fake." Texting makes that easier, too.

The worst thing about teen texting culture is that it is unacceptable to call and ask a direct question. Psychologically, the ability to talk, listen, empathize, and ask questions is the touchstone in communication. Texting is the worst possible training ground for anyone aspiring to have a mature, loving, sensitive relationship. Developing that ability requires learning how to assert yourself and be vulnerable, to say what you feel and think, to be curious about the other person's experience, to deepen connection through understanding without being so afraid of what the response will be. The net effect of texting culture is that teens silence themselves in the very relationships that offer the potential for the most critical social and emotional learning of the age. Literally and metaphorically, the mute button is on.

Facebook or Fakebook: Image, Identity, and the Empty Obsession with Presentation

Spike, a recent college graduate, tells me of his sense of confusion and betrayal after breaking up with his girlfriend, with whom he had maintained a long-distance relationship with weekend visits for a year. They had known each other since high school and much of their ongoing conversation had been through Facebook posts and texting. He had always felt they were "very compatible" in some ways, but over time he realized that his aspirations in life had evolved and on some fundamental points it was clear that they wanted different things. When he told her he wanted to break up and shared his reasons, she was devastated. She tearfully explained that the things she had said on Facebook weren't really true and that she was open to other ideas. Spike continues:

> What stunned me was when she said, "But I'm not really like that," and I'm like, "What? Why didn't you tell me that? Are you serious?" She said, "I didn't know how to—I was scared, which probably makes you even more sure you want to break up with me." She got that right! It made me really question the level of trust and intimacy. I mean, I thought I knew her, and hearing this has shaken me up. If she couldn't trust me enough to tell me these major things about herself, what does it mean to be in a relationship? All her texts and Facebook posts and time we spent online were just her posturing, being who she thought she should be.

Spike is in his early twenties and his girlfriend is a bit younger, but the Facebook persona she had maintained so carefully throughout high school and well into college was barely even an artifact of herself. It really only represented the girl she thought she needed to be. But she couldn't give it up. Eventually, she became disconnected from both her "online self" and her authentic self, so much so that when she finally let go of the online identity, she had no sense of who she really was at all.

The adolescent search for identity is not so much a hunt for the prize fully formed, but a journey that defines you through experience and insight gained. It is one thing to be an armchair traveler to faraway places you cannot visit; it is sadly another to bypass lived experience in favor of a screen life and identity composed of posts, pics, and a superficial social media portfolio. The image of the preening teen is nothing new, but for some teens, obsessive attention to their online image or identity far exceeds the everyday mirror checks or seasonal obsession we associate with a self-conscious age. I see this presentation anxiety in my practice, and recent reports on emerging patterns of depression and loneliness among heavy tech users suggest some teens spend upwards of half the day—eleven hours—tending their Facebook profiles.

"Teens are keenly aware of their online image, and they can spend hours looking at other people's photos," says Kelly Schryver, who wrote her college thesis on teenage girls on Facebook and is now an education content associate at Common Sense Media. "The natural tendency to compare ourselves to others never gets to rest. Girls especially worry about others posting ugly photos of them and often feel left out after seeing events they weren't invited to. They share photos with an audience in mind, often expecting positive feedback from friends like 'omg! u look gorgeous!' A few teens have even admitted to me they'll ask their friends to 'break the ice' and comment on their profile pictures, hoping that others will follow suit."

Chloe, seventeen, describes it as a fact of life for teens, much as checking the mirror has always been routine before going out the door. Online they feel compelled to dress for the occasion, which never ends. "Facebook gets people obsessed with appearance because what you choose to make your profile picture and what you look like, and what you're wearing, how much you're revealing, all that kind of stuff matters," says Chloe. "People don't realize how much they center their social life around Facebook and around the pictures that they're posting."

The online personas become part of a round-the-clock popularity pageant in which teens judge each other and themselves, not always kindly.

"Everyone gives their info page on Facebook a fair amount of thought," a sixteen-year-old boy tells me. "You think about how you can be yourself but not look just like everybody else," he says. "I think a lot of people worry about what other people think about them . . . I don't think anyone should ever fake it. But there's definitely a certain aspect to wanting to seem like you're cool."

Teens tell me about the hidden time they spend cultivating their identities on Facebook, Instagram, YouTube, or school e-mail groups. This includes their profiles, photos, and other personal-page content, as well as the comments they post on other people's pages. It's important to sound smart, funny, popular, snarky, sweet, sexy—or whatever earns you social capital. It is tedious and time-consuming work and, they say half-joking, it presents ethical dilemmas about truth in packaging: do you tinker with truth, or post that hip-thrusting pin-up-girl-in-the-bikini pose, or brag about a party adventure that you weren't really in on, or Photoshop your image?

In fact, for teens with eating or body image disorders life online and presentation anxiety is torture. Carly, sixteen, recovering from anorexia, remembers how she lied to her parents when they discovered she had been back on proana web sites—she told them it was "a friend who had been over"—and they have no idea now that she spends as much time as she does on Facebook comparing her thin anorexic look with her current appearance.

> It really is pathetic. I know I shouldn't do it but I can't help it. I spend hours just looking at pictures of myself in ninth grade when I was anorexic, and wishing so much I looked like that . . . I know I should take those pictures of me down but I don't. I'm always trying to like figure out, do I look cute enough in this? Or why doesn't my hipbone show anymore? I can be okay for a little bit at school, but it's hard not to take out my iPhone and check Facebook, and look at my photos . . . and just hate myself.

Girls tell me it's not unusual for them to take two hundred to three hundred photos at one event and then spend two hours ago-nizing over which three photos to post. One could argue that teens have always been preoccupied with packaging themselves to look at-tractive, and that is true. The difference today, teens tell me, is that everything you put up is public and to some degree permanent in that other people can save it and potentially use it against you. It is worrisome when you know that any "mistake" you make will be there forever for kids to tease you about, and that kids are really mean to each other. "You can hit delete but that doesn't mean anyone else does when you ask them to," one girl told me.

Privately, kids have scrolled though their photos with me and said on more than one occasion, "I'll just keep this for revenge . . . you never know when it might come in handy." This is one screen game that nobody can beat.

Nasty Is the New Norm and Porn Is the New Nasty

Reflecting cues from reality TV and lines from pop lyrics and, in-creasingly, porn videos and commercial pop-ups, social cruelty and sexual degradations have become familiar refrains in adolescent on-line banter and texting. Notes like the one Nora got from the boy she hardly knew are commonplace. In a focus group of teen boys from different high schools, the boys describe examples of typical online teen boy comments and posturing. A nerdy boy calls another boy a drug addict and the other kid called him a retard with no friends, knowing he has a brother with Down syndrome. Calling a boy a fag remains the catchall slur of choice in that realm, and is common online, although Milo, sixteen, describes a convoluted netiquette emerging around the topic at his school:

My school isn't very homophobic, but it is somewhat . . .
if you aren't out, some kids—not me, but some—feel like it's
fair game to call a gay kid a fag, or gay. We've had conversations about this—if someone is openly gay we will absolutely
not make fun of him, but if someone acts gay and isn't out, they
will call them gay. That's the mind-set of the nastiest kids in
my grade. Not me.

And just what does it mean to "act gay" and why is it okay to
harass them if they "aren't out" but might be gay? We're letting
kids down by not teaching them how dangerous this kind of thinking and ignorant behavior truly is. Adolescents are known for their
own denial of reality, even when lives are at stake. Or on the public
stockade.

Another boy explains that there is an unspoken rule about humiliating friends online—you would never do that—but others, as he
says, are "fair game."

It is common practice for kids to go on friends' Facebook accounts
and fake identity. Gino, sixteen, explains that the week before, when
some guy friends were staying over at his house, he had forgotten
to log out and then fell asleep. His friends went on his computer,
onto his Facebook page, friended about a hundred and fifty people he
didn't know, and messaged about fifty of his Facebook friends "in a
very creepy way."

My friend friends this person he found on Facebook who
lives in Spain—they message her "you are the most beautiful person I have ever seen so please just accept my friend
request cuz I need something to jack off to tonight." She
messages back a heart face smile. I'm pretty sure she wasn't
a woman.

Gino isn't upset with his pals. "My friends know it's not me—that tone isn't me, no one would actually *say that*—so I don't get mad at that stuff at all."

#Sex#, #Sexts#, and #Intimacy#

Kerry, seventeen, comes into therapy and asks, "Is this weird? My best friend sends pictures of herself naked to her boyfriend, and then he sends her pictures of himself, like with an erection, in response. Having digital sex—is that normal?"

"What do you think?" I counter.

"NO! I think it's weird, but, you know, I don't want to be a prude or anything . . . "

Kerry wants to have a boyfriend, is curious about sex, and understandably is confused by what's going on. We talk about how her longing to be "liked, in a relationship, have a boy like me, think I'm pretty and sexy" in the moment can momentarily delete inhibition, but not the painful self-consciousness that would soon follow about what you'd just done. We talk about how her friends who spend hours grooming their Facebook photos to perfection could then suddenly be willing to "bare all" and sext pictures, and even put up with abusive sexual language. When it is presented as the normal way to get a boyfriend—cool, "empowering"—and the opportunity is right in front of you, impulsivity combined with thrill and risk taking and the fantasy of a fairy-tale outcome . . . send! And then reality hits, or texts back, as it did when her other friend Gia has a picture of herself topless sent around the school and then was IM'd, "your saggy boobs disgust me" by a boy in her chemistry class.

In a study of sexting among about six hundred students at a high school in the Southwest, researchers said that nearly 20 percent of all participants reported they had sent a sexually explicit image of themselves via cell phone. Almost twice as many reported that they had received a sexually explicit picture via cell phone, and of these, more than 25 percent indicated that they had forwarded such a picture to others. Of those reporting having sent a sexually explicit cell phone picture, more than one-third did so despite believing that there could be serious legal and other consequences attached to the behavior. The researchers concluded that, "given the potential legal and psychological risks associated with sexting, it is important for adolescents, parents, school administrators, and even legislators and law enforcement to understand this behavior." I would add that we must talk about it, teach about it, and if inspired, recommend the web sites that take an activist stand against it.

Vik offers to show me a thread of conversation from a recent weekend when his friends used his account—with his identity—to connect with a girl they hadn't met but wanted to.

"It's an example of people, like, courting," Vik says, then hesitates. "Okay, this is really hard to read to you without laughing," he says as he pulls up the conversation in which his friend—let's just call him Clark—and another friend used Vik's account to contact a girl whom they knew to be a friend of a girl the two boys wanted to hook up with.

"I don't know her at all," Vik says, "but for some reason they had some interest in her." If you met these two friends, you would think they are two of the nicest kids around. Both are very polite, very funny, studious kids. And this was their idea of "courting." Be forewarned, this content contains graphic sexual language:

Clark: ok whatabout Heidi whatever, are you friends with
    Heidi?
Girl: Hey this is weird lol
Clark: Hey just tell me
Girl: Yes. I'm going to go now

Clark: Wait no . . . can you hook me up with her, just as a favor
    I'll suck your dick . . .

Vik explains to me that the girl is "a little mannish" so that comment was Clark being "kinda mean."

Girl: She is out of your league let's be real, no way you could get
    with her.
Clark: Challenge accepted. Your challenge is to go for weeks
    without dicks in all four of your holes #slut#jk# il-
    ove you#pleaseforgive me#if youdont I might cry#im
    alreadycrying#nobody loves me #blowme
Girl: Who is this? Shit Vik you're so fucking funny like those
    are really witty and shit
Clark: #jk#you know youwant it ;) but in all honesty you actu-
    ally think I cant get with her.
Girl: There is so little of you that I want. goodbye
Clark: psMy dick is the only part you want.# Come and get it . . .
    but please . . . no don't.
Girl: and it's little
Clark: digging yourself into a hole, nicely done fuck face . . . I'll
    dig my way into your hole ;)
Girl: You're like too irritating to be funny..you're like really
    too..i can't come back with a comeback, are we at war now,
    I'm gonna get you

Vik explains, to be sure I understand this is not to be taken seriously: "This is my friends being idiots, basically, with the girl not
really understanding who it is."
"So this was an example of courting?" I ask Vik.
"They were just trying to be funny. In all honesty, they were just
doing what they wish they could be doing on their Facebook page,

be a creeper, they're just kinda letting off steam . . . in some way . . . they're like experimenting."

I ask Vik why he thinks it turned so nasty, to which he replies: "It didn't turn nasty. That is the norm for our generation."

Many teenagers say that sexting is part of flirting and, as long as you trust the other person, it can be a fun part of courting. Some boys say it's a much less embarrassing way to see if a girl is interested in you.

"I sent her a photo of me with my hand in my pants, kinda taking my lead from this album cover to see how she'd respond," one boy says, "and when I got a 'cute' back, I knew she liked me, and it was, you know, safer to try to get with her." In this context, when the photos stay in the privacy of a relationship, it is part of what teens themselves describe as a courtship dance. I have also occasionally heard high school boys and girls say that sexting is harmless if you trust the other person—you can't get pregnant sexting, you can't get an STD sexting—*Hey, it's safe sex, right? Yeah, if you really trust the person.*

But here is the catch, if only in terms of risks: unlike handing off a snapshot, which of course could possibly be passed around or replicated, the medium carrying the message serves many functions, and the capacity for someone to use it to expose any photo to the world makes the trust required much greater and the risk that much higher. Parents need to help their kids see that what can be, in one context, a sign of trust, can in another be turned against you as a major violation of trust. We all take this risk when we tell another person our secrets. But a whispered secret has a much shorter half-life than online evidence in text and photos.

ELLA: "TONIGHT I'M GOING TO RUIN SOMEONE'S LIFE"

The question of "how much" is often the focus of our conversations and concerns about teens' media exposure, screen games, and online lives, and rightly so. Research shows that at every age, negative effects of media exposure and tech use tend to rise the longer children are plugged in. I see that reflected in the troubles that children of all ages bring into therapy or act out in schools where I'm called to consult. But I also see the significant impact—devastating sometimes—of seemingly insignificant exposure. In some instances, like those I've described in earlier chapters about younger children, it may be a matter of a child's first exposure to content that traumatized him or her for whatever reason—a sad or scary movie they weren't developmentally ready to process or in some cases content of a disturbing sexual or violent nature.

With teens, their accumulating life experience may put them at risk for snapping, if you will, "for no reason." A closer look, sometimes possible only in the course of ongoing therapy, typically reveals contributing factors. Teens who are emotionally neglected or overly pressured by their parents are at high risk for this. A contentious divorce or destructive family dynamics can do it. Binge drinking and other substance abuse or addictions (including tech), undiagnosed depression and anxiety, a history of social humiliation, or untreated physical or mental health concerns, sexual abuse, trauma, or loss all are common factors. Sometimes these things are known within a teen's close circle of family and friends, sometimes they are not. As we know, children—including adolescents—sometimes hide their pain, overwhelmed or ashamed by it, or afraid to share it with the adults in their life. Known or unknown, the wounded psyche is a tinderbox of trouble, and when factors come together just so, the match is lit.

Last fall, I got a frantic call from the mother of an eighteen-year-old college freshman. Her daughter, Ella, had just been suspended from the top-ranked college that had been her first choice. This mother, a tireless community volunteer, and her husband, a successful businessman, had both worked hard to raise a well-mannered daughter. Ella had grown up in a lovely but remote rural community. After much deliberation, her parents let her attend boarding school, where she excelled as captain of the squash team, a poet, and a well-liked peer. Her mother could barely describe what had transpired.

Ella had been suspended for malicious intent to harm a fellow student. She had taken another young woman to a party, plied her with alcohol, and left her, which led to a cascade of very serious consequences.

Apparently, the trouble had begun when the young woman—whom I'll call Christy—called Ella a "slut" at a party. This undid Ella. She had what she later described as a "kind of violent reaction." While stewing, Ella had watched *Gossip Girl*. Inspired by the plot of the episode, Ella took revenge by taking Christy to a fraternity party and encouraging her to drink too much.

While drunk at the party, Christy had sex with Simon. Several people saw Christy and Simon thus engaged in a frat house bedroom with the door open. The next day, Christy accused Simon of date rape and Ella of malicious intent. Simon was cleared; Christy was over the age of consent and there was no evidence of nonconsensual sex; to everyone who saw them it appeared that Christy was having "a good time" and no one observed any force. But Ella was suspended. As it turned out, just before going to the party, Ella had changed her Facebook status with a line inspired by the episode of *Gossip Girl*: "Tonight I'm going to ruin someone's life."

I am not suggesting that *Gossip Girl* was to blame for Ella's behavior or that posting her intentions on Facebook made it Facebook's fault. However, as we look at ways that media and tech sometimes become catalyzing influences in adolescent behavior, Ella's story is worth noting.

I met with Ella and her parents several times together and then continued to see Ella alone for two years. Over the course of our therapy, Ella uncovered why she had reacted so viciously to Christy labeling her a slut. It had touched a deep nerve because Ella had been sexually assaulted by an acquaintance at boarding school when she was fourteen and had never told anyone. Like the vast majority of victims of sexual abuse, she internalized the shame, felt she was somehow to blame, and was terrified to tell anyone. Nor did she tell anyone when she was sexually accosted by another boy on a community service trip abroad.

Psychologically speaking, Ella lived in the vortex of that never-discussed trauma, unaware of its power to overwhelm her. She had surprised herself by the raw intensity of her fury at Christy. What wasn't surprising, however, was that in the grip of such deep and painful emotions, this smart, creative young woman was acting out in dangerous ways, including one modeled for her by a TV show about clever, cruel, impulse-driven teenage girls. She hadn't watched *Gossip Girl* with the 700-plus SAT part of her brain; she had watched it through the lens of unresolved sexual trauma. In the grip of what amounted to a post-traumatic stress response, she followed a convenient, celebrated cultural script. A child doesn't have to have been sexually assaulted to be vulnerable this way, to be in the grip of powerful and disorienting emotions.

Gossip Girl is just one in a long list of viewing possibilities—*Pretty Little Liars, How to Rock, Shake It Up, Jane by Design, Vampire Diaries*, and all the "girls fighting" YouTube videos that echo similar themes: social climbing, girl fights, betraying your friends, breaking the rules or laws, hooking up, drinking or getting drunk, and knocking down others within a social hierarchy. Merely entertainment? We know that much younger children watch these types of shows and then emulate them on the playground and in their early online behavior. By middle school they are experimenting with creating illusions online, profiles and images of the kid they'd like to be (or think they should be), rather than who they really are, ones that are designed to impress their peers at any cost. By high school and college, kids are acting out these scripts in real life, somehow unaware that when real human beings are involved there are real consequences.

It is hard to know the role that the media and online cultures play in the incidence of teenagers with diagnosed mental illness, but the numbers and tragic reports in the news suggest we should be paying more attention. The most dangerous delusion may be to be in denial about our cultural infatuation with treating each other in such profoundly degrading, humiliating, and soul crushing ways, as if it doesn't matter.

MAUREEN AND HER MAD EX: YOUR WORST

NIGHTMARE IRL: ABUSIVE RELATIONSHIPS

FIND NEW EXPRESSION IN SOCIAL MEDIA

AND THE TECH MILIEU

Her senior year was off to a great start, and all Maureen wanted was a boyfriend to make life complete. Stefan seemed to have it all together. He was a take-charge, good-looking guy. He partied but not too much, studied but not too much, and had a sly, dry sense of humor. He started paying attention to her and his attention was wonderful. He texted frequently, showed an interest in what she was doing and with whom, showed up to surprise her when she didn't expect him. Within a month or so, though, he was sending her hundreds of texts a day, on the one hand saying incredibly sweet, saccharine things, and on the other hand increasingly threatening and controlling.

Four months into it she started to feel queasy about it, but didn't know how to get out of the relationship. Then he got mad at her for going on a ski trip with her family. He insisted that she get her family to take him along and, when they wouldn't, he punished her with constant texts belittling her and her family, making vile sexual comments and threats, and demanding she respond or he'd stop being "Mr. Nice Guy" and then she'd better watch out. He made a crude version of a popular break-up music video, using her name in it, and posted it online. *She* broke up with *him*, but he wouldn't break it off with her.

As Maureen stood firm on her decision, his abuse intensified via texts and e-mail, and in torrid, rambling online posts using pornographic, sexually graphic language—and names of her friends and relatives. He wrote and posted hate-filled commentary about her body, her looks, her personality, and aspects of herself that he knew she was extremely sensitive about. This excerpt captures the tone, though none of the most graphic and ragingly offensive sexual detail, from a note that was five times this long:

> hope your new bf doesn't mind ur face you fuck up . . . I would be more upset if I havent hooked up with 6 girls, fucked 3, and cheated on u with 2 more . . . im doing this because u need 2 be put in ur fucking place . . . ur legit a dumbass . . . cause i realize ur the biggest whore now haha.. like im ashamed tht i ever dated u . . . u r the worst person i have had sex with.. ur fat.. like im surprised i cud even get it up . . . done with being mr. niceguy . . . i dnt want to ever see u again.. i am not depressed over u whatsoever.. and i can promise u tht we will never talk again . . . dnt worry im leaving u alone like i have complete hate towards u.. i honestly wish i never ever met u.. told u not to cross me bitch . . .

Nothing could have prepared Maureen for this vicious and humiliating assault or the new waves of humiliation as she showed the letters first to her parents, then, as the court action required, to the judge, police, and school administrators. It especially pained her to show her father. She held back showing him some of the pages, she said, because she didn't want her father to think of her that way and she didn't want to break his heart any further than it was already broken.

Abusive partners have always used belittling, derisive comments to control and punish their partner, and unfortunately contemporary male culture has always included permission to openly rate and ridicule girls and women, especially in sexual terms. But the combination of media sex and violence, online pornography, texting, and other social media has normalized this kind of behavior and the impulse to broadcast it.

This boy used every possible force from tech to torment this girl. His threats and harassment continued when she left for college, and soon he was making references to events and people in her new life—he was stalking her through online connections. She eventually obtained a court-ordered restraining order against him, but this kind of story has no tidy ending. The courts can mete out consequences for the boy, but it is Maureen who truly suffers the consequences of his actions, which include the forever life of some of the material he posted online and of course the memory of the experience itself. Maureen is a bright, lovely, loving girl, who is still hurting and suffers deeply. It is hard for her to resist going online looking at Stefan's posts, which are full of party photos and endless iterations of the great times he is having at the college of his choice.

Clearly this boy is extremely troubled; most young people going through breakups would never turn that kind of vitriolic and pornographic fury on one another. But the normalization of porn in teen life, the ease with which tech and texting are used as weapons of communication, the emotional intensity of the age, and the amoral environment of the online culture make a dangerous mix in the hands of adolescents whose anger or upset is no longer child-size.

Adolescence is often associated with the eruption of serious psychological issues, a time where underlying anxiety and depression creep to the surface, where emotions and intensity become unmanageable, where early childhood traumas resurface. In some instances media and social networking have been the means by which troubled teens seek help or are identified by others who reach out to them.

Lives of teens have been saved when a friend responds to a post that reveals self-harming or suicidal behavior. Even strangers have been known to send an alert to adults who may reach the teen. A school counselor told me about a situation in which a student of his, a teen in California, read a suicidal post of a teenager in England. The California teen was able to contact appropriate grown-ups in England who were able to get to the suicidal boy, who had intentionally overdosed, and take him to the hospital in time to save his life. However, for some of today's teens who struggle with mental illness, tech has become a gateway to self-harm and further risk-taking behavior, or the medium for their message in cases of computer addiction. I remember Ross, whose tech addiction—and he was an addict, with all the life-consuming symptoms of addiction—began as he began using screen play and web browsing as a coping tool during his parents' bitter divorce. As they descended into more selfish behavior, including fighting over whether their clearly addicted son needed help, Ross moved deeper into his own world of fantasy violence, fantasy sex, and pornography.

The proliferation of web sites that kids of all ages can go to, to learn how to hurt themselves in a moment of distress, is disturbing. I have sat with teens who show me the web site where they figured out how to cut their bodies and compared their incisions with other cutters, competitively. Or girls with eating disorders who go to the proana sites, to learn new ways of starving themselves or compare themselves to others to fuel their disorder.

At the same time, tech gives troubled kids like Stefan access to material that amplifies their internal disturbance rather than helping contain them and bring them down from their dangerous inner place. And the Internet gives them access to some of the most damaging, vitriolic stuff and the ability to target others.

In the adolescent flux of identity and sexuality, disturbing media and online content coupled with the dynamics of the impersonal interpersonal communication technologies, says Jackson Katz, "are literally reshaping the nature of social interaction and the nature of what it means, if you will, to be a boy or a girl, or in a relationship. . . . In the absence of thoughtful discussions about sexuality, and thoughtful, information-focused education, the pornography industry rushes in and introduces our sons to an incredibly brutal form of men's sexuality."

Garage Band Boys: Playing the Fame Game Hits Cyber Sour Note

Fame is the do-it-yourself dream of the digital era. We're all indie artists eager for an audience on YouTube or streaming videos to viral success through social networks, so it is hardly surprising that fame is the number one value of adolescence and young adults. Tech has made the adolescent dream of fame a distinct possibility, grandiosity a reality. Teenagers love the idea of using tech to score a hit, whether that is in terms of popularity or revenge. Tech has made it easy to borrow from popular content online and many kids are adapting popular videos or music to personalize them in ways they hope will gain them popularity and fame in their own world at school or among friends. However, a moment of creative inspiration can quickly turn into a viral disaster when teens' bid for a social hit eclipses better judgment and they hit send without thinking of consequences.

Liam, Hugo, and Tim are sophomores at a suburban school known for its academic rigor, creative curriculum, and high standards for student conduct. In all the ways that tests can measure, they are well above average, but in all the ways that tests don't measure, they are happily ordinary adolescent boys. When in the company of adults they are well mannered and in the school community they are well regarded by their peers and teachers. They have good relationships with their parents, but even in a good relationship, that doesn't mean they tell their parents much of the daily grist from their lives. Not, at least, until the grist hits the fan.

One afternoon they heard from a friend of theirs that he had been rebuffed by a girl over the weekend and they did what generations of kids have done: they wrote a song about it. They got together after school, but instead of jamming on their guitars at home, they went to the school's music lab. And instead of just jamming on guitars there, they did it on Garage Band, a digital recording program for creating, recording, and uploading music to the Internet. On the school's equipment, they created an extremely clever (technologically and artistically speaking) adaptation of an extremely vulgar, sexually violent popular song in which they named the girl who had spurned their friend and they identified the fact that she went to the rival school.

Using their extraordinary creative talents and their cultural smarts about mimicking and adapting music, they did not simultaneously engage the part of their brain that had been told (1) don't make personal attacks or use identifying information online, and (2) never do anything that would put you, your family, your school, or anyone else in an embarrassing situation. Thinking it a joke, they pieced together the song, eventually including the girl's name and school, recorded it, uploaded it, and sent it to friends. By the next morning, the girl's parents had found out and were planning to take formal action, and right about then the boys realized they'd done a stupid thing.

A few weeks later we sit together in a quiet room adjacent to that of Mr. Latzer, the school principal, who had arranged the conversation with me as another opportunity for the boys to take responsibility for what they had done, reflect on what they had learned, and articulate the message they wanted to share with other students. The boys reflected on the way it had all started and how naive and unthinking they had been.

Hugo: Just like, in the heat of the moment, nobody really
    thought about any consequences that it might have. It just
    seemed like such a contained thing, like, we were just mess-
    ing around . . . it was a joke between a bunch of us . . . no
    big deal at all . . . we weren't thinking of any, you know,
    of consequences or anything like that (or) that she (might
    think) that we were personally attacking her.
Liam: We just didn't think about anything that would happen
    because of it. We just assumed that everybody else would
    like it.
Tim: That they'd like it as much as we did.
Liam: Yeah, like think it was a good joke as much as we
    thought it was.

They hadn't considered that broadcasting the demeaning song with
the girl's name in it could be construed as malicious; it never crossed
their minds. In their minds, they were being good, loyal friends to their
friend. This is how the self-absorbed adolescent brain works. They
didn't see the piece as a personal attack because, as they explained, they
didn't know the girl. They didn't see that they'd invaded her privacy
and breached the boundaries of decency and fair play. Shortly after it
made the rounds of their friends, but before word came back of the
girl's parents' ire, some friends said the girl was very upset. Yet they
let pass an opportunity to apologize to her directly and seek to make
amends. In their minds, the song was about their friend, not her.

Once the girl's parents, understandably angry, took action against
the boys, and the school administration and the lawyers came on-
board and legally accessed the Facebook page for relevant content,
the gravity of the situation became real for them.

Tim: That was literally probably the scariest moment of my
    life.

Hugo: That scared me, I was like, whoa, that is real stuff . . .
    and then it was just totally out of our control. . . . You just
    never know what happens when you put stuff out there.

But wait, I say. Parents and teachers had obviously told them that
nothing is private once it's online, no matter what the privacy settings
suggest. That when you hit send, then the stuff is out there.

Tim shrugged: "It didn't really resonate."

Indeed, Hugo's dad had warned him and had offered some excellent advice.

"My dad always says, if you think something without saying it,
then just think it. If you can say it without writing it down, just say
it and don't write it down. But like the last thing you want to do is
put something out there on the Internet that you're never going to get
back."

He'd done it anyway. And it was now part of the assembly talk
the boys made on being good digital citizens, and on how stupid
choices can lead to serious trouble. Says Hugo: "Now whenever I
post anything I literally reread it and think about whether or not I
want to do this."

For more than twenty-five years I've been in the trenches with teenagers and the refrain has been the same: *I'm sooo sorry, Mom, Dad, I
just wasn't thinking, I just wasn't thinking!* Girls and boys alike, crying
in my office because of some big mess they've gotten themselves into.
It doesn't matter where they live, what type of school they attend,
what their grades or interests are, who their parents are. This is the
adolescent brain. They are so drawn to the fun of having that party
when their parents are away, the social points that will come with it,
the idea being the "it" kid of the weekend who everyone wants to
be friends with, that they forget or ignore every conversation with

parents about trust, family, privacy, protecting the sanctity of our homes. In the moment of irresistible thrill and the stimulation of the adventure, they delete years of loving, instructive parent conversations. Teenagers have always been drawn to the romance of the risk and in many ways what these kids did reflected the age-old developmental psychological processes at work. But tech has given them a new set of keys to the car, a new kind of substance to use and abuse, a new form of independence, and the illusion of both fame and anonymity.

And as is always the case, good kids are getting into bad trouble. In these moments, as we'll see in the next chapter, how we talk to our kids and help them and the family make meaning of these events makes all the difference in the world.

Chapter 7

Scary, Crazy, and Clueless

Teens Talk about How to Be a Go-To
Parent in the Digital Age

*When you're a little kid you can kinda tell your parents
anything and you don't worry about it. But then you hear
what they say about your friend or some other kid and it's
so scary—and then you think, hmm, do I really want to tell
them stuff now? I mean in the big picture, you have your
whole life with your parents and if you tell them you made a
mistake it can really ruin things. I don't want them to look at
me badly for the rest of my life.*

—JED, FIFTEEN

I have spent the past three years interviewing more than one thousand children ages four to eighteen to get a sense of what makes a parent approachable. I have asked them to tell me what parents do that makes them feel safe and secure, able to go to them if they have a problem, need advice, or if they're worried about a friend. Trust, after all, is at the heart of the child-parent relationship from the earliest back-and-forth communication through which infant and parent learn to "read" one another.

Our children never outgrow their need to trust us. And they never stop trying to read us. As they grow and their sphere of activity expands, the subject matter of their lives grows more complex and the back-and-forth between us becomes more complex. They continue, much as they did from birth, to watch us closely for cues that tell them whether we are approachable. They come to understand how each parent will react, how Mom, Dad, stepparents, and partners differ in that regard, and under what circumstances each is likely to be welcoming, irritated, or angry. They develop a keen sense about which parent to approach with what kind of situation. Who goes ballistic over a B on a test? Who takes mistakes in stride? They learn when it's okay to interrupt a parent at work and for what reason. And they know—or believe they know—when their parents are the last people in the world to approach. This is how we earn our reputation with them as reliable and trustworthy—or not. In many ways, tech hasn't changed a thing about that bedrock feature of the parent-child relationship. But tech has heightened the need for it.

Interestingly, when I ask children what makes parents approachable, they often focus on things parents do that make them *not* want to go to them for advice or help. What fascinates me in these conversations with children, and with teachers, too, is how often three adjectives come up, again and again: Scary. Crazy. Clueless.

A sixteen-year-old boy describes how his mother took his laptop from his bed as he lay sleeping and secretly rewrote his English paper, which was due the next morning. Not knowing she had made changes, he submitted the paper electronically without opening the file again and only discovered her "editing" when he received his paper back and didn't recognize it as his own.

A teacher tells me about a fourteen-year-old boy who copied and pasted a text block from an online source without attributing it— plagiarism—and how he sat shaking uncontrollably at her desk when she confronted him about it and told him that she was obliged to inform his parents. He was so terrified of facing his father that the school had to take steps to ensure his safety. His father had routinely criticized the boy's academic performance and work ethic, and had recently upped the ante, attempting to motivate him on deadline by threatening to cancel the boy's robotics competition plans unless he got an A on the paper.

In a high school focus group, after one boy jokes about his parents' tech-free dinner rule to allow time for family conversation, a girl responds in a resigned tone that her parents show so little interest in any aspect of her life except her grades that "they're clueless about what I do online or off. At least your parents care about you."

Another says she was mortified when her mother ignored the most basic netiquette and, after seeing that a girl had posted a crude comment on her Facebook wall, posted a scolding note to the offending girl and called the girl's mother. She didn't realize that the crude comment was a phrase from a popular song the kids commonly used as shorthand for *I'm having a bad day.*

When children and families see me privately, often in a crisis or when a child's situation has become more than they can manage or ignore any longer, parents' responses are always a significant factor. What we say and how we say it matters to our children. As we'll see later in this chapter, parents' reactions can turn a disaster into an extraordinary learning opportunity for the children and an entire community. Or they can trigger an emotional backdraft—the explosive effect of a sudden whoosh of parental intensity on an already volatile situation, or one they simply don't understand. In the emotional rubble, with trust and communication broken, adults' responses are critical in repairing the damage and strengthening connections going forward.

Know that I write this not only as a therapist, but also as a mother whose two children would be the first to tell you that I have been all of the above—scary, crazy, and clueless—at memorable moments. In order to soften my potentially unhelpful reactions, it helped to have an amnesty policy with our kids: if you're in trouble and you call us first you're not going to get in as much trouble as you will if you wait. We did this because we know that the first response to a situation becomes part of the situation. And tempting as it might be to call a friend when you're in a sticky situation, as parents we're the ones with your best interests at heart and the experience to respond in the most effective way. The deal also includes this, though: as the parent, you can't ask the details—who supplied the alcohol, who bought the weed—the most important thing is your child's safety in the moment and they're calling you because they feel unsafe and as a parent you have to hold up your side of the bargain.

One of the Garage Band boys mentioned in the previous chapter told me that in that "scariest moment of my life," when he realized the damage done and more was in store from his online prank, he knew he had to tell his dad right away:

> My dad throughout my whole life has stressed honesty more than anything else. He's like, "Honestly I don't care what you did, as long as you're honest with me and tell the truth," because he doesn't really value anything more than he values trust between us. So going to him was just kind of like instinct, I guess.

Isn't that the parent we all want to be? We try. But then there we are: scary, crazy, clueless. And the question we so often ask our kids confronts us: What were we thinking? The problem is that we were most likely thinking in a panic, reactively, with emotions high and adrenaline surging. Nothing is closer to our hearts and hair-trigger reflexes than our children's safety, their health, education, and future as adults with careers and families of their own. Nothing makes us feel more vulnerable than when our children are struggling. Today's parents are involved parents. We want to be involved in our children's lives. We want our children to come to us, to seek us out with questions about how to navigate their lives, their worries, insecurities, good things, and feelings of upset. But we're not always well equipped when they do. This is a difficult, often unsettling time to be a parent and it can make us hypervigilant and sometimes overreactive. In fundamental ways, the landscape is vastly different from what it was when we were growing up.

First, we have always understood certain aspects of childhood development as givens: a baby's natural progression from nonverbal to verbal communication and from crawling to walking, and later, as our children mature, sexual and other predictable milestones. But dramatic shifts in our culture and the influence of media and tech appear to be affecting even those foundations of development in ways that are unnerving for us.

Second, some behaviors that once were generally viewed as "bad" or dangerous are now considered the norm. Children routinely go to adult concerts and access online material geared to adults. They watch adult programming on TV with content or commercials that promote drinking, recreational drug use, casual sex, screen violence, and pornography as entertainment. When we see damaging gender stereotypes and extreme body images for boys and girls being promoted as ideals, we want to protect them from harm, and yet our capacity to do so is limited. Even when we take our children to kid-friendly films, the trailers for upcoming attractions are often nothing we would want them to view, even briefly.

Third, tech has diminished parental control and parental influence. It has eroded the boundary between work and home, putting enormous pressure on parents who want to be fully present with their children but fear risking their status at work, if not their job. Facebook and other online social spaces are virtually unlimited and, with parents busier than ever and often away from home, children rely on their peers at a much earlier age. While children's move toward autonomy and independence has always included more privacy and secrecy, the uncomfortable fact is that tech allows them to hide more from us, including serious trouble, and turn instead to strangers online. That's enough to scare any parent—and sometimes scare our kids, too.

Finally, economic uncertainty makes parents deeply anxious about their children's future career and life success. College admissions have become so competitive that many parents feel compelled to act as their child's marketing strategist. In a crisis, parents may shift into crisis mode themselves, more concerned with damage control than accountability. They step in as fixers. In our scary, crazy, clueless moments, our fears speak so loudly that in all that noise a child can hardly get a word in edgewise.

Our desire to be connected to our children, coupled with chronic anxiety about the world they are joining, can lead us to behave in ways that make us unapproachable and useless to them, especially in a crisis. In "roam and learn" mode online and in their tech-connected lives, children can turn to all sorts of places and people for advice when they are in trouble. Many of the sources are good, many are not. We want to be the first responder when our children need help. And just as with a paramedic or ER team, the first moves you make and the quality of your response have a critical effect on outcomes. As parents we want to create a family where our children understand the importance of coming to us first when they are in trouble. We need to behave in ways that encourage them to do so.

So here is what our children want us to know about the things we do that keep them from coming to us when we most wish they would. Just try these on for size, see what fits and what doesn't fit. Know that everybody has done many of these things many times. I have. So has the parent you admire the most. The point is not about getting it right all the time. It is just about doing better. When we can learn from one another's mistakes and listen to the teachers and adults who care about our kids, we all benefit.

"Do What I Say, Not What I Do" Tells Your Kids They Can't Trust You

If Scary, Crazy, Clueless were a screen game, Level One would take you through territory so mundane you would never imagine that the small stuff you set in motion or traded away there could eventually cost you the game. Children are eager to describe some of these everyday patterns of behavior that undermine their trust in us.

For starters, they say, we tell them to be honest, responsible, and respectful of others. *Be a good sport, a good team player,* we admonish them when they get whiny about a class project or a soccer game. *Study hard, play fair, be polite, calm down, do what I say, don't be late, don't talk back, don't watch so much TV—and get off the computer!* Then we do the very things we've told them not to do. When what we say and what we do don't match up we become hypocrites. It's hard for children to believe in the integrity of what we say when our words and actions are out of sync. Their top three peeves:

WE BREAK RULES AND SAY IT'S OKAY

Two sisters, one ten and the other fourteen, tell me how scary it is when their father drives them to after-school activities and texts while driving—"especially on the switchbacks" in their mountain community. They have asked him to stop but he snaps at them, assuring them that he can text and drive safely. They know that is not true; they've heard the talk at school, the research that shows how dangerous texting and driving is—and they know that in their state it is also against the law. It is frightening to be in the car with him and it is frightening to see him rationalize his behavior this way. It is also deeply upsetting that he gets mad at them. Beyond the safe-driving issue, he is telling his daughters that men in power, men at the wheel, don't have to go by the rules; they don't have to respect the girls' concerns about safety. That's a dangerous model for future relationships.

Children of all ages don't like when we bend or flout the rules, cheat, or do the wrong thing. Preschoolers don't like when a parent cuts ahead in the carpool line or steps into a kids-only zone with uninvited supervision. Older kids complain of parents who ask teachers or school administrators to bend the rules for them or ask for other kinds of special treatment.

Parents tell kids that education is important, to respect teachers, and be a team player, but then forget all that when the Thanksgiving assembly or the class presentation conflicts with vacation plans. Your child may delight at the thought of beating the crowd to Disney World, but deep down the contradiction makes children uncomfortable. Most of us do this a time or two and our children learn about making exceptions and what justifies one. But when parents do it repeatedly, they turn an exception into an expectation of special privilege and send a confusing message about playing by the rules: rules are relative and made to be broken, as long as you can rationalize well enough or just not get caught. In a larger sense, ditching school this way also trains them away from investing themselves in an effort that requires commitment; they've learned that commitment is not something the family particularly values. In all of these the lesson learned is that rules are made to be broken. They'll remember this when they are old enough to break the rules themselves.

WE ARE INCONSIDERATE OR RUDE TO OTHER PEOPLE

Kids don't like when parents are rude to a teacher or to anyone—whether it's the babysitter or a grocery store clerk.

"It makes me mad when Dad complains in a really angry voice

to the waitress about something that's not her fault," says a sixteen-year-old girl. "It just makes him look like such a jerk."

"My dad heard that I was really upset and cried after my math test and he called the teacher and yelled at him," an eleventh-grade girl says. "I was so embarrassed I wanted to die."

When mom and dad are rude to each other, that is especially unsettling for children. Parents' texting fights rattle them. "My mom and dad have text wars on the way to school and it makes me so sad," an eight-year-old girl tells me. Children of divorcing parents tell me, as a twelve-year-old boy did, that the divorce is "a hundred times worse when they say awful things about each other. I mean, a lot of parents get divorced but not everyone is so immature about it." Another boy, sixteen: "OMG my parents try to look at the e-mails and texts they each send me as if it's a competition, so I never let them hold my phone. Hate it."

WE SAY MEAN THINGS ABOUT OTHER PEOPLE, INCLUDING OTHER CHILDREN

The way you talk about other families and other children affects whether or not your children confide in you. Children see us as untrustworthy when we are catty or cruel in our comments about other people. Whatever you say about another child your child will hear as a prequel to how you would judge her in the same situation. Remember, beginning in fifth grade children may test you by telling you that a classmate did something that in fact they themselves did. You establish your safety and approachability as a parent not just by how you respond to your own children, but by how generously or judgmentally you talk about other children, especially if they are in trouble.

The only time I don't tell my parents something about a kid in trouble is if it's someone I want them to like. I don't tell my parents when my friends are in trouble or messed up because they get really judgmental and I don't want them to think worse of her, so I just hide the truth and hope they don't hear from other parents.

You want your parents to hear you if you're telling them something bad or serious, and then it's like you almost want them to forget it. I hate it when they say "you told me XY and Z about that kid," and then they take it up to the next level— "don't you ever try that!" Ugh!

My dad is always saying bad things about different ethnic groups or people's religions—basically racist things, you know—which means I can't bring my friends home and I'd never even tell him about some of them.

I hate it when my mother says mean things about what her mother or her sister wore to Thanksgiving. She is just so catty. I wish she could hear herself.

If my friend does something stupid, and I KNOW he won't do it again, my parents don't listen to me, and they don't let me hang out with that kid again . . . so I don't tell them when I do. I want to be able to say, "We are kids, we need to learn from our own mistakes."

Scary, Crazy, Clueless: The Three Faces of Reactivity

What feels scary to kids is when parents feel too intense, too judgmental, too rigid, too harsh—when there is a big disconnect between what the child is feeling and the parent's assumption about what the child is feeling. When parents become aggressive, attacking, judgmental, when they throw their weight around or are emotionally out of control—that is scary for kids. When parents get so invested in their own reaction or opinion or what must happen next to a child, the child hears that the parent is unstoppable, determined to see things just one way, and determined to understand—or misunderstand— the situation in that way only, no matter what other factors may be relevant. The parent is on a rampage. Extreme judgments or cata- strophic predictions paint you as always attuned to life's catastrophic potential: *You can ruin your chances for college, for marriage, for life! You will never have that child in this house again! We would die if you were ever caught doing that! Promise me you will never . . .*

"Crazy" parents amplify the emotion or the drama in any given situation. If their child is a little upset, they are more upset. If a child thinks something was sort of unfair, the crazy parent is adamant that it was completely unfair. A twelve-year-old girl struggling with how to respond to a friend's hurtful e-mail tells me she couldn't talk to her mother about things like this because mom always had a way of ramping up the drama on everything. "She'll say 'that's horrible!' and then get started, and then I not only have my friend to deal with but my crazy mother, too." A fifteen-year-old says he is reluctant to share his e-mail notes to his teacher (students and teachers com- municate this way routinely at his school) with his father because "whenever I ask for help or something and the teacher has an away message, my dad calls the teacher personally and it's really embar- rassing."

We teach kids to let things go; crazy parents hold grudges. We teach kids to deescalate conflict; crazy parents escalate conflict. We tell kids to dial down their emotional thermometer; crazy parents get into a fever pitch of emotionality and reactivity. That parental intensity is unsettling to kids and they shut down. Sadly, they lose the chance to learn to solve their own problems with their parents as their coach. "I can't show my mom any of my friends' photos because she thinks it's her job to call everybody's mother," a sixteen-year-old girl tells me. "It's one thing for her to tell me what's right and wrong, but . . . "

Crazy parents send nasty e-mails to coaches or drama teachers when a child doesn't get chosen for the team or isn't cast in the play. They overidentify and feel too much of their child's pain or misery and, instead of coaching the child to accept what is and be resilient, they get caught up in an enmeshed reaction that makes matters worse. Meanwhile the child loses out on one of the most important lessons in relationship with authority, learning how to talk when you feel something's been unfair or there's been a misunderstanding, or you feel hurt or confused. When parents get crazy and move in to fix things, it also sends the message that they believe their child can't handle the situation. *If my mom thought I could handle it she'd let me handle it, right?*

Children describe clueless parents in somewhat pitiable terms: naive or uninformed, inept, ineffectual. A clueless parent isn't attuned to the context of a situation or certain important nuances, is gullible, or makes awkward blunders. Ryan, the fourth-grader who loved gaming, the more nasty the screen violence the better, had figured out how to manipulate his parents, so he went by their rules but played what he wished. "Sometimes when I think it might be inappropriate I ask my parents if I can watch it," he told me. "Usually they'll say, 'Have you watched anything like it before?' I say, 'Yes.' And they say, 'Whatever, I can't stop you now.'"

More specifically, parents are clueless in two different ways. Some parents are clueless in that they do not understand what's going on or they are overwhelmed by it. Their kids are so far ahead of them—technologically and as citizens of popular culture—and they don't know how to navigate the gap. Perhaps they're behind the curve because they are working such long hours to support their family, or because they are digital immigrants, or for whatever reasons are involved in taking care of so much of day-to-day family life that they feel unable to pay attention to the goings-on in their child's life.

The other way that parents are clueless is when they have abdicated their parental authority in favor of keeping the peace, being friends with their kids, or rescuing their kids from consequences. Essentially, they put on blinders to avoid what they don't want to see. Maybe in what little time they have to be with their kids, they don't want it to be unpleasant. Or they don't want to "sweat the small stuff"—a reasonable approach unless they label everything small stuff to avoid confronting any big stuff. Maybe they don't trust that they can have the kinds of guiding parental conversations about setting limits, holding kids accountable, and coaching them about handling anger, frustration, and disappointment. Maybe they truly don't care.

The father of the fifth-grade boy who had e-mailed the "four guys in a shower" YouTube video to a girl in his class (and faced disciplinary action for it) defended his son's right to "free speech" and told the school principal that the image wasn't pornography because "there was no penetration." (If you looked closely, he said, you could see that.) Many adults would consider that an absurd stand for a parent to take regarding both the content of the video and his ten-year-old son's "right" to send it to the unsuspecting girl. But this man wasn't thinking about the impact of the image on the girl, who was shocked and, in fact, traumatized by it. She didn't have an understanding of First Amendment law that might allow her school to dismiss it as inoffensive based on a legal technicality, as the boy's father claimed. Nor did she have the same casual attitude that the boy had about sending a sexually explicit video for laughs. He wasn't thinking about what his argument communicated to his son, either. The teachers had talked with the boy; he had acknowledged he'd done wrong. The father's response was foolish and disingenuous, and everyone, including his son, knew it.

Some parents back down to avoid a child's meltdown. They are at a loss when it comes to handling the ordinary everyday limit-setting and corrective guidance that is a part of responsible parenting. Their children become overempowered because they know they intimidate their parents; all they have to do is have a fit and their parents will back down.

A mother and father describe what a fight it is to get their seventh-grade son to bring his game time to a close at night, or when he has been on for far too long. He swears at them and bristles if they step in closer to him. Eventually he gets off, but it is always angrily, and he punishes them by carrying the hostility through to the dinner table or whatever family time they have. "I know I should set limits on my son's gaming, but honestly, I've tried and it's like World of Warcraft up close and personal and I can't stand it. I surrender to keep the peace."

Sometimes kids will say, "My parents want to be my best friend, so I play them, but they are clueless." A clueless parent tries too hard, misses cues, often engaging over superficial things while failing to have meaningful conversations with their child about life and values, about expectations and consequences.

An early clue in a family therapy session of a parent who may be in the clueless category is when I hear the words "My child and I are best friends." The clueless aspect of this is that as a parent you need to parent—and your child wants you to be a parent. It is wonderful to be close to your children, but friends don't have the responsibility that parents have. When parents prefer being a best friend to being a parent, they remove themselves from the essential hierarchical relationship in which an uneven and necessary distribution of power, authority, and guidance must exist. You can be close to your child *and* be a parent. Clueless parents have not learned how to live comfortably with the necessary, healthy dissonance that being in a position of authority creates.

These three labels—Scary, Crazy, Clueless—are useful as a way of seeing ourselves through our children's eyes. But in a moment of conflict or crisis, they become three roars of a single beast: reactivity. And we all know what that sounds and feels like.

"I feel so utterly incompetent," says a partner in a law firm. "I vacillate from being trusting to practically paranoid about what my two teenagers are doing online. They tell me they are doing homework and yet there are constant pop-ups and YouTube videos and suddenly I'm a raving lunatic, yelling at them—and I hate my tone of voice—I sound like I am questioning them on the witness stand, when they are hardly criminals. I just can't get it right!"

GILLIAN'S STORY: REACTIVITY, REDEMPTION, AND RESTART

Jessie and George, both in their early forties, had tried to make their home the perfect hangout for their four children and their children's friends. They wanted their kids to have active social lives; they just wanted the socializing to be in their own home so they could keep an eye on things. All four children—their daughter, Gillian, fourteen, and their three sons, ages eight, ten, and sixteen—enjoyed the full array of personal tech. Cell phones, laptops, wireless, and open access to online activity were a given. But they had pulled the plug on all of it for Gillian when they called me for an urgent first consultation. Gillian had gotten into trouble with an online friend, they said, and now tech was at the center of a disciplinary showdown that was going from bad to worse.

George does most of the talking in the initial part of our first conversation, Jessie nodding thoughtfully as he tells the story. He describes his daughter as "hardheaded, strong-willed, and tough." She'd met a boy her age named David through a friend a few months earlier. David lived a few towns over so it was hard for them to get together in person. Instead, they posted on one another's Facebook walls and through the site's private messaging option. They also video chatted frequently.

George and Jessie say they had had ongoing and what felt to them to be open conversations with their children about sexual intimacy, trying to instill in them their own values. "We have been very direct with our children about sex," George says. "We have always said it is a sacred thing and they should wait until they meet somebody they love."

They hoped that talking about these subjects would help steer their children away from risky and needless experimentation.

Recent events involving teens and drinking, several sexting incidents involving students at the school, and, most disturbing, the date rape of the daughter of a family friend made their children's world verifiably scary to George and Jessie. As we talk, I also learn that Jessie, as a child, had been sexually abused, which George knew and which understandably made them both all the more fiercely protective of Gillian.

Then came "the incident." George and Jessie discovered that Gillian had met up with David once again in person and given him, in common parlance, a hand job—brought him to orgasm with her hand. And had experimented with oral sex. David, being fourteen, with access to the Internet and no thought to consequences, e-mailed two of his friends about his new experience. The rumor mill churned quickly and eventually someone approached Gillian's older brother and said, "Heard about your sister. Sorry, dude." Gillian's brother, embarrassed, then furious at being embarrassed, and upset about his sister, told his parents. They were livid.

Meanwhile, Gillian had forgiven David this indiscretion—he'd confessed what had happened as soon as he realized that the whole thing had blown up. He had apologized for not thinking about the implications for her. He genuinely cared about her—he had shown that many times before this happened and they both wanted to keep seeing each other. Upon learning of their daughter's sexual foray, however, George and Jessie reacted by yanking all of Gillian's tech privileges, basically cutting her off from her cell phone, Facebook, IM, and video-chatting with anyone. They ordered her not to see or talk to David again.

In the six weeks since, Gillian had become withdrawn and uncommunicative with them. She was not sleeping or eating well, and seemed depressed to them. Now, in addition to their anger and disappointment in her, they were worried about the downward spiral they observed.

George, Jessie, and I met several times in a short period of time before I asked to meet Gillian. I wanted to understand their side of the story thoroughly and we needed the sessions for them to reach a calmer and more approachable place in their thinking so they could have productive conversations with Gillian. Understandably, they had had very strong reactions upon hearing their daughter was experimenting sexually in ninth grade. It seemed too young to them. They both felt that Gillian had compromised their family values, that she was no longer safe in the world, and that their trust had been broken. Nor were they ready yet to hear Gillian's side of things, they were so entrenched in their own personal reactions to the situation. It had triggered traumatic memories for Jessie and that only brought out George's protective stance all the more. After several extended sessions in which they each worked on their own reactions to Gillian, they were ready to work in a calm and constructive way with their daughter. It was time to hear from her.

Gillian was relieved to meet me; she knew she needed help dealing with this unexpected big mess. At fourteen, she was a clear-eyed girl who was mature and assertive in communicating her feelings. She tells me that she didn't feel badly about what she had done with David, but she did feel badly about the damage it was causing in her family. She really liked David and felt comfortable with him. They had talked about whether she was ready to get involved sexually and he had not pressured her. She hadn't felt ready to do anything more than what they did, but she had been willing to try that—she was naturally curious after all, and she cared for him. Further, they had agreed they would stop at any point if she felt uncomfortable about it, and when that did indeed happen, they stopped. She thought David had been stupid to brag to his friends but she also felt he'd been honest afterward and sincerely sorry. "He gets what he did wrong," she says, "and he learned from it. He's not going to do it again."

Most upsetting to Gillian was the explosive reaction from her parents—their anger and accusations, their assumptions and their tone, and of course the restrictions they had placed on her.

"I'm never going to be able to go to see David again," she says, starting to cry. "They assume the worst in him. They always assume the worst in people. And now I have to lie to my parents to keep talking to him. I have to go on my friends' Facebook accounts just so I can be in touch with him. It's not good at fourteen to feel this way. I used to be so close to my parents and I thought they would let me be myself, make my own choices. I chose this and I don't feel bad about what I did."

But now she was lying to her parents, something she had never imagined herself doing and something she did feel bad about. She was despondent over her parents' reaction over the break in their relationship, and over being cut off from all her friends and her life online. She acknowledged that she was not eating as she should, partly out of loss of appetite from feeling depressed, but also because she felt furious and powerless, and refusing food seemed one way to protest.

In our work together for the next four months, George and Jessie were able to sort out the tangle of issues and feelings that had triggered their extreme reaction at the time and which had grown only more complicated in the ensuing weeks. Part of this was that they had lost sight of what it means to be a parent of an adolescent—that you can't control your child at this age. They are, to borrow from Kahlil Gibran, "sons and daughters of life" who "come through you but are not you." Gillian was not being a bad girl or a bad daughter or a bad person by discovering herself in a sexually intimate and relatively safe situation with a boy who actually cared about her. This was not an impulsive act and they were more mature than many kids are, including their discussion that they would stop if she felt uncomfortable, and the fact that they did so.

This was a comparatively good relationship that was significant for these two kids, even though it had existed in large part online and through Skype. They had seen each other several times. They had hung out and talked, talked a lot, in fact, and a quality of connection existed between them that George and Jessie couldn't see. All they saw was that Gillian was sexually active in the ninth grade and they were bereft about that.

In our conversations they were able to look more objectively at the situation and their reactions to it. George was a doting dad with a romanticized view of his daughter as something of a fairy-tale princess whom he would walk down the aisle to marry her prince someday. He was clueless about his daughter as a fourteen-year-old real girl, fully developed, fully of her generation, and at a large high school where she hung around with other ninth-graders who were having parties and doing things in ninth grade that he probably hadn't done until eleventh or twelfth grade. He struggled to reconcile those disparate realities in his mind—that it could be possible for his fourteen-year-old daughter to have a situation like this in the context of a thoughtful, respectful, safe, and positive relationship with a boy. As he and Jessie processed all of that in our conversations, they were able to see that they had done a very good job with Gillian, evident in that she had been able to say "I'm comfortable with this and not that," that she had not done this with a stranger, that she hadn't downed vodka first, that it was not impulsive, and that David really did care about her. He had not pressured her but had asked respectfully for what he wanted and she had felt comfortable with the level of physical intimacy they had shared.

What had felt so awful to Gillian was that first of all her parents made it sound so awful—that she had done something inherently bad. She rejected that idea. She didn't feel ashamed of herself and didn't think she should. She was indeed on a depressive slide and a hunger strike from being unilaterally cut off and disconnected from all her girlfriends, all her friends who were boys, as well as from David, to whom she was used to texting or talking or Skyping daily. Her parents had felt so frightened at feeling out of control that they had taken complete control in a way that was scary to her. Their punishment—that she surrender her phone, shut down Facebook, and have no life online for six months—seemed so irrational to her and they seemed so clueless about so much that she couldn't see a future for herself with this harsh discipline.

In continued conversations in sessions with all three together, they were able to communicate authentically, listen to one another's points of view respectfully, and work together to create a deeper understanding for navigating the teen years ahead. Gillian came to appreciate and agree with her parents' concerns about pacing herself in being sexual with boys. Gillian's parents realized that they had overreacted and they apologized for that and eventually they let her resume her relationship with David. David wrote her parents a letter of apology and eventually he came to visit at the house. It was awkward. And then things got better.

When something awful happens, or something we think is awful happens, it can push us over the top. What I loved about this family and what was true about this family throughout, was the family ethos—the value they placed on talking about things and working through relationship impasses to deepened family connections. What could have been a longer and deepening family crisis became a family crisis that strengthened everyone's understanding and connection to each other. However serious the situation might be, we all have that chance with our children. Everyone can listen, process, grow, and grow closer.

The opportunity to meet a surprising moment with equanimity starts early, even with sexual material. Parents have told me about their upset upon finding their young child confronted with a steamy love scene—or even porn—as they went online for more innocent diversions. Just clicking on the names of children's characters like "gumby" or "my little pony" can now lead to pornographic sites. So we have to be prepared for that because while web browsers like Mobicip help block against porn, chances are at some point it's going to find its way to your child's screen. You're going to need to calmly say, "Let's move on—pictures like that are not meant for children. Thanks for showing me that. Now let's find your game." Nothing complicated. No big lectures or drama. You want your child to know that it's always safe to show or tell you anything, and that you simply will not lose your cool.

Scary in the Worst Way: When Adults Bully Children

A school counselor described how parents often respond to instances of children's social cruelty by openly demanding the harshest punishment for the accused child—as long as the child isn't their own: "They don't want you to deal with that child—they want you to vaporize that child," she said. "Their attitude is, 'No, he just shouldn't exist anymore, get him out of here.'"

Clearly there are times when a child has behaved so badly that, in order for a school to continue as a safe place, the child needs to leave. But too often parents act as if getting rid of the problem child is the only solution, teaching kids that, metaphorically at least, annihilation is an option in life as well as in screen games. Having sat through many conversations debating discipline options including suspension and expulsion, I appreciate the emotional heat on all sides. These are

often complicated situations, and legal issues sometimes make it difficult if not impossible to be transparent with all involved when, in
general, transparency is best for all.

Marty, a single dad, called me in a panic when he received a call
one afternoon from the police in the suburban community where he
lived with his two sons, nine and eleven years old. Rich, the elder,
was being accused of bullying a boy, a former classmate who had
moved to a neighboring community the year before. Marty knew
there had been some friction at one point between the two boys, but
it had been nothing serious, just the petty rivalry that is so typical of
ten-year-old boys jousting for their place in the boy pecking order. As
it turned out, Rich was on a group e-mail list that still included the
former classmate and one day another student circulated an e-mail
in which he made a comment that dredged up something from that
past. Thinking he was sending a snarky postscript to the boy alone,
Rich had written "I hate you & your sneakers suck." Suddenly he
realized it was a Listserv—going out to all on the list—and he knew
instantly that he'd been wrong to send it even to the boy alone. He
told his dad and e-mailed the boy directly to apologize for the comment. He also replied to the group again, saying to ignore his earlier
message, that he'd sent it by mistake. He had never heard back from
the boy. It had been two months.

The police called to tell Marty they were investigating a complaint
about a three-year history of bullying by Rich of this boy. Marty was
dumbfounded. Rich was terrified. He had already admitted to his
dad that he'd written the one insulting note, but except for his e-mail
apology there had been no other communication between him and
this boy—only the group e-mails that occasionally circulated, and he
had never talked about the boy in those before.

"The officer talked to me as if I were harboring a criminal," Marty says. "She said she was investigating this history of bullying by
Rich, and when I said there was no such history, she said, 'Aren't

you worried about this? I mean, if he's thinking and writing these thoughts at this point, what could he be doing in five or ten years?' Her tone was very accusatory, as in 'there's something wrong with your kid, here.' It was not meant to be objective. She had already made up her mind about Rich—already decided he was this terrible kid guilty of bullying—apparently after taking a complaint by this boy's parents."

Marty immediately called Rich's school to tell them about the police call and to ask the principal and Rich's teachers from that period if they had seen anything that could be construed as bullying by Rich. If it was true, he wanted to know. There had been no bullying, they said. Furthermore, the police had already called them and had been told the same thing.

Marty called the officer to suggest a meeting at the school with the boy and his family to discuss their concerns. He also asked the officer if she knew that Rich had written the apology to the boy after the one note. The officer told him to send a copy of the apology. They never heard from the police again. Marty heard indirectly that for whatever reason, the boy's parents had chosen to respond to Rich's one e-mail to their son by reporting a history of bullying where there was none. They had never contacted Marty to discuss any concerns and afterward they never apologized for reporting Rich to the police.

We are all adapting to the ways in which tech has altered the communication patterns between everyone. As new laws try to catch up with the ever-changing tech effect, it is inevitable that some will be mismanaged and cause harm rather than protect. In this case, the accusing parents became so reactive that they ignored all first-responder moves—call the school and maybe the parent(s) of the boy and get both sides of the story, since a ten-year-old's report of a three-year history may not be the full story. Or perhaps the parents ramped up the details for impact—who knows? Whatever the reason in this case, it was not the truth. When adults—in this case the accusing parents and the police officer—are so quick to accuse a child of a crime, we have to wonder: are we acting in the best interests of our children?

Each time I hear a story like this of bullyish adult behavior—and unfortunately it is not an isolated case—I am reminded how vulnerable everyone is to fear and anxiety about the negative impact of tech on their children or the children in their community.

Sadly, we know of incidents in which social networking has been used as an antisocial weapon with tragic outcomes, including the suicide of victims of extreme cyberbullying that included harassment or public humiliation. While well intended, zero-tolerance policies and some other antibullying attempts to protect children have had mixed results. So we see a first-grader suspended from school for singing a popular rap lyric, while some of the worst offenders go under the radar and create such fear among their peers that no one will report them to a teacher.

Sometimes adults overreact in efforts to police and punish what amounts to common social aggression that is part and parcel of the learning curve of childhood. The mean, teasing, thoughtless, impulsive behavior that is a part of daily life for kids requires not just an antibullying approach, but a comprehensive prosocial social-and-emotional learning curriculum and an approach to education that reinforces the development of character traits that we know lead to both academic and social success.

As Paul Tough writes in *How Children Succeed*, we need to think much more about the role of school in giving children the tools they need not just in reading, writing, and math, but in interpersonal relationships and character traits that lead to socially and personally responsible kids. These are the skills required to effectively prevent any one of us—child or adult—from bullying. In my work with schools and parents, I see all of us—educators, parents, and children alike—engaging with these issues. We need to make sure these social-emotional skills are now in the center of what children are learning year after year at school, and that these lessons also make it home to parents.

As we work to develop a more intelligent way to address aggression in our culture and among our children, we need to take an honest look within, too. Sometimes parents are quick to try to protect their own children from disciplinary consequences for aggressive behavior, but want no such immunity for other children. It is easy for errors to escalate via tech and social media, and the stakes are high for our children. As the grown-ups, we have to work more diligently to sort out our own fears and projections or whatever the psychological forces at work, to see these situations clearly, and to respond reasonably. It is becoming increasingly clear that schools and parents not only need to help children of all ages resist antisocial cultural norms but to develop the character and humanity to change these norms for the better.

The Garage Band Boys: Turning a
Sour Tune into a Teachable Moment

The opposite of scary is approachable. The opposite of crazy anxious is calm. The opposite of clueless is informed and realistic. And the opposite of reactive is responsive. The Garage Band story in the previous chapter is a model of adults responding effectively and being approachable, competent, and guiding. It shows how each response becomes part of a larger one that creates a valuable learning experience and ultimately strengthens communication between kids and their parents.

If we look and learn from what went right, here it is: The principal remained calm, gathered all evidence, and wasn't scary. Unlike the police officer who was envisioning a future sociopathic criminal, he assumed these were children of character on a common learning curve. He knew these were good kids who did a bad thing, that they had made a mistake, and that he would hold them accountable. He empathized with the girl and her parents and listened to her story and the impact on her, her friends, and her school. He talked several times with the principal of her school. Of course, there would be a disciplinary committee meeting, but true to his vision as an educator, he wanted to help these boys learn and grow from what they had done. He would talk to all parties involved—and there were many: each boy and each boy's parents, the girl and her parents, the other school, the students in his school, his board, the school newspapers, and the local press.

His plan for discipline allowed the boys to account for what they had done and remain whole, not shamed. He cautioned faculty and parents from overstigmatizing the boys, reminding all in the community that learning from mistakes is how we build character, that children are not their mistakes, and that they shouldn't label children as their mistakes. In addition to being suspended for a few days, the boys were required to research, write, and make a school presentation about the impact of the media on teen risk-taking behavior and what they had learned from their mistake. They also did community service teaching Internet safety to younger kids. He met with each boy and his parents and made it clear that this situation had a beginning, middle, and end. Honesty, transparency, and accountability were important to model for children. The nature of their error, the discipline, and their subsequent actions to take responsibility and set things to right would all be part of the record; this would not, as parents often fear, "ruin their chances" for college, their lives, or their futures.

When the principal learned that I was writing this book, he invited me to interview these boys and their classmates, knowing that it would help them process the experience, gain new insight from it, and share the lesson with a wider audience. This school, known for its academic rigor, also created through adult leadership and mentorship a psychologically safe place to learn not just from classroom curriculum but the hard lessons from life.

Liam told me later how the adults' responses to the situation made him appreciate them and their wisdom all the more and brought him and his parents closer:

> Before, I wasn't really the most open kid, so they'd ask the usual "what happened at school today?" and I'd be like "Oh nothing." So now they're definitely a little more interested in what I'm doing, and what's going on. It helps the relationship. I used to hide a lot of stuff and just keep my personal life personal, and not concern them with it. It helped me learn a big lesson about being upfront and honest about everything with my parents.

When I spoke with the boys who created the offensive song and released it online to friends, they talked about how the responses by all involved had turned their personal worst-case scenario into a hard and valuable learning experience for them and for the larger community. Their teachers and the school principal used the teachable moment to hold the boys accountable while creating a way for them to make amends to the victim of their prank, and to expand the lesson to others through assembly presentations. Ultimately the boys emerged wiser, feeling stronger in their sense of self, grateful to their school, and closer to their parents.

Speaking of Sex: Taking the "Scary, Crazy, and Clueless" Out of It

Michael Thompson described an eighth-grade health class workshop in which fourteen boys talked about sex and sexuality. When he asked how many had had meaningful conversations with their parents about sexuality, three said they had; eleven had had none. "That's very typical," he told me. "These are fourteen-year-old boys, and their parents had not stepped up to the plate. There is all this handwringing about video games . . . but meanwhile, their boys are up there watching an enormous amount of porn. The only thing you can do is to start teaching early, in a non-inflammatory way. Because the problem is, if you start too late and you're too hysterical, you lose all credibility."

You may feel out of your league in talking about sex with your child. Many, if not most, parents do. I am confident you know that you should and that most likely the big sticking point is how to do it. Many years ago, when I was a school psychologist at Phillips Academy Andover, I taught a course for eleventh- and twelfth-graders called "The Psychology of Love and Human Sexuality." In this class students could talk openly and thoughtfully about sexual intimacy, and how the culture represents a full range of relationships. We talked about good love and bad love, sexuality in all its fluidity, about gender and culture and politics, all in a way that helped students clarify for themselves their own values and what they want for themselves in this important part of their lives. It's ironic and extremely unfortunate that at this moment in time educators in many schools are forbidden from having conversations like this with a class, in a learning environment, while at the same time kids are surrounded by media and entertainment that include gratuitous and violent sex, pornography, and unhealthy relationhips. Despite this discouraging landscape, teens tell me today, just like the students I worked with in the past, that they want good love, true love. They know that a real relationship is better than hooking up.

I've had conversations like the ones that follow, in small groups and in packed auditoriums with teens of all ages. Whole books have been devoted to the why and the how of talking with kids about sex, and I've listed a few in the bibliography. In the spirit of this chapter, however, I want to share a script—actually two, one to use with girls and one to use with boys. These are the basic versions of those I've offered when parents call me in a panic or feel mounting concern over sexual content their child may be encountering. Maybe it's porn. Maybe it's a popular movie or TV show in which casual sex is a through-line. Maybe it's a conversation you overheard. You fill in that blank—I'm using FWB as the starter here.

These suggestions are geared toward an older tween or teen. At any age, the important thing always is to take your child's unique self, experience, social milieu, and readiness into account; also bring your own values, reasonable concerns, and assumption of good intentions (your child's) into your conversation. In this instance, some key points include healthy values; cultural messages; sex that is connected to love and intimacy, in the context of a relationship, and disconnected sex; the matter of "hooking up" or friends with benefits, where there is no commitment to a relationship; and the extreme level of objectifying kinky or violent images of porn. While these examples follow the more common heterosexual scripts, with boys using girls, the same dynamics also occur with girls using boys—*boys are my toys*—and in same-sex relationships.

You may want to have conversations like this in the car, my favorite place to hold my children hostage when I need to download a talk with them. I start these conversations with "I need to tell you some things and I need you to listen, you don't have to say anything back, but I want you to listen . . . "

THE BOY TALK . . .

"I've been hearing stories of boys convincing girls to give them blow jobs—this friends with benefits thing, and I need you to hear me out on this. First of all, any girl, for any reason, even if she agreed to do this one day, can turn around on a dime and accuse you of raping her. Any girl can say you forced her, even if you didn't, and you can be accused of rape. How scary is that? And when parents learn that their baby is having oral sex with you, they might just go ballistic and go after you even if the girl doesn't want them to. So understand that there are legal issues here that you have very little control over.

"Okay. So that's the legal stuff. But the other piece is that you are growing up in a culture where it is considered cool for men to use women as their sexual plaything without any sense of caring, without any responsibility for her experience, and that is really damaging to the girls but also to you. If you think of women as there for your pleasure, you stop thinking about the person, that's what objectifying means, and sex is no longer about love, it's about you and your gratification. That kind of sex doesn't last, it doesn't work for real relationships.

"You guys see so much sex that is disconnected from romance and love, and I want you to have good loving relationships. When you watch violent porn and all this stuff with threesomes and kinky and violent sex, it can really mess up your ability to have a loving sexual relationship with a partner you care about. Do you really think most girls that you would like would ever do that stuff with you? Do that kind of role-playing or whatever? I hate to break it to you—if you watch that stuff and that's what you get used to, you are going to have a real difficult time with real women.

"So if you want to have really great sex and a long-lasting love life, think about what kinds of porn you watch, don't ever, ever force a girl to do anything, don't get her drunk, don't go there. Understand that you really need to know who you are being sexual with. The more you discover your sexuality in a good relationship where you really like and care about each other, the better and longer your sex life will be."

You can close with a simple, all-purpose: "So how was that?"

THE GIRL TALK . . .

"I've been hearing that some boys are pressuring girls to be friends with benefits—for blow jobs and sex with no strings attached—and

I need you to hear me out and help me understand what you think about it. Who benefits from friends with benefits? (Usually they laugh, and they know the answer to that: boys. Girls often are more willing to talk than boys are, so wait for a response but be ready to move ahead if you don't get one.)

"Here's the deal: Friends with benefits gives boys permission to use girls and it's a tricky term because that's not really something a boy who's your friend would ask you to do. It might get your hopes up that the boy really does care about you. But the whole point of friends with benefits is it gets him off the hook for having to care about you. No boy who really cares about a girl would ask her to be a friend with benefits.

"Another thing: It's a really bad introduction to sex for a girl because there's no relationship. There's no commitment. There's no caring. When you're young and figuring yourself out sexually, it's so important to do that in a real relationship, where the person really cares about you. And friends with benefits isn't really about satisfying the girl, right? It doesn't go both ways. (Expect an *eeew* response.) Friends with benefits teaches girls that it's your job to take care of boys' sex needs and desires and to forget about your needs and desires. It teaches you that who you are and what you care about and what you want and are comfortable with sexually doesn't matter. Girls really want romance, love, want to feel special, want to feel they really matter to the boy and vice versa. In other words, girls want a real relationship. And that's a really good thing to want when it comes to being sexual.

"Look, good boys are getting a lot of bad messages about how to treat girls when it comes to sex—they're getting the message that it's cool for men to force women to do sexual things or use you, use women, as their sexual plaything without any sense of caring. So the guy comes on like he desperately needs you to jerk him off and then all the guys brag to each other about what they got a girl to do. They may broadcast that online without any thought to what that's like for you and how it might affect you. That is really damaging to girls, to you. It teaches you that you don't deserve more from a boy. No caring. No commitment. But you *do* deserve more. When it comes to being sexually active, you want to wait till you are in a good relationship, with someone you know really well and they care about you, so that you can have fun learning about all that together. You don't want to feel used or scared or hurt by hooking up with boys who don't really care about you, doing stuff you aren't comfortable doing.

"Sex is many things—it can be wonderful or creepy, good or bad, messy and fun—and the best way to protect yourself and your discovery of your sexuality is to learn about this part of you with someone who cares about you. Oral sex can be a wonderful thing, too, when you are old enough and ready for it and it's part of a mutual, caring experience—when you're being intimate with someone you know so well and love so much that it is enjoyable—for both of you—it goes both ways. That's just not something most kids your age find *mutually* appealing—and that's the point. When you do things too young it can scare you and spoil something that might be something you'd enjoy in a really special way later with the right partner, under the right circumstance.

"Listen, you all are growing up in a culture that has disconnected sex from romance, disconnected sex from intimacy, and I want you to have good loving relationships. And as you get older, you'll find that there are good men out there who want to have good loving relationships, too. That's worth the wait."

In either case, some parents get a minimalist reply, and that's okay. Others tell me that their kids respond with tales from the front or with questions. If that happens, you take a deep breath, stay calm, and follow your child's lead.

Effective sex education involves much more than simply talking about biology and health. It involves talking about love, values, decency, integrity, gender roles, sexual fluidity; honoring sexual orientations; transcending cultural roles; being better than the media; understanding the risks of sexual objectification, connected sex, and disconnected sex (to intimacy); getting the connection between alcohol and drugs and sex; honoring your own body; understanding that our body, mind, and spirit are interwoven; and realizing that we can heal from hurts but want to prevent them whenever possible. As soon as you hear you kids using terms like "friends with benefits," "booty call," or "hooking up," enter the conversation with them. Encourage your children's schools to have these conversations. Keep it calm and focused. It may leave them speechless, but you'll have delivered the message—without the drama.

Choosing the Parent You Want to Be: Approachable, Calm, Informed, Realistic

When our children are young, it's possible to dream that our current closeness will never change, that we'll always be the lap they climb into when upset, at least metaphorically. That is not a guarantee. We have to cultivate that relationship over time and in the everyday moments that establish patterns and expectations of trust, honesty, and closeness. Just as schools are working hard to teach digital citizenship and netiquette to students, parents, too, need to have this be part of the daily dialogues at home.

If you listen to other people describe lists of behaviors or transgressions of other children and think to yourself *No way—not my child!* then you may still have the illusion that it is possible to control your children and control how the world responds to them. The sooner we understand that they will make their own choices, or will have regrettable experiences, the sooner we can turn to finding ways to be helpful—open, calm, and approachable.

Given the world they are growing up in and the culture that they are destined to join, it is all the more important that our children be able to turn to us for advice, let us into their world to see what they are truly doing, and tell us the truth when they are in trouble. For them to be able to turn to us isn't just a question of whether we allow it or expect it; they must first feel they can do so safely. *Watch what you say. Don't add to the drama. Practice, practice, practice.*

Chapter 8

The Sustainable Family

Turning Tech into an Ally for
Closeness, Creativity, and Community

*Maybe there is a "legacy" kind of training on how to just be
with each other as we are in the moment. That might be "the
pearl of great price" . . . the treasure getting lost in the digital
age.*

—LARA, SEVENTY-TWO

A friend described her nine-year-old son parked at the kitchen table
late one afternoon watching a *Star Trek* DVD on his laptop while
doing his homework as she fixed dinner and the rest of the family
trailed in from the day. First one older sister, then the other ambled
through, each popping her cell phone into a charger in a clutter of
electronic paraphernalia around the computer station at the far end
of the counter. My friend had plugged in her Kindle to recharge,
propped by her husband's iPad, which was tethered to an adjacent
port. At one point her son glanced up from the spaceship *Enterprise*
on his screen, cast a quizzical look at the lineup of digital gadgetry,
and quipped, "Cool! We're like a docking station!"

Not always so cheerfully, that is what parents tell me they worry
that family life is becoming: everyone in orbit, docking for dinner just
long enough to refuel and swap consumer notes about their newest

upgrades and acquisitions; a staging area where tech is the organizing theme of activity and conversation—and relationship. Facebook replacing face time. iChat replacing us-chat. There is no longer any real generational divide in this regard. Digital natives or immigrants, we all love our screens and digital devices. So how do we embrace the interface without losing ourselves in the matrix? How can we sustain family connectivity, in vivo as well as through tech, in the digital age?

Sustainability has been described as "the ability of an ecosystem to hold, endure, or bear the weight of a wide variety of social and natural forces which could compromise its healthy operation." It is what the philosopher and author Bryan G. Norton calls a "philosophy of adaptive ecosystem management." As a child and family therapist, it is a worldview I hold dear in my life and my work, which is so often about replenishing depleted resources, letting hurtful behaviors lie fallow, and thoughtfully nourishing new growth. What new seeds of action need planting? I try to clear the space for new growth, new skills, new insights to flourish, tending as a partner until someone is ready to till his or her own field. Each season inevitably brings new growth and often surprising development.

Parenting can feel like endless weeding, so repetitive and tedious. Gardening requires slow consistent attention. My years as a gardener remind me over and over again that our best intentions often don't produce what we envisioned. We have to check and recheck the soil and add new nutrients. Sometimes we have to learn not to fuss so much. The garden teaches us to persevere and remain hopeful, knowing full well we can't control the weather.

A family *is* an ecosystem. Hardy. Diverse. Resilient. Fragile. Whatever the relationship dynamics, family members create and share an environment that is uniquely theirs; they are interconnected. Each of us wants our family to grow and thrive, to endure in the best sense of that word. We also recognize that we live at a time and in a culture in which "a wide variety of social and natural forces which could compromise its healthy operation" have indeed merged. Facebook, texting, screen games, and the brain's dopamine pleasure response—all these intersect in powerful ways that can compromise the well-being of our children and families.

The good news is that we have everything we need to create sustainable families—loving, thriving human ecosystems. I say this having seen on an intimate scale the ways families fail or the way a family misses the mark with this child or that one. I am endlessly moved by the way a family can adapt and grow and flourish. Families respond to loving attention. They are the original renewable resource. It is never too late to turn a nurturing eye to family and in the process to update attitudes or patterns that aren't working as you'd like. In my work with happy families and with struggling ones, I have seen certain shared qualities that exemplify what I call "the sustainable family." These attributes transcend politics, religion, education, income, and ethnicity. They offer a universal operating system, so to speak. You can be a liberal or conservative, religious or free spirit, urban or rural family and they will help nourish your family. I'll turn to them separately in a moment, but in short, here is what distinguishes the sustainable family.

The sustainable family is a family that has created a fabric of connectivity that is strong and many layered. It can deal with a crisis with elasticity, without unraveling. It is flexible, not brittle, and has high tensile strength forged by spending time together. It values family life above life online and has the wherewithal to understand that you cannot create a sustainable quality of family togetherness unless you make it a priority. In sustainable families, tech can be used in a wide range of excellent ways, but the primacy of being mindful, attuned to each other with all your senses, and fully present for one another without a media interface is the foundation for the humanizing connection. Sustainability is about cherishing the finite time you have with your children to create your family, not taking for granted that you or they will always be there, open and willing to be with you. This means stepping up to manage the media and tech, remove it when necessary, not to exploit it or be exploited by it. It means not to numb out or avoid family engagement. At the heart, a sustainable family is child centered in ways that provide the most loving, supportive, and uniquely human context for healthy growth and development. That can seem a daunting task, but so much of it comes down to day-to-day choices we make in the moment with our children. That is not to suggest they are always easy choices to make or implement, but they are essential ones if we care about our children's futures.

Neil Postman concluded his 1982 book *The Disappearance of Childhood* by entreating parents to conceive of parenting "as an act of rebellion" against the dehumanizing aspects of the emerging tech and popular culture. "It is not conceivable that our culture will forget that it needs children," he wrote. "But it is halfway toward forgetting that children need childhood. Those who insist on remembering shall perform a noble service."

This requires diligence on many fronts, he wrote, including teaching children delayed gratification, responsible sexuality, self-restraint and manners, civil behavior, language, and literacy. "But most rebellious of all is the attempt to control the media's access to one's children," he wrote more than thirty years ago. This call to action is only more urgent today as we seek to manage our children's access to ever more pervasive and sophisticated media.

Whether you envision yourself as a parenting rebel or a sustainable family activist, you want your family culture to be a counterculture to negative aspects of the dominant media and online culture. *What values do I want my children to end up with as adults, and am I living the lifestyle and teaching the lessons that embody these values?* Just as it is possible to be in denial about the long-term consequences to the environment of shortsighted decisions, we can be in denial about the downside of too much texting, not enough true quality time, the so-called educational games, and unquestioned assumptions about tech use in our families, deluding ourselves that more and faster are necessarily better. It is true that some of the best computer games teach life skills—strategic thinking and collaborative skills, for instance. But not all life's lessons can be taught online. Artificial intelligence is getting smarter every day, but nothing can duplicate the distinctly human wisdom encoded in life's deeper experiences.

Technological innovation will always outpace the research on its effects, but we already know that independent of any benefits the hyperconnected tech-mediated life may offer, it comes at a real cost to children's psychological health and well-being. Martin Seligman, a psychologist, father of the field of positive psychology (the study of well-being), and author of *Flourish*, identifies five elements that research has concluded are highly significant contributors to well-being. They are: positive emotion, engagement, positive relationships, meaning, and accomplishment. When you think about these five elements in the context of our relationship with technology— the dopamine hits, the fun of social networking or screen games that reward growing competence, and the instant gratification of getting what you want online (whether it's world news or world nudes)— it would explain why all that connectivity can feel like well-being. And why, in moderation, it can be. It also may explain, however, why excessive texting, Facebooking, and other tech connectivity that replaces authentic, embodied human connection erodes the basic foundation for well-being.

As with environmental sustainability, the threat to sustainable families is in not adapting, not opening the circle for new people and new ideas, generative connections and new technology. Tech can be a tool that strengthens family connection, or it can dilute family connection. When family members "fly solo" too much and spend too much time pursuing their singular lives online with their out-of-family social networks, family cohesion erodes. Family ties loosen. Today's family must develop a relationship with technology without losing sight of the primacy of family relationships, because it is in protecting and cultivating these relationships that we make a family sustainable.

Plugged In/Unplugged: A Story of Two Families

Purists can be blind to the wisdom of a middle path. At times I've heard the reasonable *how much tech is okay* conversation devolve into a moralistic debate between someone who smells the roses and someone who is allergic to them. There is no need for that. I want to share briefly the stories of two families with very different philosophies about living with (or without) tech. I've known the dads, adult brothers Eli and Ivan, since they were seven and nine, respectively, in elementary school. Now in their mid-forties, both with wonderful parenting partners, they have taken dramatically different approaches to the role of tech and media in their families. Each is fully literate in the world of tech; their careers involve sophisticated applications in science and technology, so their difference is philosophical, not based on understanding or access. Both were in demanding corporate jobs when they began to be aware of the extent to which tech was keeping them from family engagement and the impact of that on them and their families. Lest you think this came easily to these dads, I can tell you it did not. They both struggled with the role of tech in their work-family balance, and it took each of them a few years to grasp the cost-benefit ratio and figure out how to bring those into balance, to make the shifts and the sacrifices in a way that felt right.

Eli, a scientist and consultant, traveled internationally for his work and was on the road constantly when his first child was born. He describes himself now as having been addicted to his phone and computer, using quick games and connectivity as a stress-buster from his demanding work and a way to avoid the tugs on his heart from being away from his family.

"Ultimately addiction is about misery stabilization," he says. "It's about mitigating your anxieties for something or other, and in each little three-second nugget when I did that, an awful lot of them were, 'okay, I'm going to check in on something—my e-mail, my stock profile, or something else—because that's going to be a more pleasant experience than the one I'm having at the moment.'" He started to connect the dots of those three-second fixes and was disturbed by what he saw. "I started saying, 'You know what? This is not living life as it is. This is trying to create some more happy reality than I have, and that's not a good way to live,'" he says. "I needed to allow myself to have the terrible emotion that I was having and not try and snuff it out with something more pleasant. I needed to say, 'Yes, this feels cruddy,' or 'Yes, I am in pain,' or whatever it is, and allow that thing to be. I think ultimately all addictive processes are about numbing pain, and even the teeny little ones that are all fueled by the Black-Berry experience are still about that. It's trying to look for something to be more happy than it is."

Older brother Ivan was an early tech adapter, a teen tech wizard who skipped college inspired by the likes of Steve Jobs and others in the inner circle of architects of the new realm. He was at home in the tech industry and he brought the tech industry home with him from work each day. He was (and still is) a total gadget freak. His home office was a mecca of high tech, new tech, and experimental tech, and he was never unplugged. He couldn't pull himself away. A few years into parenthood he began to realize that in his constant multitasking and distraction with tech he was absenting himself from the fuller real-life human dimension of family life with young children—the part that tech couldn't touch, or shouldn't.

"The shift for me was mostly due to just getting older, seeing time pass—it was simple as that," he says. His two children, ages one and four, would be starting school in a few years, and he was struck at how quickly the first few years of their lives were going. "I realized, 'I've got just five more years of this and then they're going to be gone,'" he says, referring to his youngest starting school. His own father had been absent from the family beginning when he was young, Ivan says, "so I really don't want to pass that experience on to my kids."

Exhilarated by their epiphanies, the two brothers set out to become more actively and thoughtfully engaged fathers. Getting a grip on tech was important to them both. Both were partnered with women who cared deeply about connecting directly through family conversation and activity and parenting to reflect that. The brothers' vision and strategies were quite different, though, and eventually the digital footprint in their homes and the house rules each family developed in regard to children's access to tech proved to be a study in contrasts.

Eli's children have extremely limited access to tech, screens, and entertainment media. The family desktop computer has a child-safety monitor on it. Sienna, seven, and Sarita, four, use the computer almost exclusively for schoolwork and very little beyond that. Their brother Matt, twelve, is not allowed to do chat, play computer games, or surf the Internet. He can search Wikipedia and some other sites, but the security block screens for any questionable content. The family has a laptop but it is not Web connected. Eli and his wife, Jocina, each have a smartphone and they have a family TV, but no cable, so they basically don't get any channels and watch very little—typically about half an hour a week. They use the TV to watch DVDs and when Matt turned twelve he got a Wii with a couple of games on it for Christmas. He gets an allowance of thirty minutes a day of Wii, which he can save and pool for a longer playtime.

Ivan's children live in a highly wired household full of tech and electronics, including a big flat-screen TV on which they play all kinds of parent-approved games. Call to Duty is the most violent one. The family has iPads and each boy has a laptop and a cell phone with limited texting. They have a family computer in the kitchen and the kids are clear on the rule about no screens or electronics in the bedrooms. They have to ask permission before getting a new anything—whether it's a movie, app, video, or game. Rather than having fixed time limits Ivan and Carmen have a more flexible approach based on how the kids are using the rest of their time. *Is homework done? Are chores done? When was the last time you played that game? Do we need to have a family conversation about something?* Generally weekends are the times when the kids go on screen, but they can be on e-mail and talk with friends when they're on their computers doing homework.

When the two families get together for vacation twice a year, the brothers have had some tense moments when they argue about tech rules for the cousins during visits. Originally each dad thought the other should be more accommodating and honor the visiting family's parental preferences. That presented problems enough, but then what about destinations away from both their homes? It hasn't been easy to figure out, but they continue to work it out year to year. Their love and commitment to each other, to family, and to the rituals of shared vacations helped them find common ground and compromise.

From this longitudinal but casual and completely nonscientific study of two related but very different families, the good-news finding is that both models work. All five children are thriving. They are creative, kind, thoughtful, generous, intelligent kids, as so many children are with and without iPads, iTouch, widescreen TV, single or multiplayer games, cell phones, texting, the latest movies, and so forth. These cousins adore each other and play hours on end together when they visit twice a year. Although the two families' approaches to the use and role of tech are dramatically different, both are strong, loving examples of what I think of when I think about sustainable families. This also mirrors what I have found consistently, family by family, child by child for nearly three decades as media and tech have transformed home turf.

From looking at Eli's and Ivan's family experiences—which are at opposite ends of the tech spectrum—and the experiences of many, many families in between, seven qualities or attributes emerge as a practical guide for the sustainable family in the digital age. These are essential but not precise; they are open to interpretation because that is the nature, the work, and the play of family life.

1

The sustainable family recognizes the pervasive presence of tech in today's world and develops a family philosophy about using it that reflects and supports the family's values and well-being. The family has its own ways—tech and nontech—of hanging out, messing around, and geeking out.

The most vibrant, healthful relationship with tech begins with none. Ultimately, tech works best in the context of the safe, secure, loving environment we first create for our children face-to-face, unplugged and tech-free. That's because no matter how expansive media and tech become in our family life and no matter how much we use it and enjoy it, our children's capacity to engage with it without disengaging from family comes from their deeper connection to us.

So the sustainable family starts there, building deep connections of presence and emotional attunement. Children need to feel safe and secure, and their first and most powerful experience of that is in relation to us. They "learn us," reading our reliability and approachability through the years in the way we are with them. This includes our tech habits, since so many young children consider screens competition for their parents' attention and, as they grow older, our tech habits establish a baseline norm for them. It's important that we save multitasking for tasks and resist including our children in that category, just as we hope they don't relegate us to a to-do-list status when they are older. Face-to-face, words of love matter. They are one way that children—at any age—hear how much they matter to you. Our daily and nightly *I love you* and the reassuring *we're here to protect you* or *we're here to help you* mean a lot to a child. Just knowing that there are grown-ups who want to protect them is reassuring and critical. It also gives us credibility later when we say we're setting limits (on computers, say) or setting curfew for their protection. Reasonable limits, discipline, and consequences help children understand that there are rules they need to follow to be safe; it's a kind of protection and a sign that they matter to us.

Tech talk is everybody's second language today, so the sustainable family uses it at home. Whether you keep a no-tech, low-tech, or tech-rich home, your children are of the world and they need to know their way around. They're going to see things out there and you want them to feel they can tell you about it. In addition, each family needs to develop its own house rules, an actual contract, about the use of tech—in an ongoing way because it will continually change with each stage of child development and each new tech innovation. You want your child to be inquisitive, thinking, and comfortable chatting among friends about how things work at your house—and why. In the largest way, no matter how you feel about media and tech, establishing house rules for your family gives you a chance to (a) acknowledge different realities and value systems outside your home and family without demonizing others, (b) explain why you are making the choices you are for your family, without demonizing others, and (c) practice responding to your child without showing yourself to be scary, crazy, or clueless. The conversations you have with your six-year-old about why he must ask permission to watch a TV show or go on the computer preps the soil for conversations with him as a teen about limit-setting, responsibility, and consequences when he is exploring far more adventurous and risky territory.

Ongoing conversations may address matters of life online, family tech use, school and legal rules, netiquette, etiquette (yes, the old-fashioned kind that helps our children learn that there are times when tradition trumps tech), and digital citizenship. Think of tech talk as part of your family's "media diet," even if you don't allow much media in the house. Talking about it is an entirely different matter and opens the conversational channels for thoughtful discussions about practically anything.

Strategic thinking sounds complicated but it is not and it is a lot simpler than crisis intervention. At home and at school, a responsible-use contract regarding media, computer, and tech use, a contract that everyone signs, clarifies expectations for all and makes it easier to set limits without seeming arbitrary when your child's friends come over—and easier for your kids to set limits with their friends.

However fixed or flexible you make the contract, you are somewhere on the responsible parenting continuum because you are setting limits, establishing consequences, and modeling responsibility.

Some tech-specific principles for the sustainable family:

- Parents take responsibility and engage actively with children. This means installing protective software, blocks, or apps to limit their access to levels of content they are not developmentally ready for or which you simply don't want them to access.
- Children earn privileges by showing responsibility and establishing trust. Why wait for a $500 credit card bill for Miss Kitty app accessories that your six-year-old clicked to buy as she played on your smartphone on the family vacation? Establish clear rules about access and downloading *anything* and other rules that keep tech safe and fun for your child at any age.

- Mistakes are treated as teachable moments. Children are encouraged to fess up and they are allowed to learn from mistakes. Shaming them is not an option. Hearing them out, talking it through, adding to their understanding, and holding them accountable is.
- *Family members have shared expectations re communication via tech. You're entitled to say, If I text about a date, or a plan, please text back—like now. If I text and say I'm going to leave a voice message, know that you should listen to it—like now. If I call and let it ring three times, it's important, pick up. If you said you will text so I know you are safe, DO! There is a shared sense of responsibility to the person at the other end who loves you. It can change daily, but changes need to be agreed upon and acknowledged.*

Here are some other helpful reminders to consider when you give your child a phone with texting privileges. No need to be an ogre spelling these things out; by giving your child a phone, you're expressing trust in her readiness to be responsible. Be encouraging—but clear:

- Having this phone is a privilege that can be revoked. Like getting the keys to the car. It's not really a gift. *I'm giving you this to use with the understanding that I expect you to use it in the ways we've discussed. If you take it to school, you follow school rules—no exceptions. Don't use this phone to initiate or forward mean stuff or sexy stuff. No porn, no illicit anything. And if anything happens to the phone—if you lose it or break it—you're responsible for replacing it.*
- Parents should always have the password to the phone. *I reserve the right to scroll through your phone if I'm ever worried about your safety.*
- This phone doesn't replace real conversations with your

family and friends: *I know a phone isn't just a phone any-more, it's a minicomputer, but remember that important con-versations should happen face-to-face. Don't use this to hide or escape from what's going on around you. Do not live your life on your phone. Don't let your phone become your life.*

- Safety, health, and good habits come first and always: *No using the phone when you're crossing streets. No sleeping with your phone. The phone stays off during homework and family meals. It's easy to get dependent on all this tech—don't get caught up in it like that.*

  I have seen so many parents and kids build resentments, agonize for hours, or burst into tears because of mixed mes-sages through mixed media. Family means we work harder to get it right as much of the time as we can.

- Common sense and manners matter and must prevail in tech-assisted communications. A philosophy of courtesy teaches your children that. Resist the fast-twitch urge to text or per-haps even e-mail anything with serious emotional content. Big news, bad news, sad news, sometimes even the happiest news can be sensitive communications. In many families, there are still many people who feel hurt when they read about a death or engagement online, rather than hearing it and sharing their felt reaction with you. Is it okay in your family to e-mail a condolence note? Maybe, maybe not. Check it out. It's family.

2

The sustainable family encourages play and plays together.

In this "crazy busy" world families need to play together. In in-terviewing children for this book, I was surprised by how often kids of all ages talked about how meaningful it is to them to have fun as a family. Some would talk about "the time when . . . " and launch into tales of the simplest, sweetest, goofiest times spent with a parent or sibling or cousin. Throwing snowballs at a tree; skipping stones across a creek; playing charades, cards, or board games; baking cookies or a "mystery casserole" for a potluck—it doesn't have to be fancy. Children from diverse ethnic and socioeconomic backgrounds often share similar stories of delight at the simple things that bring their family members together to play, laugh, and just hang out and be together.

Kids also talk about how tech often gets in the way of family time. The letdown of vacations consumed by a parent online, a weekend where a parent got gridlocked on a computer and reneged on a fun plan, small details—they were all playing cards except for the one who had to be online. Broken promises can lead to broken hearts. It is so easy to forget what it is like to be a child. Children want their parents to want to play with them at all ages, whether it's a board game, Wii, or the school picnic. They feel proud and special when their parents delight in their company. Kids know that you enjoy them when you actively choose to be with them, when you take turns and let them choose what would be fun to do, when you co-create fun times, and when you don't back out of playdates. All kids understand that there are times that work interferes, that "duty calls" parents and they must answer. But as with most things in relationships, there is a tipping point. If you aren't big on follow-through or have too many excuses, you and your child lose this vital connection of play and shared company.

The father of young children tells me: "A lot of how we connect is just going walking, shopping, going to movies, you know, taking the dog for a walk, watching TV, reading, whatever—just hanging out together. Nothing really big and fancy, it's just being together."

The mother of a teen tells me: "A lot of our connecting is around the media. So we watch *Modern Family* together. We watch *Glee* together. We watch *30 Rock* together. And it's in those conversations watching TV where I hear about her life."

It is this "just being together"—really *together*—with parents and family that gives children confidence, pride, and security. They feel they belong, they feel the connection, and they are more likely to talk about things that matter to them in that setting than at any other time. You can use this family ethos of creative play to make tech-free, unstructured, imaginative play alone and with friends a priority for your child—and for yourself. In addition to family play, a child's solo and peer play nurtures curiosity, grit, and zest and a host of social and emotional learning closely linked to well-being and success in school and life. Play is where children discover their own talents and inspiration. It is where they practice concentration and how to work through frustration. Play is the best fertilizer for growing kids.

There are two kinds of play that are really important for families to nourish in kids. One is the capacity and opportunity to play alone, to enjoy dabbling in their own imagination, finding their way from boredom to something that grabs their attention, learning how to enjoy their own company for extended periods of time. And the way you do this is to create numerous situations in which you are in a place or setting at home or on vacation or traveling where it's the old-fashioned equivalent of a rest hour or downtime. Whether that's with a sketch pad, modeling clay, Legos, a book, or Jenga blocks, it's all about discovering "me, myself, and I."

The other type of play is as a family—not just with peers or siblings—but to be a family that knows how to play together across generations, siblings, cousins, extended family, and family friends.

Whether playing alone in the backyard or with others in the living room, the point is you bring your imagination. It's no different from everybody sitting around the living room with their computers except it's unplugged. Bake something. Make something. Invent something new.

Creativity comes in infinite operating systems in a child's mind. For example, Eli's son Matt is writing a book by hand, pen to paper, illustrating it with maps and magical characters. Ivan's son Luca, whose passion for tech is as strong as his tech-wizard dad's was as a teen, is developing a computer game, "geeking around," learning how to write code. Neither original artistry on paper nor original artistry on the screen is better or worse than another. Each boy is using his medium to express his own authentic deep creativity and imagination. This is levels above commercial screen games that replace the prerogative of a child's imagination with a commercial concept, script, and action. Indeed, when kids are making up games, helping each other understand the rules, and improving on the rules and strategy, this kind of play can teach cooperation, empathy, patience, brainstorming and co-creating, teamwork—all leadership skills.

Further, when a child's family life is grounded in the primacy of family and good values, in daily conversations about who we are, and in understanding each other, what we are working on, where we need encouragement or help, and where we feel safe, connected, and optimistic, the impact of negative online or screen play content or accidental exposure to crass or upsetting images or words can be processed and, in that conversation and deepening understanding, become yet more good compost for the garden of life's lessons.

If you're stumped for nonscreen play ideas, turn to your child. Children are expert play consultants. Enjoy the opportunity to let them be the experts. They'll appreciate the leadership role, and it also allows you to learn from your child, or each of your children, what kind of play appeals to them. And if you're both stumped, then together you can use screens for help—one of the great things about tech is that you can go online and find anything. Get yourself some books about play. There are wonderful books available for parents about all kinds of age-appropriate, nontech play.

Play is a window into your child's experience. We talk about all the benefits that children gain from play, but we know from research and play therapy that it also offers parents a way to hear and understand aspects of our children's life experience that otherwise are hidden or unclear to us. In playing with them or watching them at play, we witness their orientation to learning, creating, problem solving, curiosity, and resilience. One child may love tactile play, another wordplay, another rough-and-tumble play. Children use play to process experience and emotions with which they may be struggling. Think of play as a way to see your child's inner life expressed. Enjoy getting to know your child in play!

That includes teens. In both high school and college today, drinking has so permeated the teen and young adult culture that drinking games are some of the most commonly played games, period. A lot of kids don't like it but go along with it rather than look uncool or sit home alone. Others do stay home or hide out in their dorm rooms because it can be difficult to find same-minded friends to hang out with or places to go where drinking isn't what it's all about.

While some would never say directly to their parents, "I really don't enjoy all that," teenagers who don't like the party social-drinking scene are often at a loss for what to do on the weekend. While they may seem busily engaged with their computer, they tell me they often wish their parents would suggest something fun for them all to do together. For these kids weekend nights can be a lonely time. Many struggle with finding friends who know how to play, unplugged, and are easygoing. In focus groups, teens tell me they love to play charades and some of the same old games people have loved for years because they're so much fun. If you have teens who seem alone at home on weekend nights, occupying themselves by watching movies or being on the computer, think about inviting them to play some games with you or do something that's interactive and fun. It helps to enlist college kids and older cousins to introduce them to the games because they'll think it's more cool.

Most of all, it's important to know that you don't have to wait for them to ask to hang out with you. Think of something they would enjoy and see if they want to do it with you. If not a movie, then maybe a card game. Invite them to fix something, make something for someone, cook with somebody. Enlist them to help you with something that needs doing or planning, for the family or in the community. Try to be thoughtful about eliciting their help, aiming for something that can be shared and is enjoyable.

3

A sustainable family nourishes meaningful connection and thoughtful conversation that shares feelings, values, expectations, and optimism.

Family is the language lab of the digital age. Children's tech-connected socializing has taken them out of face-to-face conversations and limited their opportunities to build the basic skills for live dialogue and that entire dimension of interpersonal communication. It is essential that families create ways of coming together and talking about all kinds of issues, matters of the heart, fights, plans for the weekend—the family equivalent of circle time in school that can offer an opportunity for thoughtful conversations and a process by which they can talk about the things that are really important to them, feel heard, respected, and helped. And in a way that nurtures not only individual growth but the strength of the family as an entity that is there for each of them. As wonderful as video chatting is, humans learn the nuanced art of conversation face-to-face. Given how much of kids' so-called conversation is with other kids, most of whom rely on tech to communicate, the role of parents and teachers as conversation mentors is more important than ever. It is in their live conversations with us where they are continuing to learn the art of personal communication and how to be part of a dialogue or group conversation. Families are the safest place to learn how to work through the hard stuff of relationships and learn to talk to one another.

These conversations are a place where everybody in the family— parents and children alike—can practice positive engagement, self-control, problem solving, listening, being curious, and sharing ideas and insights. Two kinds of conversation with your child have particular potential for great or dismal outcomes. One is the much-vaunted dinnertime conversation. The other is the conflict resolution conversation, which is part and parcel of life with children at any age. To borrow from Benjamin Franklin's observation about death and taxes as the only sure things in life, I would add that the only things certain in family life are eating and arguing—sometimes at the same time.

Dinner first. We hear all this research about the family dinner and why it is so important. But we all know: family dinner can be horrible! Dinner should not be conversations about tests. Dinner should not be about stressful things. Dinner is not about fighting; dinner is not where siblings get to be mean to each other or you get to nag. The research on why dinner is so important has to do with how families connect at dinner. Not all families can eat together. If you can't eat together, you can create this context at other times, so don't worry.

One of the most important qualities of dinner together is that it's a time when you can be curious about your kids and they can be curious about you. So the kinds of conversations you want to have at dinner are about what makes them tick. *What did you do? How did you come up with that? What did you do next?* It's a time to invite each other to brainstorm solutions to problems at work or at school. If you can think of a dilemma you had in your workday when you didn't know what to do, share it with your children. This is great role modeling for children that you, too, face moments of insecurity or confusion, not knowing what to do. Bring your kids into your moments of confusion and talk about how they would solve the problem. Update them another day. *Remember that advice you gave me?* Share how it worked. That teaches problem-solving skills, enhances their sense of efficacy and, I believe, nurtures their motivation and confidence for action. If you want to strengthen their belief and their ability to solve the world's bigger problems, invite them to help you solve the little and big problems you are dealing with. It makes them feel empowered. Kids hate it when they are criticized and critiqued and shut down for being naive, or stupid, or too young to know better. Leave that off the menu.

When families see so little of each other in a busy day, it's tempting to use dinner to preach rather than discuss, tempting to raise stressful topics. Better to do that away from the table so kids don't inhale their food and run. At its best, dinner is about sharing stories, solving problems, no pressure, no meanness, no putdowns, no sarcasm—and no tech distractions. If you want your children to feel like they really matter to you and you're really curious about them, show them with your undivided attention. Some families keep a digital tablet at the table to answer questions and check facts. Use it to serve the conversation—not break it.

Now from eating, let's move to arguing.

Principles of conflict resolution and mediation translate well to the home front, whether the conflict is over a teen's party plans or an eight-year-old's playdate meltdown. Eli, whose work includes mediation, shared these steps he used in a sibling meltdown with his young daughters:

- First, extract yourself from everything else to make yourself fully available. Cooking or whatever else, it can wait.
- With younger children, get down on one or both knees at eye level with them and ask whoever seems to be most hurt by this: what do they think just happened? Then ask the

other person, what was their experience of this? With older kids get everybody on the same level, sitting in sofas and chairs—literally level the playing field.

- Establish your neutrality and helpful intention. Say that you're there as a third party, just helping people say what they need to say and helping them feel like they've been heard; to make sure each person has heard and understood the other, and then had a chance to say how all that sits with them.

- Apologies are part of the process, but not an end in themselves. This is as much about helping each child feel heard about whatever wasn't working. Accountability must be there, especially for the purpose of laying a foundation for future interactions. You want them to think: *What does it mean that somebody's feelings were hurt? What's my role in that, do I want to do anything differently, and what might that look like?*

More than ever, family is the training ground for the art of social conversation. Ivan and Carmen made a concerted effort to teach their sons from an early age how to participate in a conversation and simple courtesies. "We've taught them that when we are sitting here at lunch it is a reciprocal process; you need to know how to have a conversation. I ask you a question and you need to answer and ask one back. Sitting there in silence or offering one-syllable answers are not acceptable. Please and thank you, good manners," Carmen says.

Finally, we can even fine-tune the way we talk about our children's successes, to make our comments more than gushy praise. Rather than jump to *you deserved it!* be curious about each step along the way, each choice point and why they chose (well) to act as they did, and invite them to let you walk with them on their path to success. We boost their sense of efficacy and well-being, optimism, and enthusiasm. The simple act of naming three things in your day,

whether at bedtime or at the dinner table, and describing in detail
what you did, step by step, to create that outcome can alter your
child's sense of self, confidence, optimism, and well-being. And your
partner's. Yours, too!

4

**In the sustainable family, members understand the unique-
ness of each person, encourage independence and individual interests, and foster their independence in the context
of family.**

Each member of your family has his or her unique wiring, temperament, ways of self-soothing, sense of humor, ways of being creative, vulnerabilities, and neediness. What motivates one child shuts
another one down. What one finds challenging, another finds overwhelming. Music to one is fingernails on a chalkboard to another.
And so on. The more we understand and genuinely appreciate each
member of our family, the safer and stronger our family is for each
and all. For parents, this is about loving and raising the child you
have, not the one you thought you'd have or perhaps wish you'd had.
Being family does not mean that we are all alike, that we all love to
read or work out or hike. It may turn out that way, but don't expect
it—and don't press for it. In a sustainable family, people are allowed
to be who they are. What makes a family family is that at baseline the
members accept, respect, and care for each other in spite of frustrating differences. Siblings fight but can learn to work at getting along.
Extended family often includes people you would never choose as
friends, but still . . . they are family.

Years ago I worked with the family of four boys with a huge
commitment to family travel, often including service-work vaca-

tions that involved the whole family or sending kids at a particular age on a special away trip or camp adventure. This was great in theory and almost everybody in the family seemed to be gung ho about the next trip. However, time and time again the third son, now age fifteen, would become "difficult" just before leaving, cry, and not want to participate in anything after two or three days. Eventually it was his turn to go on a celebrated rite of passage in this family—a summer travel experience after ninth grade. The first two sons had come back glowing from their trips. The third son was so distraught and homesick that his mother had to go get him. He couldn't make it through the session. None of them could understand it. "But we *all* love to travel!!" his mother said, echoing everyone's sentiments. "This is what we do as a family—*this is our thing.*"

After one meeting with Zak, now seventeen, it was clear to me that he had been having full-blown panic attacks on the trip and had a long history of social anxiety and intense separation anxiety; he hated leaving his home, even with his family, to go on vacation. He also struck me as a very smart boy with some attention and processing issues, which turned out to be significant when evaluated. It took everyone time to realize that he had a completely different kind of wiring and temperament from the rest of them—different, not flawed or less-than.

Now they work diligently as a family to make it okay for him to be himself and to develop strategies so he can handle family trips. One big compromise involved tech. The parents had strict rules about no cell phones on family trips, at dinner, and at family events. Zak has a long-standing very good friend, a girl who really gets him and can calm him down. Now his parents let him bring his phone on trips because they know it stabilizes him to talk with his friend or at least know that he can. This makes it easier for him to travel with them, which they all prefer to his staying at his grandparents'. When you reinforce your love and appreciation and acceptance for your kids in these flexible ways, it makes them feel that you not only accept them but you believe in them and are proud of them for being their quirky self.

Circumstances and people change and a child who has "always been . . . " (fill in the blank with the identity he or she has had in the family) can present a self that is new to us but no less authentic to the child. This could be anything—coming out as gay, pursuing interests we never imagined and which break the mold of the child we thought we knew, suffering an injury that deletes a dream of following a particular passion, or discovering a serious physical or mental illness. All of these can test a family's mettle. Fortunately, acceptance and flexibility are inner resources that are always ours to give and to practice, and these moments offer all of us the chance to see ourselves anew and appreciate this new dimension in our family.

One summer when Eli's eleven-year-old son Matt was visiting his cousins in Seattle, his uncle noticed a lump on the back of his neck. The diagnosis of cancer came quickly. Lara, his grandmother, stepped up to help get Matt and the whole family—two parents with outside jobs and two sisters with busy school lives—through the nine months of chemo and radiation.

Lara took charge of Matt's care, making the two-hour trip to Seattle weekly during the course of his treatment. She was quite familiar with all the side effects of chemo and radiation, having been through it herself and having helped other friends through the challenge. Lara would sleep over and stay as long as it took, each round, to tend to the full range of Matt's physical reactions. But this wise granny also knew that it would take more than physical care to get Matt through this harsh course. So she and Matt, with his parents' blessing, embarked on a simultaneous journey: they watched the six *Star Wars* movies over and over, throughout the nine months of treatment. In this low-tech family, suddenly not only did Matt have an iPad, he had it in his bed! Lara told me later that she knew that if they could keep talking about "destiny, who gets this assignment, who gets that one, forces of healing and destruction, and Luke Skywalker's mythic fight for his life," that those conversations would help this eleven-year-old child make meaning of his hard assignment. Together, this eleven-year-old and his seventy-two-year-old grandmother rode the waves of nausea, of waiting for results, of too-weak-to-sip-soup, and of feeling better and preparing for the next big battle with the iPad and these movies as good medicine, a force for healing.

5

A sustainable family has built-in mechanisms for healthy disagreement. Parents set limits, act thoughtfully with parental authority, and do the hard parenting work of demonstrating accountability, authority, openness, transparency, and not *just trust me* but *here's why.*

A teenage girl once remarked to me that she wished her mother wouldn't be so flexible sometimes about accommodating her requests (or even her demands) for exceptions to family rules about "important stuff." Sometimes when she felt pressured by her friends to join in something she had reservations about or simply didn't want to do with them, she would say she had to check with her mom. They would all gather around the smartphone screen to watch the text exchange, confident her mom would cave because she always did, and the girl's heart would sink a little when her mother would text back her predictable okay (with smiley face).

"It's just really hard for me to say no to my friends, and I could use the help . . . " she told me.

Kids want their parents to set limits, be clear, transparent, and flexible but not endlessly so. They need to hear at every age that privileges and responsibility go together, and that no matter how new the attraction, they earn access to it the old-fashioned way: at an age-appropriate time by showing they are responsible and trustworthy and can handle it. Otherwise, our lack of clear limits can send the message that they're entitled—to material things, to free rein regarding tech time and usage, to power on demand.

Ideally, our parental intentions in setting limits are first to protect our children from influences or situations that present a threat to them or simply are developmentally inappropriate for them; then to guide them until it is clear they are able to make smart, safe choices for themselves. Responsible limit setting is not about micromanaging or a parent's controlling behavior. Most children welcome a parent's steadying hand as they learn to ride a bike, and would never demand that you let go so they can pedal into a busy intersection. This is not the case when they want something—the candy bowl, the computer password, the keys to the car—and you and your limits stand in the way. This power struggle is the source of much of the everyday disagreement between parents and children and the underlying parenting job is to help a child develop self-regulation and practice self-control. Kids need a process for that, and they need parents to provide that structure until they can internalize it and show they can manage new freedom and responsibility.

One of the most challenging aspects of this is that unfortunately when our kids are arguing with us or out of control, pushing against our limits, our first reaction is to push back, verbally at least. As we saw in the preceding chapter, we sometimes react in ways that kids experience as scary, crazy, or clueless is not helpful. So once you get yourself in control, the most effective way to teach your child self-control is, first, to be consistent about limits and consequences and, second, to shift the conversation from correcting to connecting—use your words to help your child get in touch with his or her own "inner boss" (executive functioning) and listen to it. This shifts the locus of control back to the child's internal process, off a dependency on an outside source and a reflexive resistance to it.

Here is how that sounds:

- Corrective (parental boss): *Are you on Facebook?! Shouldn't you be doing your homework?*
- Connective (inner boss): *You were going to write that paper before the weekend so you'd have a good weekend. Are you gonna stand by that decision you made? Or Are you following through on the commitment you made to yourself to do your homework?*

You want to help your kids make plans, take control of their lives, and rather than say, *You are not doing what you should do,* you're saying, *You are not sticking to the plan you wanted for yourself.* It's a different point of entry. For perfectionist kids, who are tyrannized by a tough inner boss, learning how to say no can be just as important. *I hear you say you are exhausted and can't stop adding to your collage poster. Maybe you need to go to sleep and trust that it's good enough? What do you think?* You want to help them turn the volume up on the voice inside of them that is about controlling themselves, rather than you being the external voice for self-control.

When you and your child disagree, channeling your inner mediator can help you stay calm and constructive: I have often heard Eli say to his kids, "You know, this is as far as I'm really willing to go in this direction. Is there some other way that we could make this kind of a thing work?"

6

The sustainable family has values, wisdom, a link to past and future, and some common language that they share with family and friends.

"It was a three-generational gathering for an old lady's birthday, rather Proustian in the tone of voices, the way the kids and old people were connected in levels of play, talk and silence. And there was no electronic machine in view. No TV. No iPad. No laptops. No reference, even, to contemporary media. What we did was hang out with each other from late afternoon to long past sunset and just be in the heart of connection in the moment; all of what needed to happen had space and time to emerge. Not at all pretentious or studied, just plain present with each other at a birthday party, full enjoyment of each other without media," Lara would write later about her birthday party.

Lara was turning seventy-two and we were gathered for dinner to celebrate her birthday. There were nine of us there, three generations of family and friends. They included her son Eli and his three children. It was a divine evening of easygoing togetherness. The girls cut out paper hearts to hand to everyone present, making sure no one was left out. Then they would leave to work on a play or tumble quietly. About three times throughout the evening, when they felt the need to play more energetically, they asked their father to play "She'll Be Coming 'Round the Mountain" on his guitar and we all sang while they galloped around the living room till they were spent.

Later they played and listened quietly. Not once did a child interrupt or have a meltdown. Everyone was okay to be who and what they were in the moment. Matt drifted upstairs to read and draw and the girls nestled into any adult's side or lap when they wanted closeness and affection. They would get up without fanfare when they wanted space.

These are the kind of ordinary summer nights that settle a child's soul, nurture creativity in the company of family and extended family, deepen their security and a sense of belonging in the world and being held in a community family of loving good adults who so clearly care about them. This could not happen with the TV on, with cell phones ringing, or with anyone texting. It was cheese, crackers, wine, and sparkling water and then on to a potluck dinner. Seamless.

A few days later, Lara reflected on the mingling of memory and the moment in the timeless quality of that evening. "What is old and familiar about Saturday night's birthday party is my own memory bank from my early summers—all during World War II and up, all the fifties," she said. She recalled an almost surreal sense of timelessness when they were together this way, despite difficult circumstances. "It's the sense of 'slow time no time and always enough time' even though there were air-raid tests and ration coupons—and a massive unspoken fear. The grown-ups were always there, talking and reading and knitting and sitting and watching and commenting on our plays. There wasn't any television, really, until my folks got one much later and put it in the basement. They rationed us to one show a week—Ed Sullivan on Sunday evening with them present."

She remembered playing Scrabble and charades—the prescreen first-generation versions of those games—and inventing plays, performing songs from musicals, reading out loud, and memorizing poems to recite at those gatherings—that was the main event. The kids dressed up and presented plays to the grown-ups. They performed on invisible instruments and had contests and spelling bees. They'd fight over who got which old hat for their parts in the shows.

"The absence of media gave us our own imaginations," Lara said. "We trusted our imaginations and let them run the moment of play . . . the stories emerged from the playing, not from the TV. The adults just sat all around the kids and the playing—as if those big people were the permeable circle through which the small people could enter, leave and return. It's a body memory of no media anywhere except a radio that was for listening to the war news or a children's hour. And only connection . . . of running over for a hug and just listening to the drone of their talk while they aren't listening to us—the most important thing was that we were all there, whoever the 'we' was at that moment."

Years later, bringing up her own sons, and now with their children, her grandchildren, what she treasured was the same sense of timelessness: "Slow time no time always enough time."

When Fred and I were young parents, we worked to preserve a taste of that for our own children, despite everyone's much more scheduled and work-pressured lives. Summers in a tiny cabin on a tidal cove in Maine were the venue. With kids mostly grown just a few years ago we finally added a big-screen TV for movies, and Wi-Fi. But throughout their childhood, the mainstay was games of Sardines, Acey Deucey/Rummy 500, and Monopoly, shooting rockets, and playing *Little House on the Prairie* in the cove, lying on the picnic table and counting shooting stars, and the business of sitting around the fire and making up stories were compelling as well as steadying, solidifying, and bonding in an ancient kind of storytelling tradition. Old friends, new friends, family, and community, summer after summer, we all created a culture that was a counterculture to the crazy busy, plugged-in rest of the year.

As families age and develop together, as children become adults with partners, sustaining a family becomes a cocreation; a move toward interdependence, a shared sense of responsibility, an understanding that how we talk about other people in our family—including those dearly and not-so-dearly departed—shapes how we evolve as a family. We may make family out of friends to share ritual and tradition. With every word we are creating and re-creating, again and again, the story of our family, the narrative in which everyone gets a part and no one gets written off or written out.

The sustainable family is open, adaptive, inclusive, tolerant, and flexible. It is also always a work in progress. Sustainability is not a static thing; it's not a measure or a score you can achieve and bask in your accomplishment. It is a practice. When challenges present themselves, when mistakes happen, when family members disappoint or offend one another, or when any of the many kinds of difficult family situations arise, so does the opportunity to practice.

7

Sustainable families provide experiences off-line in which children can experience and cultivate an inner life, solitude, and connection to nature.

Solitude, deep thinking, stillness, a full sense of empty contentment, and a soulful feeling of gratitude cannot be found online. We have to cultivate this capacity for peace and nourish this appetite for spirituality and the search for big meaning. Whether in the sanctuary of the forest, a holy place, or a poem, sustainable families provide opportunities to encounter and engage this part of our selves. Instead of plugging in the earbuds, listen to your self, find your inner GPS, Google search your own life experience, plug into your soul. Children need time off-line to develop this internal relationship with their selves, with deep reflective thinking, to learn that being alone is not necessarily boring, lonely, or scary.

Nature, in particular, also reminds us that we share the planet with everybody. We are ultimately all one family, all sharing the same planet. When we call her Mother Earth, we remind ourselves that we are one family. Like any family we face difficult challenges, major conflicts about getting along and sharing resources. In the political discourse of war and terrorism, we forget the great mother; being in nature also reminds us that we can have a politic in which we work for peace, sustainability, and sharing what we have. Tending the garden, locally and globally. Whether your kids play in tidal pools or watch planet Earth, or get to do both, in today's world we can't afford to disconnect our own family from everyone else's or the global family.

For many of us, it is in nature that we discover and recover our deepest connection to our selves. The wilderness teaches us to tread lightly, to coexist, and to be gentle with each other. In the peaceful moments of quietly sitting in a canoe or on a cliff's edge, peace itself seems attainable. There is a reason children love to tell you about their summer times, the magic of fireflies and fishing. In nature we get to inhabit our natural selves, our natural bodies. It can be a powerful even if not always pleasant connection.

TV shows and documentaries can show us in vivid detail extraordinary facets of nature. Screens can bring nature into our living rooms, and that is good; not everyone can take a hike when they wish. That said, we can get directions to the trail on our GPS but we cannot experience the hike and the inner journey in the same way. A young man I know, who grew up in the woods of New Hampshire and now attends an urban college, likes to start his day on his computer, "walking" a trail filmed precisely to give the viewer the behind-the-camera view as the trail unfolds. His years of being in nature—walking unmediated along trails like those—inform his online virtual hike. It is his embodied experience of nature and the wild that connects him to the visual image in a meaningful way.

Nature teaches us a different kind of strategy game for survival. It is the best classroom I know in which to learn lessons about the dangers of excessive force, clear cutting, gratuitous violence, hoarding, and exploitation. And about how small we are. Contrary to the "unlimited potential" of artificial intelligence, being in nature teaches us nature's lessons of limitations and sharing; that we are all fundamentally interconnected, not just technologically.

New York Times columnist Nicholas Kristof wrote about hiking the Pacific Crest Trail with his daughter, noting how, "In short, the wilderness humbled us, and that's why it is indispensable."

Tech does not teach humility; quite the opposite.

"In our modern society, we have structured the world to obey us; we can often use a keyboard or remote to alter our surroundings," Kristof wrote. "Yet all this gadgetry focused on our comfort doesn't always leave us more content or grounded. It is striking how often people who are feeling bewildered or troubled seek remedy in the wilderness."

In a father-son talk about online basics, Ivan explained to his seven-year-old son Peter the concept of "the cloud," the great repository of data in cyberspace where private information shared online goes and why you shouldn't put your name or home address on a game online. As they took a hike that afternoon, Peter gazed upward into the heavens. "Daddy, does God live in the tech cloud? Is God in charge of tech?" It is so often the youngest voices that ask the biggest questions.

With all technology's power we tend to relate to it as a higher power and think of it as unlimited, the great force in the cyber sky that unites us all. Tech is a tool, and when used well, it can serve all of humanity in thrilling ways. But we have seen the dark side, with too many examples of the harm that it enables in the hands of malevolent, or even simply careless, people. We can't let the sway of new tools distract us from old truths: as a species the human spirit thrives in the context of good relationships and a sense that we are all fundamentally connected to each other and part of something larger than ourselves. Something wondrous.

When we all get caught in our screens, caught in the World Wide Web, and we disconnect from an ethical culture, the humanistic values that all religions share, we forget about the other forces that connect and ground us. We forget to be grateful; we forget what it feels like to be grateful. The sustainable family recognizes that creating a sustainable future for the next generation of families requires we give children a sense of stewardship. As Kristof wrote, "To guarantee wilderness in the long run, we first need to ensure a constituency for it." We could say the same for creating a constituency for a compassionate culture, one that protects childhood and supports children and families. That is the job of parents and families.

This is not a simple time and the big questions about how we use media and tech are not simple. The answers are nuanced and we have to be willing to hold the complexities and think deeply about their implications, resist facile, fast-twitch answers that insist "the kids are all right." The kids are not all right. Not completely. The World Wide Web makes it possible for us to go online and see, hear, and feel how we are interconnected in ways never before imaginable. These times require us to hold the reality that we are all connected in so many deep and invisible ways and to recognize that tech makes the invisible visible in ways that can be challenging for children and ourselves.

Sustainability is about pace and sequence—for our children, it's about the pace and sequence of growing up. For humans, sustainability is about the long-term maintenance of well-being, which has environmental, economic, social, and spiritual dimensions. Ultimately, the sustainable family is about stewardship. As parents we are stewards of our children, and our children are stewards of the future.

Our challenge as technology continues to open new worlds of possibility is to not let new opportunities and new apps obliterate old truths. Children need our attention; children flourish in families that work hard at the hard work of being a family. We have not yet proven that we can work as the large global family we so desperately need to be. Fortunately, we can bring humanity and technology together on a smaller scale in our own homes and our own families and we can teach our children how to be in this new world. There we can deepen connections, cultivate closeness, and push *pause* more often to savor the gift of time and the primacy of family.

Slow time no time always enough time.

Acknowledgments

This book is about connections and how we establish and maintain the kinds of relationships that will help us all live up to our potential in the digital age. The project has been for me its own case study in the alchemy of chance meetings; connecting with teachers, parents, and students all around the country; reconnecting with old friends and colleagues; making new online friends; and discovering new technological applications that have transformed the process of research and writing.

It began with a fortuitous meeting at HarperCollins, in which Lisa Sharkey invited me to have a conversation about writing a book, and asked Gail Winston, executive editor, to join us. From the moment I met Gail, I was struck by her integrity and acuity. Since then, I have come to cherish her warmth, humor, and wisdom. Through Gail I connected with my agent, Kim Witherspoon, whose tenacity and belief in the book have been unwavering. At each stage in the process of writing this book, these two extraordinary women have been my polestars, keeping me balanced, guiding me brilliantly through this project.

At HarperCollins, I am also very grateful to Maya Ziv, editor, for her meticulous attention to detail during every stage of this process, and Robin Bilardello for creating the perfect book jacket. I hope that everyone judges the book by your brilliant cover! I thank Stephen Wagley for his diligent work as copy editor, and at InkWell Management I thank William Callahan for keeping everyone connected.

I will be forever indebted to my friend, fellow psychologist and prolific writer Michael Thompson, who helped in so many ways, especially in connecting me to Teresa Barker, who became my invaluable partner in this process. I could not imagine a more dedicated, intelligent, and delightful person to work with! I always looked forward to "going to work" on our smartphones, headsets, Evernote, and joinme.com with such a calm, upbeat, meticulous, and utterly professional colleague. My collaboration with Teresa, whom I have yet to meet IRL, is a glorious example of how technology can facilitate thrilling working relationships between two people who sit down to work together thousands of miles apart.

For her part, Teresa extends thanks to her family—husband, Steve Weiner; son, Aaron, and daughter-in-law, Lauren; daughters, Rachel and Rebecca; beloved mentor and mother-in-law, Dolly Joern; sister (and schoolteacher), Holly; and parents, George and Maxine, for their invaluable contributions—and to soul sisters Sue, Margaret, Kathy, Leslie, and the Elizabeths, whose friendships, begun the old-fashioned way, have flourished across years and screens. And to Michael, who e-mailed one December day to introduce a friend.

The research and networking for this book took me into more than thirty schools, also thousands of miles apart. Most were independent schools; public schools welcomed us, too. There, and in focus groups coast to coast, I interviewed more than 1,000 children ages four to eighteen, more than 500 parents, and more than 500 teachers, representing a diverse range of backgrounds. The one thing they all had in common was an eagerness to share from their own discovery, delight, and difficulty with the impact of technology on their lives.

I was able to do much of the research for this book thanks to my long-standing consulting work and close collegial relationships with many heads of schools, principals, faculty, parents, and students. Educators understand and hold dear the concept of the teachable moment: that growing up is full of missteps, and that an essential pathway to learning life's big lessons is to learn from our mistakes. This is a book full of stories about really good kids making big and little mistakes. It's about parents doing the best they can and learning along with their children. It's about school principals and teachers discovering what it means to educate children in the age of technology. I am immeasurably grateful to each of you for your trust in allowing me to come into your schools to interview students, teachers, and parents. I wish I could name each of you who have been so helpful in this way, but of course to protect your privacy, I cannot.

Thanks to all the children, parents, teachers, school counselors, and therapists I have known over many years who, upon hearing about this project, volunteered to be part of it. I had several confidential conversations with college students and other young adults, as well as grandparents, to include their perspectives. You all know who you are, and I thank each and every one of you. This is a topic close to all of our hearts, and I was consistently moved by the willingness of all those involved to tell me their stories—the voices in this book range from age two to eighty-nine.

Except for experts interviewed, all the names mentioned in this book have been changed to ensure privacy. Identifying details have also been changed for privacy, and many stories are composite stories, when several people shared similar information. I have chosen not to list the thirty schools that participated in this project in order to counter the temptation by anyone to guess where this or that happened, and to underscore that any story in this book could have happened in any number of places (indeed, many did happen in more than one school, one community, or one life!). We are all equally vulnerable, and rather than point fingers, let's support each other and learn from one another.

In addition to interviews conducted confidentially, I would like to thank the following experts who generously contributed to this book through personal interviews: Craig Anderson, Mark Bertin, Ellen Birnbaum, Tina Payne Bryson, Dimitri Christakis, Gene Cohen, JoAnn Deak, Ned Hallowell, Mimi Ito, Jackson Katz, Mike Langlois, Madeleine Levine, Liz Perle, Denise Pope, Harvey L. Rich, Michael Rich, Kelly Schryver, Nancy Schulman, Robin Shapiro, Daniel Siegel, Lydia Soifer, Michael Thompson, Yalda Uhls, Donna Wick, and Maryanne Wolf. In their various and different ways, these professionals are all deeply committed to protecting childhood and family relationships, and I was honored by their willingness to contribute to the book.

I also want to thank my dear friend and inspiring colleague Janice Toben, a visionary and pioneer in developing social and emotional learning curricula. Janice is my partner in much of the SEL schoolwork that I do. Huge appreciation to her and our talented, dedicated, and delightful colleagues Rush Sabiston Frank, Nick Haisman, and Elizabeth McLeod—for all the different ways we work together, at the InstituteforSEL.org and in schools, and for your help with this book.

My thanks to Tony Dopazo and his gang at Metro-Tech Services in Boston for their professional help with all things digital, all the time.

As a therapist, I am continuously awed by the depth and resilience of the people with whom I have the pleasure of working. I extend my deep appreciation to those of you who gave me permission to share aspects of our work together in this book. Two others deserve my thanks: my high school- and college-age research assistants who contributed to this book in numerous ways. Since one of these ways was to share confidential stories with me, you must remain nameless. You were each a total pleasure to work with!

I am so lucky to have such dear friends and extended family,and thank you for all the phone calls, stories, e-mails, links to YouTube, TED talks, texts, blogs, Instagrams, Facebook posts, and research articles that you took the time to send my way to help me stay current in this ever-changing field. Bits and pieces of so many of your lives are woven throughout this book that I am protecting your privacy in not naming names. However, please know how deeply I cherish your presence in my life and especially all the ways, technological and IRL, that you stayed connected and patient while I was so pre-occupied with this project. Thank you for not pushing delete on me!

A special thanks to Carolyn Peter, Melanie Gideon, and Alexandra Merrill, all extremely accomplished wise women who so generously read the entire manuscript for me, each of whom brought her professional expertise and critical eye to the work.

And a shout out to my painting pals. Thank you for the pleasure of your friendship, artistic inspiration, and your patience with my many no-shows through the three years of my work on this book. Our studying, painting, and traveling together are the best new "lightest highlight" in my life.

This book is about sustaining family relationships, and I am so thankful to have the unending support of my family. In my first family, my parents Rosalind and Lee Steiner instilled in each of their daughters many of the traits and values that are described in this book (grit, optimism, resilience, curiosity) long before the field of SEL existed. Big love to my sisters Terry Steiner and Nancy Steiner, with whom I shared our childhood and first learned what it means to be family. To both my sisters' families, all my nieces and nephews, and to all the Adair families down South, thanks for being such an engaged, helpful, and encouraging family. A most special thanks to Nancy and her husband, author David Michaelis, whose generosity, enthusiasm, sage advice, and unwavering support from start to finish have been invaluable.

To my other NYC family, my heartfelt thanks for your extraordinary gifts of time and hospitality; for the endless supply of "kitchen confidential" conversations and the warmth, wisdom, and humor they provided and that permeate this work. Your remarkable "sustainable family" graces these pages in so many ways, I cannot imagine having written this book without you. On the Boston homefront, my enduring love and gratitude to Margaret O'Neil, who is an essential part of my life and our family and without whom I could never do the work I do. To my goddaughters and their partners, Elizabeth Atterbury and Joe Kievett and Emily Atterbury and Diné Butler, thank you for the depth of our connection and your inspiration as artists, educators, and activists. To Eric Allon, old friend and longtime legal counsel and adviser: you are always there when I need you and you are always right!

Finally and foremost in my heart, my deepest love and gratitude go to my husband, Fred, and our children, Daniel and Lily. For thirty years Fred has wholeheartedly supported each new endeavor I take on, and the writing of this book was no exception. Fred, thank you for your utter and unquestioning belief in my work. I am in awe of your ability to help me hold on to my vision when I couldn't see where I was going, your ability to ask the right and hard questions to help me get to the next level, and your patience with this process. Thank you forever for the depth of your love and intelligence that come through in our daily morning sessions with the newspapers, our evening dinners, and my meltdowns! Your unwavering commitment to this project, including reading (and sometimes rereading) every page, has been a great gift. Most of all, you have been my dearest partner in creating and sustaining our family. Everything I cherish about love, family, and connections is anchored by our partnership.

Long before apps and texts and tags, I called Fred my PPD—my "permanent prom date"—with love and amazement that we found each other. Just when I thought it wasn't possible to love anyone or anything as much, along came our children. Daniel and Lily, it is through loving you that I have learned most deeply what it means to be a parent. I am profoundly grateful for all you have taught me about love, families, and growing up in the digital age. I am so touched by the consistency with which you each would ask "How's the book, Mom?" and then offer some fresh insight or story, always lightening my load. I thank you for your humor, wisdom, your sharp readers' eyes, and technological savvy. I am eternally grateful for your presence in my life, and for each and every way that we connect.

Notes

Introduction: The Revolution in the Living Room

1   "All the wisdom in the world": S. Fraiberg, *The Magic Years: Understanding and Handling the Problems of Early Childhood* (New York: Fireside, 1996).

4   Kids between the ages of eight and eighteen: V. J. Rideout, U. G. Foehr, and D. F. Roberts, *Generation M2: Media in the Lives of 8- to 18-Year-Olds* (Menlo Park, CA: Henry J. Kaiser Family Foundation, January 2010); S. Kessler, "Children's Consumption of Digital Media on the Rise," Mashable Social Media, 14 March 2011, http://mashable.com/2011/03/14/children-internet-stats/.

4   Stimulation, hyperconnectivity, and interactivity: Interview with Gene Cohen, director of the Center on Aging, Health, and Humanities at George Washington University, December 2000; G. D. Cohen, *The Mature Mind: The Positive Power of the Aging Brain* (New York: Basic Books, 2005) and *The Creative Age: Awakening Potential in the Second Half of Life* (New York: HarperCollins, 2001).

5   We use the language of addiction: M. H. Keung, "Internet Addiction and Antisocial Internet Behavior of Adolescents," *Scientific World Journal* 3 (November 2011): 2187–96; M. Choliz and C. Marco, "Patterns of Video Game Use and Dependence in Children and Adolescents," *Anales de Psicologia* 27, no. 2 (2011): 418–26; R. Kittinger, C. J. Correia, and J. G. Irons, "Relationship Between Facebook Use and Problematic Internet Use among College Students," *Cyberpsychology, Behavior, and Social Networking* 15, no. 6 (2012): 324–27; S. Goldberg, "Parents Using Smartphone to Entertain Bored Kids," *CNN Technology*, 26 April 2010; P. Flores, *Addiction as an Attachment Disorder* (Oxford, UK: Rowman and Littlefield, 2004); W. Powers, *Hamlet's BlackBerry: A Practical Philosophy for Building a Good Life in the Digital Age* (New York: HarperCollins, 2010); L. Rosen, *iDisorder: Understanding Our Obsession with Technology and Overcoming Its Hold on Us* (New York: Palgrave Macmillan, 2012); J. Palfrey and U. Gasser, *Born Digital: Understanding the First Generation of Digital Natives* (New York: Basic Books, 2008); S. Turkle, *Alone Together: Why We Expect More from Technology and Less from Each Other* (New York: Basic Books, 2011); J. Steyer, *Talking Back to Facebook: The Common Sense Guide to Raising Kids in the Digital Age* (New York: Scribner, 2012); N. Carr, *The Shallows: What the Internet Is*

Doing to Our Brains (New York: W. W. Norton, 2010); K. Kelly, *What Technology Wants* (New York: Viking, 2010).

5 Pregnant women who regularly use cell phones: "Cell Phone Use in Pregnancy May Cause Behavioral Disorders in Offspring, Mouse Study Suggests," *Science Daily*, 15 March 2012, http://www.sciencedaily.com/releases/2012/03/120315110138 .htm#.UHriJP43q98.email; "Cell Phone Use May Reduce Male Fertility, Austrian-Canadian Study Suggests,"*Science Daily*, May 19, 2011, http://www.sciencedaily .com/releases/2011/05/110519113022.htm#.UHrhAoPFtaE.email; B. Rochman, "Pediatricians Say Cell Phone Radiation Standards Need Another Look," *Time*, Healthland, 20 July 2012, http://healthland.time.com/2012/07/20/pediatricians-call-on-the-fcc-to-reconsider-cell-phone-radiation-standards/; O. P. Gandhi, L. L. Morgan, A. A. de Salles, Y. Y. R. B. Herberman, and D. L. Davis, "Exposure Limits: The Underestimation of Absorbed Cell Phone Radiation, Especially in Children," *Electromagnetic Biology and Medicine* (2011): 1–18.

5 We joke, but the truth is that research: Keung, "Internet Addiction"; B. E. Wexler, *Brain and Culture: Neurobiology, Ideology, and Social Change* (Cambridge, MA: MIT Press, 2006).

5 As adults we may choose to mess with our minds: K. Murphy, "Cellphone Radiation May Alter Your Brain. Let's Talk," *New York Times*, 30 March 2011, http://www.ny times.com/2011/03/31/technology/personaltech/31basics.html?_r=1&gwh=52BDB3 D2D3CF0EACA3E80E27C61CD719; "Exposure to Mobile Phones before and after Birth Linked to Kids' Behavioral Problems," *BMJ–British Medical Journal*, 7 December 2011, http://www.sciencedaily.com/releases/2010/12/101206201242.htm; S. Byun, C. Ruffini, J. Mills, A. Douglas, M. Niang, S. Stepchenkova, S. K. Lee, J. Loutfi, J. K. Lee, M. Atallah, and M. Blanton, "Internet Addiction: Metasynthesis of 1996–2006 Quantitative Research," *Cyberpsychology, Behavior, and Social Networking* 12, no. 2 (2010): 203–7; H. W. Lee, J. S. Choi, Y. C. Shin, J. Y. Lee, H. Y. Jung, and J. S. Kwon, "Impulsivity in Internet Addiction: A Comparison with Pathological Gambling," *Cyberpsychology, Behavior, and Social Networking* 15, no. 7 (July 2012): 373–77; H. Xiuqin, Z. Huimin, L. Mengchen, W. Jinan, Z. Ying, and T. Ran, "Mental Health, Personality, and Parental Rearing Styles of Adolescents with Internet Addiction Disorder," *Cyberpsychology, Behavior, and Social Networking* 13, no. 4 (2010): 401–406.

6 Doctors and researchers, from local emergency rooms: B. Worthen, "The Perils of Texting while Parenting," *Wall Street Journal*, 29 September 2012, C1, http://online .wsj.com/article; N. N. Borse, J. Gilchrist, A. M. Dellinger, R. A. Rudd, M. F. Ballesteros, and D. A. Sleet, *Childhood Injury Report: Patterns of Unintentional Injuries among 0–19 Year Olds in the United States, 2000–2006*, Centers for Disease Control and Prevention, National Center for Injury Prevention and Control (Atlanta: Centers for Disease Control, 2008); B. Keim,"Is Multitasking Bad for Us?" NOVA science-NOW, 4 October 2012, www.pbs.org/wgbh/nova/body/is-multitasking-bad.html; PBS interview with Clifford Nass, 1 December 2009, www.pbs.org/wgbh/pages/ frontline/digitalnation/interviews/nass.html; F. Zimmerman and D. Christakis, "Associations between Content Types of Early Media Exposure and Subsequent Attentional Problems," *Pediatrics* 120, no. 5 (2007): 986–92; D. J. Siegel, *The Mindful Brain: Reflection and Attunement in the Cultivation of Well-Being* (New York: W. W. Norton, 2007).

16   Parents' chronic distraction: S. Luthar and B. Becker, "Privileged but Pressured? A Study of Affluent Youth," *Child Development* 73, no. 5 (September/October 2002): 1593–1610; M. Levine, *The Price of Privilege: How Parental Pressure and Material Advantage Are Creating a Generation of Disconnected and Unhappy Kids* (New York: HarperCollins, 2006); P. Flores, *Addiction as an Attachment Disorder* (Oxford, UK: Rowman and Littlefield, 2004); E. Aboujaoude, *Virtually You: The Dangerous Powers of the E-Personality* (New York: W. W. Norton, 2011); J. Bowlby, *The Making and Breaking of Affectional Bonds* (London: Tavistock, 1979).

16   In addition to the issue of distracted supervision: Rosen, *iDisorder*; Levine, *The Price of Privilege*; Aboujaude, *Virtually You*.

17   Apple creator Steve Jobs famously said: Steve Jobs, "You've Got to Find What You Love," Stanford University Commencement Speech, 12 June 2005, http://news.stanford.edu/news/2005/june15/jobs–061505.html.

18   Parents have lost their job: R. Taffel,"Decline and Fall of Parental Authority," *AlterNet*, 22 February 2012, http://www.alternet.org/story/154249/the_decline_and_fall_of_parental_authority?page=entire; Interview with Dimitri Christakis, February 2012.

19   We live today in what the media theorist Henry Jenkins: H. Jenkins, *Convergence Culture: Where Old and New Media Collide* (New York: New York University Press, 2006).

19   Facebook, which invites us to carefully craft: Rideout et al., *Generation M2*; A. Smith, L. Segall, and S. Cowley, "Facebook Reaches One Billion Users," CNNMoney, 4 October 2012, http://money.cnn.com/2012/10/04/technology/facebook-billion-users/index.html?hpt=hp_t3; Kittinger et al., "Relationship between Facebook Use and Problematic Internet Use among College Students."

20   Tech has altered our social discourse so rapidly: C. Murphy, *Are We Rome? The Fall of an Empire and the Fate of America* (New York: Houghton Mifflin, 2007); S. Biegler and D. Boyd, "Risky Behaviors and Online Safety: A 2010 Literature Review," Harvard University, Berkman Center for Internet and Society, 4 November 2010, http://cyber.law.harvard.edu/research/youthandmedia/digitalnatives; Turkle, *Alone Together*; Palfrey and Gasser, *Born Digital*; G. Paton, "Twitter and Facebook 'Harming Children's Development,'" Independent.ie.com, 20 October 2012, http://www.independent.ie/business/technology/twitter-and-facebook-harming-childrens-development–3266055.html.

21   Experts suggest that much of our tech connection: Interview with Dan Siegel, February 2012; Wexler, *Brain and Culture*.

22   The impulse to respond instantly begins to *feel* necessary: D. J. Siegel, *Mindsight: The New Science of Personal Transformation* (New York: Bantam Books, 2011).

22   Some say that people haven't changed: L. Fogg, *Facebook for Parents* (New York: Wiley, 2012); L. Kutner and C. Olson, *Grand Theft Childhood: The Surprising Truth about Violent Video Games and What Parents Can Do* (New York: Simon and Schuster, 2008); Siegel, *The Mindful Brain*.

23   Children and parents are showing signs: E. Hallowell, *Connect: 12 Vital Ties That Open Your Heart, Lengthen Your Life, and Deepen Your Soul* (New York: Pocket Books, 1999); Levine, *The Price of Privilege*; R. Taffel, *Parenting by Heart: How to Stay Connected to Your Child in a Disconnected World* (Cambridge, MA: Perseus Publishing, 2011); Turkle, *Alone Together*; D. J. Siegel and M. Hartzell, *Parenting from*

the Inside Out: How a Deeper Self-Understanding Can Help You Raise Children Who Thrive (New York: Jeremy P. Tarcher/Penguin, 2004).

23 Research confirms what we sense is true: Turkle, *Alone Together*; Kessler, "Children's Consumption of Digital Media on the Rise."

24 Researchers at the University of California's Annenberg Center: S. Gaudin, "Families Spending More Time on Social Networks, Less Time Together," *Computerworld*, 16 June 2009; *Special Report: America at the Digital Turning Point*, University of Southern California Annenberg School Center for the Digital Future, 2012, http://www.digitalcenter.org/.

25 Or to recognize the potential danger: "Cyberbullying: One in Two Victims Suffer from Distribution of Embarrassing Photos and Videos," *ScienceDaily*, 25 July 2012, www.sciencedaily.com/releases/2012/07/120725090048.htm; S. Clifford, "Teaching Teenagers about Harassment," *New York Times*, 26 January 2009; E. Bazelon, "Amanda Todd Was Stalked Before She Was Bullied," *Slate*, 18 October 2012, www.slate.com/blogs/xx_factor/2012/10/18/suicide_victim_amanda_todd_stalked_before_she_was_bullied.html; J. Steinhauer, "Verdict in MySpace Suicide Case," *New York Times*, 27 November 2008; P. Aftab, *The Parents' Guide to Protecting Your Children in Cyberspace* (New York: McGraw-Hill, 2000); S. Bauman, *Cyberbullying: What Counselors Need to Know* (Alexandria, VA: American Counseling Association, 2011); R. Kowalski, S. Limber, and P. Agatston, *Cyber Bullying: Bullying in the Digital Age* (Malden, MA: Blackwell Publishing, 2008); Rosen, *iDisorder*.

25 Research already points to serious concerns: E. A. Vandewater et al., "Digital Childhood: Electronic Media and Technology Use among Infants, Toddlers and Preschoolers," *Pediatrics* 119, no. 5 (2007): e1006–15; S. Goldberg, "Parents Using Smartphone to Entertain Bored Kids."

26 In 1982, the educator, theorist, and media culture critic: N. Postman, *The Disappearance of Childhood* (New York: Delacorte Press, 1982; New York: First Vintage Books, 1994); "The Machine of the Year 1982: The Computer Moves In," *Time*, 5 October 1983, http://www.time.com/time/magazine/article/0,9171,952176,00.html.

27 New research offers unprecedented views: P. Kuhl, "The Linguistic Genius of Babies," TEDxRainier, October 2010.

27 "I read something the other day": N. Jackson, "More Kids Can Work Smartphones Than Can Tie Their Own Shoes," *Atlantic*, 24 January 2011; G. Perna, "More Young Kids Can Use Technology Than Tie Shoes," *International Business Times*, 20 January 2011, http://www.ibtimes.com/articles/103217/20110120/young-kids-technology-study-play-computer-game-operate-a-smartphone.htm.

28 She is recalling "Cat's in the Cradle": Harry Chapin, "Cat's in the Cradle," Elektra Records, 1974.

29 Winifred Gallagher, in her book *Rapt*: W. Gallagher, *Rapt: Attention and the Focused Life* (New York: Penguin Press, 2009).

Chapter 1: Lost in Connection

33 "Stimulation has replaced connection: Interview with Ned Hallowell, March 2011.

36 Technology has transformed the ways: A. S. B. Weiner and J. W. Hannum, "Differ-

ences in the Quantity and Efficacy of Social Support between Geographically Close and Long-Distance Friendships." *Journal of Personal and Social Relationships* (in press).

36 Yet we know the darker side is there: D. A. Christakis and F. J. Zimmerman, "Violent Television during Preschool Is Associated with Antisocial Behavior during School Age," *Pediatrics* 120, no. 5 (2007): 993–99; D. A. Christakis, "The Effects of Fast-Paced Cartoons," *Pediatrics* 128, no. 4 (2011): 772–74; T. Robinson, "Reducing Children's Television Viewing to Prevent Obesity," *JAMA* 282, no. 16 (1999): 1561–67; H. L. Burdette and R. C. Whitaker, "A National Study of Neighborhood Safety, Outdoor Play, Television Viewing, and Obesity in Preschool Children," *Pediatrics* 116, no. 3 (2005): 657–62; N. A. Conners-Burrow, L. M. McKelvey, and J. J. Fussell, "Social Outcomes Associated with Media Viewing Habits of Low-Income Preschool Children," *Early Education and Development* 22, no. 2 (2011): 256–73; S. M. Coyne, L. A. Stockdale, D. A. Nelson, and A. Fraser, "Profanity in Media Associated with Attitudes and Behavior Regarding Profanity Use and Aggression," *Pediatrics* 128, no. 5 (2011): 867–72.

38 We delete from memory the steady flow: B. Keim, "Is Multitasking Bad for Us?" NOVA scienceNOW, 4 October 2012, www.pbs.org/wgbh/nova/body/is-multitasking-bad.html; PBS interview with Clifford Nass, 1 December 2009, www.pbs.org/wgbh/pages/frontline/digitalnation/interviews/nass.html; S. S. Miller, "Survey: Teens 3 Times Likely to Text and Drive," *Dayton Daily News*, 2 May 2012, http://www.daytondailynews.com/lifestyle/survey-teens-3-times-as-likely-to-text-and-drive-1369411.html; M. Madden and A. Lenhart, "Teens and Distracted Driving: Texting, Talking, and Other Uses of the Cell Phone behind the Wheel," Pew Research Center's Internet and American Life Project, 16 November 2009, http://pewinternet.org/Reports/2009/Teens-and-Distracted-Driving.aspx; National Safety Council, *Compilation of Research Comparing Handheld and Hands-Free Devices*, December 2009; National Safety Council, Attributable Risk Estimate Model, December 2009; Nationwide Insurance, May 2008, http://www.nationwide.com/news room/press-release-almost-all-americans-believe-they-are-safe-drivers-2008.jsp; F. M. Drews, M. Pasupathi, and D. L. Strayer, "Passenger and Cell Phone Conversations in Simulated Driving," *Journal of Experimental Psychology* 14, no. 2 (2008): 392–400.

39 "Family creates our first experience of ourselves": H. Rich and T. Barker, *In the Moment: Celebrating the Everyday* (New York: HarperCollins, 2002).

39 The psychologist Selma Fraiberg: S. Fraiberg, *The Magic Years: Understanding and Handling the Problems of Early Childhood* (New York: Fireside, 1996).

40 The content is more powerful: F. B. Evans III, *Harry Stack Sullivan* (London and New York: Routledge, 1996).

41 "There are a lot of minimoments": Interview with Liz Perle, cofounder and editor in chief at Common Sense Media, April 2011.

42 Children are no longer sheltered in this way: W. Buckleitner, "Tablets for Children, Including Apps," *New York Times*, 28 March 2012, http://www.nytimes.com/2012/03/29/technology/personaltech/tablets-for-children-including-apps.html; A. Brasel and J. Gips, "Media Multitasking Behavior: Concurrent Television and Computer Usage," *Cyberpsychology, Behavior, and Social Networking*, 15 March 2011, http://www.liebertonline.com/doi/pdfplus/10.1089/cyber.2010.0350; S. J. Lee

and Y. J. Chae, "Balancing Participation and Risks in Children's Internet Use: The Role of Internet Literacy and Parental Mediation," *Cyberpsychology, Behavior, and Social Networking* 15, no. 5 (2011): 257–62.

42   But plenty of it is intentional: L. Guernsey, B. Worthen, H. Kirkorian, and L. Perle, "Touch-Screen Devices and Very Young Children," *Diane Rehm Show,* 23 May 2012, http://thedianerehmshow.org/shows/2012–05–23/touch-screen-devices-and-very-young-children?page=1; L. Guernsey, *Screen Time: How Electronic Media—From Baby Videos to Educational Software Affects Your Young Child* (New York: Basic Books, 2007); S. Linn, *Consuming Kids: Protecting Our Children from the Onslaught of Marketing and Advertising* (New York: Anchor Books/Random House, 2004); J. Schor, *Born to Buy: The Commercialized Child and the New Consumer Culture* (New York: Scribner, 2004).

42   The same cynicism, cruel humor: Linn, *Consuming Kids*; Schor, *Born to Buy.*

45   Social networking has switched out: S. Dominus, "A Facebook Movement against Mom and Dad," *New York Times,* 16 January 2010, A14; M. Sedensky, "Dad's Techxecution Hits a Nerve," *Sunday Oregonian,* Associated Press, 19 February 2012; B. Worthen, "The Perils of Texting While Parenting," *Wall Street Journal,* 29 September 2012, C1, http://onlin.wsj.com/article.

45   As is often the case with discussions: D. Levin and J. Kilbourne, *So Sexy So Soon: The New Sexualized Childhood and What Parents Can Do to Protect Their Kids* (New York: Ballantine Books, 2008); E. Bazelon, "Amanda Todd Was Stalked before She Was Bullied," *Slate,* 18 October 2012, www.slate.com/blogs/xx_factor/2012/10/18/suicide_victim_amanda_todd_stalked_before_she_was_bullied.html; J. Farley-Gillispie and J. Gackenbach, *cyber.rules: Negotiating Healthy Internet Use* (New York: W.W. Norton, 2007).

48   Kids want to—and need to—own their identity: M. McConville, *Adolescence: Psychotherapy, and the Emergent Self* (San Francisco: Jossey-Bass, 1995); B. E. Wexler, *Brain and Culture: Neurobiology, Ideology, and Social Change* (Cambridge, MA: MIT Press, 2006); Bazelon, "Amanda Todd Was Stalked."

50   It is about us, about our cultural crisis: C. Hedges, *Empire of Illusion: The End of Literacy and the Triumph of Spectacle* (New York: Nation Books, 2009).

50   Incident by incident we put out the fires: W. E. Copeland, D. Wolke, A. Angold, E. J. Costello, "Adult Psychiatric Outcomes of Bullying and Being Bullied by Peers in Childhood and Adolescence," *JAMA Psychiatry.* Published online first: 20 February 2013; ():1-8. doi:10.1001/jamapsychiatry.2013.504.

50   Those pathways expand and deepen with experience: D. J. Siegel, *Mindsight:The New Science of Personal Transformation* (New York: Bantam Books, 2011); D. J. Siegel and T. P. Bryson, *The Whole-Brain Child: 12 Revolutionary Strategies to Nurture Your Child's Developing Mind, Survive Everyday Parenting Struggles, and Help Your Family Thrive* (New York: Delacorte Press, 2011); D. J. Siegel, *The Developing Mind,* 2nd ed., *How Relationships and the Brain Interact to Shape Who We Are* (New York: Guilford Press, 2012); D. J. Siegel, "Understanding Your Child's Attachment Style," *Psychalive,* http://www.psychalive.org/2011/10/identifying-your-childs-attachment-style/; D. J. Siegel, "An Interpersonal Neurobiology Approach to Psychotherapy: How Awareness, Mirror Neurons and Neural Plasticity Contribute to the Development of Well-Being," *Psychiatric Annals* 36, no. 4 (2006): 248–58.

51   "The brain is what it does": C. Davidson, *Now You See It: How the Brain Science of Attention Will Transform the Way We Live, Work, and Learn* (New York: Viking Press, 2011).

51   What it does in this regard: M. Levine, *The Price of Privilege: How Parental Pressure and Material Advantage Are Creating a Generation of Disconnected and Unhappy Kids* (New York: HarperCollins, 2006); E. Aboujaoude, *Virtually You: The Dangerous Powers of the E-Personality* (New York: W. W. Norton, 2011); R. Kowalski, S. Limber, and P. Agatston, *Cyber Bullying: Bullying in the Digital Age* (Malden, MA: Blackwell, 2008); G. Steffgen, M. S. Konig, J. Pfetsch, and A. Melzer, "Are Cyberbullies Less Empathic? Adolescents' Cyberbullying Behavior and Empathic Responsiveness," *Cyberpsychology, Behavior, and Social Networking* 14, no. 11 (2011): 643–48, posted online 9 May 2011, http://www.liebertonline.com/doi/abs/10.1089/cyber.2010.0445; M. Wolf, *Proust and the Squid: The Story and Science of the Reading Brain* (New York: HarperCollins, 2007); D. Christakis, "Effect of Block Play on Language Acquisition and Attention in Toddlers: A Pilot Randomized Controlled Trial," *Archives of Pediatric and Adolescent Medicine* 161, no. 10 (2007): 967–71; F. Zimmerman and D. Christakis, "Associations between Content Types of Early Media Exposure and Subsequent Attentional Problems," *Pediatrics* 120, no. 5 (2007): 986–92; S. Turkle, *Alone Together: Why We Expect More from Technology and Less from Each Other* (New York: Basic Books, 2012); C. Steiner-Adair and L. Sjostrom, *Full of Ourselves: A Wellness Program to Advance Girl Power, Health, and Leadership* (New York: Teacher's College Press, 2006); S. Lamb, L. Brown, and M. Tappen, *Packaging Boyhood: Saving Our Sons from Superheros, Slackers, and Other Media Stereotypes* (New York: St. Martin's Press, 2009); S. Lamb and L. M. Brown, *Packaging Girlhood: Rescuing Our Daughters from Marketers' Schemes* (New York: St. Martin's Press, 2006); Levin and Kilbourne, *So Sexy So Soon.*

51   Neurologically speaking, empathy takes time: Wolf, *Proust and the Squid.*

51   We each have what Wolf describes: Interview with Maryanne Wolf, March 2011; Wolf, *Proust and the Squid.*

52   At the same time, a Stanford University review: "Empathy: College Students Don't Have as Much as They Used to, Study Finds" *Science Daily*, 29 May 2012, http://www.sciencedaily.com/releases/2010/05/100528081434.htm; Z. Bielski, "Today's College Kids Are 40-Per-Cent Less Empathetic, Study Finds," *Globe and Mail* (Toronto), 1 June 2010, http://www.theglobeandmail.com/life/work/todays-college-kids-are-40-per-cent-less-empathetic-study-finds/article1587609/; S. Konrath, E. O'Brian, and C. Hsing, "Changes in Dispositional Empathy in American College Students over Time: A Meta-Analysis," *Personality and Social Psychology Review* 15, no. 2 (2011): 180–98.

52   In his wonderful description of family's role: R. Taffel, *Parenting by Heart: How to Stay Connected to Your Child in a Disconnected World* (Cambridge, MA: Perseus Publishing, 2011).

53   The American Academy of Pediatrics has for years: Council on Communications and Media and A. Brown, "Media Use by Children Younger Than 2 Years," *Pediatrics* 128, no. 5 (2011): 1040–45; B. Carey, "Parents Urged Again to Limit TV for Youngest," *New York Times*, 18 October 2011, http://www.nytimes.com/2011/10/19/health/19babies.html; R. J. Hancox, B. J. Milne, and R. Poulton, "Association of

Television during Childhood with Poor Educational Achievement," *Archives of Pediatric and Adolescent Medicine* 159, no. 7 (2005): 614–18.

53 Nonetheless, recent research shows: D. Elkind, *The Power of Play: Learning What Comes Naturally* (Philadelphia: Da Capo Press, 2007); R. Louv, *Last Child in the Woods: Saving Our Children from Nature Deficit Disorder* (New York: Algonquin Books, 2008); D. J. Siegel and M. Hartzell, *Parenting from the Inside Out: How a Deeper Self-Understanding Can Help You Raise Children Who Thrive* (New York: Jeremy P. Tarcher/Penguin, 2004): S. Linn, *The Case for Make Believe: Saving Play in a Commercialized World* (New York: New Press/W. W. Norton, 2008); H. Stout, "Effort to Restore Children's Play Gains Momentum," *New York Times*, 5 January 2011, http://www.nytimes.com/2011/01/06/garden/06play.html?pagewanted=all&_r=0; L. Winerman, "Playtime in Peril," *American Psychological Association* 40, no. 8 (2009): 50; T. Baranowski, D. Abdelsamad, J. Baranowski, M. T. O'Connor, D. Thompson, A. Barnett, E. Cerin, and T. A. Chen, "Impact of an Active Video Game on Healthy Children's Physical Activity," *Pediatrics* 129 (2012): e636–42; S. Kaplan, "The Restorative Benefits of Nature: Toward an Integrative Framework," *Journal of Environmental Psychology* 15 (1995): 169–82; A. D. Pellegrini and C. M. Bohn, "The Role of Recess in Children's Cognitive Performance and School Adjustment," *Educational Researcher* 34, no. 1 (2005): 13–19; A. L. Philips, "A Walk in the Woods—Evidence Builds That Time Spent in the Natural World Benefits Human Health," *American Scientist* 99, no. 4 (2001): 301.

53 But adding hours of TV and computer play: V. Rideout and E. Hamel, *The Media Family: Electronic Media in the Lives of Infants, Toddlers, Preschoolers and Their Parents* (Menlo Park, CA: Henry J. Kaiser Family Foundation, May 2006).

54 On the bright side: D. A. Gentile, P. Lynch, J. Linder, and D. Walsh, "The Effects of Violent Video Game Habits on Adolescent Hostility, Aggressive Behaviors, and School Performance," *Journal of Adolescence* 27 (2004): 5–22.

54 So let me be clear: "Best Websites for Kids," Common Sense Media, http://www.commonsensemedia.org/website-lists.

55 Michael Rich, a pediatrician and director: M. Rich, "Boy, Mediated: Effects of Entertainment Media on Adolescent Male Health," *Adolescent Medicine State of the Art Reviews* 14 (2003): 691–713; M. Rich and M. Bar-on, "Child Health in the Information Age: Media Education of Pediatricians," *Pediatrics* 107, no. 1 (2001): 156–62.

55 "Computers are the new playground": M. Thompson, *Homesick and Happy: How Time Away from Parents Can Help a Child Grow* (Boston: Ballantine Books, 2012).

57 A 2006 survey by the Kaiser Family Foundation: Rideout and Hamel, *The Media Family*; Center on Media and Child Health, Children's Hospital Boston, "The Effects of Electronic Media on Children Ages Zero to Six: A History of Research," Henry J. Kaiser Family Foundation Issue Brief (Menlo Park, CA: Henry J. Kaiser Family Foundation, January 2005); American Academy of Pediatrics, Committee on Public Education, "Children, Adolescents, and Television," *Pediatrics* 107, no. 2 (2001):423–26.

57 "Children's rooms are now almost *pathogenic*": M. A. McNally, D. Crocetti, M. E. Mahone, M. B. Denckla, S. J. Suskauer, and S. H. Mostofsky, "Corpus Callosum Segment Circumference Is Associated with Response Control in Children with

Attention-Deficit Hyperactivity Disorder (ADHD)," *Journal of Child Neurology* 25, no. 4 (2010): 453–62.

58   Scientists don't yet know how screens and media use: Zimmerman and Christakis, "Associations between Content Types"; D. Stober, "Multitasking May Harm the Social and Emotional Development of Tweenage Girls, but Face-to-Face Talks Could Save the Day, Say Stanford Researchers," *Stanford News* 25 (January 2012), news .stanford.edu/pr/2012/pr-tweenage-girls-multitasking–012512.html; "Will Hyperconnected Millennials Suffer Cognitive Consequences?" *Daily Circuit*, Minnesota Public Radio (audio), PewInternet, Pew Internet and American Life Project, 1 March 2012, http://www.pewinternet.org/Media-Mentions/2012/MPR-hyperconnected-millennials.aspx; J. Anderson, "Millennials Will Benefit and Suffer Due to Their Hyperconnected Lives," *PewInternet*, Pew Internet and American Life Project, 29 February 2012, http://www.pewinternet.org/Reports/2012/Hyperconnected-lives .aspx; J. Y. Yen, C. F. Yen, C. S. Chen, T. C. Tang, and C. H. Ko, "The Association between Adult ADHD Symptoms and Internet Addiction among College Students: The Gender Difference," *Cyberpsychology, Behavior, and Social Networking* 12, no. 2 (2009): 187–91; E. Hallowell and J. Ratey, *Driven to Distraction: Recognizing and Coping with Attention Deficit Disorder from Childhood through Adulthood* (New York: Anchor Books, 1994).

58   However, the experts also predicted: *Daily Circuit*, "Will Hyperconnected Millennials Suffer Cognitive Consequences?"; Anderson, "Millennials Will Benefit and Suffer."

58   "It's not really that you multitask": Interview with Dimitri Christakis, February 2012.

59   Based on studies of highway accidents: A. Lenhart, R. Ling, S. Campbell, and K. Purcell, "Teens and Mobile Phones," Pew Internet, Pew Internet and American Life Project, 20 April 2012, http://pewinternet.org/Reports/2010/Teens-and-Mobile-Phones/Summary-of-findings/Findings.aspx; "Driver Electronic Device Use in 2010," *Traffic Safety Facts*, National Highway Traffic Safety Administration, December 2011; "Cellphones and Driving," Insurance Information Institute, October 2008; "Distractions Challenge Teen Drivers," *USA Today*, 26 January 2007; B. Worthen, "The Perils of Texting while Parenting," *Wall Street Journal*, 29 September 2012, C1, http://online.wsj.com/article; *Safety + Health*; "Teens Warn Peers Against Texting and Driving," www.nsc.org/safetyhealth/Pages/teens_warn_peers_against_text ing_and_driving.aspx#.UIHdmhjbAhc; Madden and Lenhart, "Teens and Distracted Driving."

59   We have to be concerned about tech habits: Hallowell and Ratey, *Driven to Distraction*; J. Meisner, "Heads Up: Cell Phones Add to Risk When Crossing Street, Study Shows," *Mac News World*, 27 January 2009, http://www.macnewsworld.com/sto ry/65963.html.

60   This despite public health warnings: Lenhart, Ling, Campbell, and Purcell, "Teens and Mobile Phones"; "Distracted Driving and Driver, Roadway, and Environmental Factors," National Highway Traffic Safety Administration, September 2010; "Most U.S. Drivers Engage in 'Distracted' Driving Behaviors," USAToday.com, December 1, 2011.

61   In focus groups with more than six hundred teens: Lenhart, Ling, Campbell, and

Purcell, "Teens and Mobile Phones"; A. Halsey III, "Teen Drivers Are Texting, Just Like Their Parents," *Washington Post*, 13 May 2012, http://www.washing tonpost.com/local/trafficandcommuting/teen-drivers-are-texting-just-like-their-parents/2012/05/13/gIQA8raQNU_story.html; Worthen, "The Perils of Texting While Parenting."

62   In studies of infants' responses to their mother's voice: P. Leach, *Your Baby and Child from Birth to Age 5* (New York: Alfred A. Knopf, 2003); T. B. Brazelton, *Touchpoints: The Essential Reference: Your Child's Emotional and Behavioral Development* (Reading, MA: Perseus Books, 1992); T., Field, D. Sandberg, R. Garcia, N. Vega-Lahr, S. Goldstein, and L. Guy, "Pregnancy Problems, Postpartum Depression, and Early Mother-Infant Interactions," *Developmental Psychology* 21, no. 6 (1985): 1152–56; J. F. Cohn, R. Matias, E. Z. Tronick, D. Connell, and K. Lyons-Ruth, "Face-to-Face Interactions of Depressed Mothers and Their Infants," *New Directions for Child and Adolescent Development* 1986, no. 34 (February 2006): 31–45.

62   This example of "embodied cognition": N. Angier, "Abstract Thoughts? The Body Takes Them Literally," *New York Times*, 2 February 2010, http://www.nytimes .com/2010/02/02/science/02angier.html?pagewanted=all&_r=0; L. Miles, L. Nink, and C. N. Macrae, "Moving through Time," *Psychological Science* 21, no. 2 (February 2012): 222–23; L. E. Williams, "Experiencing Physical Warmth Promotes Interpersonal Warmth," *Science* 322 (2008): 606–607; N. Jostmann, D. Lakens, and T. Schubert, "Weight as an Embodiment of Importance," *Psychological Science* 20, no. 9 (February 2009): 1169–74.

64   Closeness counts: E. Hallowell, *Connect: 12 Vital Ties That Open Your Heart, Lengthen Your Life, and Deepen Your Soul* (New York: Pocket Books, 1999); M. Pipher, *Shelter of Our Families: Rebuilding Our Families* (New York: Ballantine Books, 1996); Taffel, *Parenting by Heart*; Siegel and Hartzell, *Parenting from the Inside Out*; J. Steyer, *The Other Parent: The Inside Story of the Media's Effect on Our Children* (New York: Atria, 2002).

65   The more we learn about the architecture: G. Schlaug, M. Forgeard, L. Zhu, A. Norton, and E. Winner, "Training-Induced Neuroplasticity in Young Children," *Annals of the New York Academy of Sciences* 1169 (2009): 205–8; S. J. Lane and R. C. Schaaf, "Examining the Neuroscience Evidence for Sensory-Driven Neuroplasticity: Implications for Sensory-Based Occupational Therapy for Children and Adolescents," *American Journal of Occupational Therapy* 64, no. 3 (May/June 2010): 375–90; D. Thompson, "The Atlantic: Kids Are Changing, Neuroplasticity Is Real, and Education Needs a Revolution," *Brain Power* 2, March 2012, http://www.brainpowerinitiative.com/2012/03/the-atlantic-kids-are-changing-neuroplasticity-is-real-and-education-will-need-to-change/; Wexler, *Brain and Culture*; G. D. Cohen, *The Creative Age: Awakening Human Potential in the Second Half of Life* (New York: HarperCollins, 2000); G. D. Cohen, *The Mature Mind: The Positive Power of the Aging Brain* (New York: Basic Books, 2006); E. A. Maguire, R. Frackowiak, and C. Frith, "Recalling Routes around London: Activation of the Right Hippocampus in Taxi Drivers," *Journal of Neuroscience* 17, no. 18 (1997): 7103–10.

Chapter 2: The Brilliant Baby Brain

66 "Look, the brain of the child": Interview with Dan Siegel, February 2012.

68 Babies come equipped with their own bonding instinct: W. Farrell, *Father and Child Reunion: How to Bring the Dads We Need to the Children We Love* (New York: Penguin Putnam, 2001); P. Leach, *Your Baby and Child from Birth to Age 5* (New York: Alfred A. Knopf, 2003); T. B. Brazelton, *Touchpoints: The Essential Reference: Your Child's Emotional and Behavioral Development* (Reading, MA: Perseus Books, 1992); Child Development Institute Parenting Today, *Ages & Stages*, http://childdevelopmentinfo.com/ages-stages.shtml; J. Bowlby, *The Making and Breaking of Affectional Bonds* (London: Tavistock, 1979); T. R. Insel and L. J. Young "The Neurobiology of Attachment," *Nature Reviews Neuroscience* 2 (2001): 129–36; P. M. Crittenden, *Raising Parents: Attachment, Parenting and Child Safety* (Milton Park, Oxfordshire, UK: Willan Publishing, 2008); D. J. Siegel, *Mindsight: The New Science of Personal Transformation* (New York: Bantam Books, 2011); American Medical Association, *American Medical Association Complete Guide to Your Children's Health* (New York: Random House, 1999); M. Diamond and J. Hopson, *Magic Trees of the Mind: How to Nurture Your Child's Intelligence, Creativity, and Healthy Emotions from Birth through Adolescence* (New York: Plume, 1999).

69 The mirroring exchange that occurs: J. Blanchard and T. Moore, "The Digital World of Young Children: Impact on Emergent Literacy," Pearson Foundation, 1 March 2010; Leach, *Your Baby and Child From Birth to Age 5*; Brazelton, *Touchpoints*; *American Medical Association Complete Guide to Your Children's Health*; Diamond and Hopson, *Magic Trees of the Mind*.

69 The baby brain comes hardwired: B. E. Wexler, *Brain and Culture: Neurobiology, Ideology, and Social Change* (Cambridge, MA: MIT Press, 2006); Leach, *Your Baby and Child From Birth to Age 5*; Brazelton, *Touchpoints*; *American Medical Association Complete Guide to Your Children's Health*; Diamond and Hopson, *Magic Trees of the Mind*.

70 There is really no way to know what it means to Henry: L. Schulz, "Intelligence, Commonsense, and Cognitive Development," MIT Department of Brain and Cognitive Sciences Early Cognition Lab, n.d.; C. Cook, N. Goodman, and L. E. Schulz, "Where Science Starts: Spontaneous Experiments in Preschoolers' Exploratory Play," *Cognition* 120, no. 3 (2011): 341–49; C. Johnson, "Researchers Study How Babies Think," *Boston Globe*, 28 March 2011, http://www.boston.com/lifestyle/health/articles/2011/03/28/researchers_study_how_babies_think/?rss_id=Boston.com+—+Latest+news; P. Kuhl, "The Linguistic Genius of Babies," TEDxRainier, October 2010.

71 Studies show that they are especially distressed: Kuhl, "The Linguistic Genius of Babies."

71 More recent studies using brain imaging scans: "Infants Do Not Appear to Learn Words from Educational DVDs," Science Centric, 15 March 2012, http://www.sciencecentric.com/news/10031578-infants-do-not-appear-learn-words-from-educational-dvds.html; A. Park, "Baby Wordsworth Babies: Not Exactly Wordy,"

Time, 2 March 2010. http://www.time.com/time/health/article/0,8599,1968874,00 .html; Kuhl, "The Linguistic Genius of Babies."

71   We also know that our child's fascination: Wexler, *Brain and Culture*.

72   Although she cannot know what Henry is thinking: Siegel, *Mindsight*; D. J. Siegel, "Understanding Your Child's Attachment Style," *Psychalive*, http://www.psychalive .org/2011/10/identifying-your-childs-attachment-style/.

73   Donna Wick, a clinical and developmental psychologist: Interview with Donna Wick, November 2011.

74   It is rare anymore that screens are completely absent: V. Rideout and E. Hamel, *The Media Family: Electronic Media in the Lives of Infants, Toddlers, Preschoolers, and Their Parents* (Menlo Park, CA: Henry J. Kaiser Family Foundation, May 2006; pbs.org/parents/childrenandmedia/article-faq.html#.UPLj-ww5qww; P. Ravichandran and B. France de Bravo, "Children and Screen Time," ChildWise, 2010, http://boysdevelopmentproject.org.uk/index.html; P. Ravichandran and B. France de Bravo, "Pre-School Children," ChildWise, 2009, http://boysdevelopmentpro ject.org.uk/index.html; P. Ravichandran and B. France de Bravo, "Special Report Digital Lives," ChildWise, 2010, http://boysdevelopmentproject.org.uk/index .html; D. Feng, D. B. Reed, M. C. Esperat, and M. Uchida, "Effects of TV in the Bedroom on Young Hispanic Children," *American Journal of Health Promotion* 25, no. 5 (2011): 310–18; E. A. Vandewater, D. S. Bickham, J. H. Lee, H. M. Cummings, E. A. Wartella, and V. J. Rideout, "When the Television Is Always On: Heavy Television Exposure and Young Children's Development," *American Behavioral Scientist* 48 (2005): 562–77.

74   A 2011 survey by *Parenting* magazine: S. Kessler, "Children's Consumption of Digital Media on the Rise," Mashable Social Media, 14 March 2011, http://mashable .com/2011/03/14/children-internet-stats/; Rideout and Hamel, *The Media Family*.

74   Another study sponsored by the Sesame Street Workshop: A. L. Gutnick, "Always Connected: The New Digital Media Habits of Young Children," Joan Ganz Cooney Center at Sesame Workshop, March 2011, http://joanganzcooneycenter.org/Re ports-28.html.

76   Immersed as we may be in a tender: Interview with Dan Siegel, February 2012; Siegel, *Mindsight*; D. J. Siegel, *The Mindful Brain: Reflection and Attunement in the Cultivation of Well-Being* (New York: W. W. Norton, 2007).

76   In the hyperconnected digital culture: E. Hallowell, *Connect: 12 Vital Ties That Open Your Heart, Lengthen Your Life, and Deepen Your Soul* (New York: Pocket Books, 1999); E. Hallowell, *The Childhood Roots of Adult Happiness: Five Steps to Help Kids Create and Sustain Lifelong Joy* (New York: Ballantine, 2002); Farrell, *Father and Child Reunion*; S. Fraiberg, *The Magic Years: Understanding and Handling the Problems of Early Childhood* (New York: Fireside, 1996); "Will Hyperconnected Millennials Suffer Cognitive Consequences?" *Daily Circuit*, Minnesota Public Radio (audio), Pew Internet: Pew Internet and American Life Project, 1 March 2012, http://www.pewinternet.org/Media-Mentions/2012/MPR-hyperconnected-millennials.aspx; J. Anderson, "Millennials Will Benefit and Suffer Due to Their Hyperconnected Lives," Pew Internet, Pew Internet and American Life Project, 29 February 2012, http://www.pewinternet.org/Reports/2012/Hyperconnected-lives .aspx; C. Nass and C. Yen, *The Man Who Lied to His Laptop: What Machines Teach*

Us about Human Relationships (New York: Penguin, 2010); B. Keim, "Is Multitasking Bad for Us?" NOVA scienceNOW, 4 October 2012, www.pbs.org/wgbh/nova /body/is-multitasking-bad.html; Bowlby, *The Making and Breaking of Affectional Bonds.*

78 Tremendous brain growth occurs: Child Development Institute Parenting Today, *Ages & Stages,* http://childdevelopmentinfo.com/ages-stages.shtml; Wexler, *Brain and Culture*; B. Worthen, "What Happens When Toddlers Zone Out with an iPad," *Wall Street Journal,* 22 May 2012, http://online.wsj.com/article/SB100014240527023 04363104577391813961853988.html.

78 "The brain was designed to develop": J. Blanchard and T. Moore, "The Digital World of Young Children: Impact on Emergent Literacy," Pearson Foundation, 1 March 2010; Interview with JoAnn Deak, Wonderplay presentation at the 92nd Street Y, New York City, January 2012; J. Deak, *Your Fantastic Elastic Brain* (Belvedere, CA: Little Pickle Press, 2011); Wexler, *Brain and Culture*; Worthen, "What Happens When Toddlers Zone Out."

78 In her "baby brain bootcamp" talks: Interview with JoAnn Deak, January 2012.

79 Those neural pathways and networks: Interview with Maryanne Wolf, March 2011; M. Wolf, *Proust and the Squid: The Story and Science of the Reading Brain* (New York: HarperCollins, 2007).

79 Tech can interrupt or weaken that connection: L. Rosen, *iDisorder: Understanding Our Obsession with Technology and Overcoming Its Hold on Us* (New York: Palgrave Macmillan, 2012); Nass and Yen, *The Man Who Lied to His Laptop*; J. Palfrey and U. Gasser, *Born Digital: Understanding the First Generation of Digital Natives* (New York: Basic Books, 2008); D. J. Siegel and T. P. Bryson, *The Whole-Brain Child: 12 Revolutionary Strategies to Nurture Your Child's Developing Mind, Survive Everyday Parenting Struggles, and Help Your Family Thrive* (New York: Delacorte Press, 2011); N. Carr, *The Shallows: What the Internet Is Doing to Our Brains* (New York: W. W. Norton, 2010); Courtney Hutchison, "Watching SpongeBob SquarePants Makes Preschoolers Slower Thinkers, Study Finds," ABC NightTime News, 12 September 2011, http://abcnews.go.com/Health /Wellness/watching-spongebob-makes-preschoolers-slower-thinkers-study -finds/story?id=14482447#.UHr0kRj0RaE; S. Turkle, *Alone Together: Why We Expect More from Technology and Less from Each Other* (New York: Basic Books, 2011); E. Aboujaoude, *Virtually You: The Dangerous Powers of the E-Personality* (New York: W. W. Norton, 2011); "Infants Do Not Appear to Learn Words from Educational DVD," Science Centric; "'Your Baby Can Read': No, He Can't," Editorial, *Boston Globe,* 26 July 2012.

80 Research by Patricia Kuhl and others: Kuhl, "The Linguistic Genius of Babies"; L. Soifer, "Development of Oral Language and Its Relationship to Literacy," in *Multi-Sensory Teaching of Basic Language Skills,* ed. Judith Birsh (Baltimore: Paul H. Brooks, 1999); Interview with Lydia Soifer, November 2011.

80 Neurobiologist Wolf says that although the science: Interview with Maryanne Wolf, March 2011.

81 "Online language programs are just tools": Interview with Lydia Soifer, March 2011.

82 Extensive research since the 1970s: D. A. Christakis and F. Zimmerman. *The Elephant in the Living Room: Make Television Work for Your Kid* (Emmaus, PA: Rodale

Books, 2006); E. J. Paavonen, M. Pennonen, and M. Roine, "Passive Exposure to TV Linked to Sleep Problems in Children," *Journal of Sleep Research* 15 (2006): 154–61; B. Wilson, "Media and Children's Aggression, Fear and Altruism," *The Future of Children* 18, no. 1 (Spring 2008), http://futureofchildren.org/publications/journals/article/index.xml?journalid=32&articleid=58&sectionid=270; V. J. Rideout, U. G. Foehr, and D. F. Roberts, *Generation M2: Media in the Lives of 8- to 18-Year-Olds* (Menlo Park, CA: Henry J. Kaiser Family Foundation, 10 January 2010); "How Teens Use Media," Nielsen, 2009; Ravichandran and France de Bravo, "Children and Screen Time"; Ravichandran and France de Bravo, "Pre-School Children"; Ravichandran and France de Bravo, "Special Report Digital Lives"; D. Walsh, "Interactive Violence and Children," testimony submitted to the Committee on Commerce, Science, and Transportation, United States Senate, 21 March 2000; D. A. Gentile, P. Lynch, J. Linder, and D. Walsh, "The Effects of Violent Video Game Habits on Adolescent Hostility, Aggressive Behaviors, and School Performance," *Journal of Adolescence* 27 (2004): 5–22.

82   All the research underscores: Council on Communications and Media and A. Brown, "Media Use by Children Younger Than 2 Years," *Pediatrics* 128, no. 5 (2011): 1040–45; "Ads Touting 'Your Baby Can Read' Were Deceptive, FTC Complaint Alleges Two of Three Defendants Settle Charges for Claims Made about Children's Learning Program; Order Bans Use of Product Name," Federal Trade Commission, 28 August 2012, http://ftc.gov/opa/2012/08/babyread.shtm; "'Your Baby Can Read': No, He Can't," *Boston Globe*; "France Pulls Plug on TV Shows Aimed at Babies," CBC News, 20 August 2008, http://www.cbc.ca/world/story/2008/08/20/french-baby.html; D. S. Bickham and M. Rich, "Is Television Viewing Associated with Social Isolation?" *Archives of Pediatrics and Adolescent Medicine* 160, no. 4 (2006): 387–92.

82   Experts have said most of this for decades: "Infants Do Not Appear to Learn Words from Educational DVDs," Science Centric; R. J. Hancox, B. J. Milne, and R. Poulton, "Association of Television during Childhood with Poor Educational Achievement," *Archives of Pediatric and Adolescent Medicine* 159, no. 7 (2005): 614–18; S. Linn, *Consuming Kids: Protecting Our Children from the Onslaught of Marketing and Advertising* (New York: Anchor Books/Random House, 2004); S. Linn, *The Case for Make Believe: Saving Play in a Commercialized World* (New York: New Press/W. W. Norton, 2008).

84   One reason TV and tech are especially risky: Siegel and Bryson Payne, *The Whole-Brain Child*; Wolf, *Proust and the Squid*; Zimmerman and Christakis, "Associations between Content Types of Early Media Exposure"; Center on Media and Child Health, Children's Hospital Boston, "The Effects of Electronic Media on Children Ages Zero to Six: A History of Research," Henry J. Kaiser Family Foundation Issue Brief (Menlo Park, CA: Henry J. Kaiser Family Foundation, January 2005); N. Christakis and J. Fowler, *Connected: How Your Friends' Friends' Friends Affect Everything You Feel, Think, and Do* (New York: Back Bay Books/Little Brown, 2009).

85   There's a big difference between *I feel angry*: Siegel, *The Whole-Brain Child*.

87   These are the roots of the resiliency: Hallowell, *Connect*; Hallowell, *The Childhood Roots of Adult Happiness*; M. Seligman, *Flourish: A Visionary New Understanding of Happiness and Well-Being* (New York: Free Press, 2011); M. Seligman, *The Optimistic*

Child: Proven Program to Safeguard Children from Depression & Build Lifelong Resilience (New York: Houghton Mifflin, 1995); P. Tough, *How Children Succeed: Grit, Curiosity, and the Hidden Power of Character* (New York: Houghton Mifflin Harcourt, 2012); J. P. Robinson and S. Martin, "What Do Happy People Do?" *Journal of Social Indicators Research* 89 (2008): 565–71; C. Steiner-Adair, "Got Grit? The Call to Educate Smart, Savvy and Socially Intelligent Kids," *Independent School* (magazine of the National Association of Independent Schools), Winter 2013.

88　The American Academy of Pediatrics, which has been firm: Council on Communications and Media and A. Brown, "Media Use by Children Younger Than 2 Years"; J. Steyer, *Talking Back to Facebook: The Common Sense Guide to Raising Kids in the Digital Age* (New York: Scribner, 2012).

88　Choose shows that teach them about the world: Best Websites for Kids, Common Sense Media, http://www.commonsensemedia.org/website-lists.

89　What about sharing story time on a Kindle: K. J. Dell'Antonia, "Why Books Are Better Than e-Books for Children," *New York Times*, 28 December 2011, http://parenting.blogs.nytimes.com/2011/12/28/why-books-are-better-than-e-books-for-children/?partner=rssnyt&emc=rss; Interview with Lydia Soifer, November 2011; Wolf, *Proust and the Squid*; Interview with Maryanne Wolf, March 2011.

89　For our infants and toddlers, though: L. Guernsey, "Why eReading with Your Kid Can Impede Learning," *Time*, 20 December 2011, http://ideas.time.com/2011/12/20/why-ereading-with-your-kid-can-impede-learning/.

89　Research also suggests that the direct screen-viewing: P. Moretz, "Traditional Books Provide More Positive Parent-Child Interaction According to Temple, Erikson Researchers," *Temple Times*, 9 November 2006, http://www.temple.edu/temple_times/november06/Traditionalbooks.html; Wolf, *Proust and the Squid*; T. Mossle, M. Kleimann, F. Rehbein, and C. Pfeiffer, "Media Use and School Achievement—Boys at Risk?" *British Journal of Developmental Psychology* 28, no. 3 (2010): 699–725.

90　If parents absent themselves and outsource reading: Interview with Maryanne Wolf, March 2011.

90　Advertising and marketing campaigns: "Children, Adolescents and Television: American Academy of Pediatrics, Committee on Public Education," *Pediatrics* 107, no. 2 (2001): 423–26; "Children, Adolescents and Advertising: Committee on Communications, American Academy of Pediatrics," *Pediatrics* 118, no. 6 (2006): 2562–69; J. Schor, *Born to Buy: The Commercialized Child and the New Consumer Culture* (New York: Scribner, 2004); "No Einstein in Your Crib? Get a Refund," *New York Times*, 23 October 2009, http://www.nytimes.com/2009/10/24/education/24baby.html?_r=1; W. Haskins, "Do Educational DVDs Make Babies Blockheads?" Technology News World: Science, 8 August 2007, http://www.technewsworld.com/story/58735.html.

91　Or the Federal Trade Commission's 2012 action: "Ads Touting 'Your Baby Can Read'"; "'Your Baby Can Read': No, He Can't," *Boston Globe*; "France Pulls Plug on TV Shows Aimed at Babies," CBC News.

91　Even among those which experts say: Common Sense Media, http://www.commonsensemedia.org/.

92　"the bigger the touch screen, the better": Wilson Rothman, "The Best iPad Apps for Babies, Toddlers and Sanity-Loving Parents," May 6, 2010, http://gizmodo

.com/5532261/the-best-ipad-apps-for-babies-toddlers-and-sanity+loving-parents.

92   Many of these toys will do more: *Tecca*, "Today in Tech," April 11, 2012, http://news
      .yahoo.com/blogs/technology-blog/7-smart-toys-today-connected-kids–030406909
      .html.

93   "It's a culture of stimulation": Interview with Ned Hallowell, March 2011; E. Hallow-
      ell and J. Ratey, *Driven to Distraction: Recognizing and Coping with Attention Deficit
      Disorder from Childhood through Adulthood* (New York: Anchor Books, 1994).

Chapter 3: Mary Had a Little iPad

99   "There is an inner life and an imaginal self": Interview with Janice Toben, director of
      the Institute for Social and Emotional Learning, February 2012.

101   "Children learn by touching": Interview with Ellen Birnbaum, August 2012; N.
      Schulman and E. Birnbaum, *Practical Wisdom for Parents: Raising Self-Confident
      Children in Preschool Years* (New York: Random House, 2007).

102   The years from birth to five: J. Blanchard and T. Moore, "The Digital World of
      Young Children: Impact on Emergent Literacy," Pearson Foundation, 1 March 2010;
      S. Fraiberg, *The Magic Years: Understanding and Handling the Problems of Early
      Childhood* (New York: Fireside, 1996); P. Leach, *Your Baby and Child from Birth to
      Age 5* (New York: Alfred A. Knopf, 2003); T. B. Brazelton, *Touchpoints: The Essential
      Reference: Your Child's Emotional and Behavioral Development* (Reading, MA: Per-
      seus Books, 1992).

102   The progression from magical thinking: W. Hartmann and G. Brougere, "Toy Cul-
      ture in Preschool Education and Children's Toy Preferences," in *Toys, Games, and
      Media*, edited by J. Goldstein, D. Buckingham, and G. Brougere (Mahwah, NJ: Erl-
      baum, 2004), 37–53; Brazelton, *Touchpoints*.

102   "The child's capacity for social interaction: S. Greenspan and S. Shanker, *The First
      Idea: How Symbols, Language, and Intelligence Evolved from Our Primate Ancestors to
      Modern Humans* (Cambridge, MA: Da Capo Press, 2004).

106   New research on school readiness: C. Wood, *Yardsticks: Children in the Classroom
      Ages 4–14*, 3rd ed. (Turner Falls, MA: Northeast Foundation for Children, 2007);
      L. S. Pagani, M. A. Fitzpatrick, T. A. Barnett, and E. Dubow, "Prospective Asso-
      ciations between Early Childhood Television Exposure and Academic, Psychosocial,
      and Physical Well-Being by Middle Childhood," *Archives of Pediatric and Adolescent
      Medicine* 164, no. 5 (2010): 425–31; S. A. Denham and R. P. Weissberg, "Social-
      Emotional Learning in Early Childhood: What We Know and Where to Go from
      Here," in *A Blueprint for the Promotion of Prosocial Behavior in Early Childhood*, ed-
      ited by E. Chesebrough, P. King, T. P. Gullotta, and M. Bloom (New York: Kluwer
      Academic/ Plenum, 2004), 13–50; F. Din and J. Calao, "The Effects of Playing Edu-
      cational Video Games on Kindergarten Achievement," *Child Study Journal* 31, no. 2
      (2001): 95–102; K. Mistry, C. Minkovitz, D. Strobino, and D. Borzekowski, "Chil-
      dren's Television Exposure and Behavioral and Social Outcomes at 5.5 Years: Does
      Timing of Exposure Matter?" *Pediatrics* 120, no. 4 (2007): 762–69; L. Takeuchi,
      "Families Matter: Designing Media for a Digital Age," Joan Ganz Cooney at Sesame
      Workshop, June 2011; M. Sadowski, "School Readiness Gap," *Harvard Education*

Letter, 2006; L. B. Ames, *Is Your Child in the Wrong Grade?* (New Haven, CT: Gesell Institute, 1966).

107 "I think the foundation of all young children": Interview with Robin Shapiro, January 2011.

107 Nor do the skills acquired: Wood, *Yardsticks*; "Infants Do Not Appear to Learn Words from Educational DVDs," Science Centric, 15 March 2012, http://www .sciencecentric.com/news/10031578-infants-do-not-appear-learn-words-from-educational-dvds.html; W. Haskins, "Do Educational DVDs Make Babies Block-heads?" Technology News World: Science, 8 August 2007, http://www.technews world.com/story/58735.html.

107 As Greenspan and Shanker wrote: Greenspan and Shanker, *The First Idea*.

108 By age three more than one-third of children: Hartmann and Brougere, "Toy Cul-ture in Preschool Education"; *Social and Demographic Trends: Millenials: A Portrait of Generation Next*, Pew Research Center, February 2010; H. L. Burdette and R. C. Whitaker, "A National Study of Neighborhood Safety, Outdoor Play, Television Viewing, and Obesity in Preschool Children," *Pediatrics* 116, no. 3 (2005): 657–62; P. Ravichandran and B. France de Bravo, *Children and Screen Time (Television, DVD's, Computer)*, 2010, boysdevelopmentproject.org.uk.

108 In these years from three to five: Interview with Janice Toben, February 2012; Inter-view with Robin Shapiro, January 2011; L. B. Ames and F. Ilg, *Your Three-Year-Old: Friend or Enemy* (New York: Dell Trade Paperback, 1982); Ames and Ilg, *Your Four-Year-Old: Wild and Wonderful* (New York: Dell Trade Paperback, 1982); Ames and Ilg, *Your Five-Year-Old: Sunny and Serene* (New York: Dell Trade Paperback, 1982).

111 According to a CBS report: C. Lagorio, "Resources: Marketing to Kids," CBS News, 11 February 2009; J. Schor, *Born to Buy: The Commercialized Child and the New Con-sumer Culture* (New York: Scribner, 2004); J. Steyer, *Talking Back to Facebook: The Common Sense Guide to Raising Kids in the Digital Age* (New York: Scribner, 2012).

111 Advertisers target the very young: S. Linn, *Consuming Kids: Protecting Our Children from the Onslaught of Marketing and Advertising* (New York: Anchor Books/Random House, 2004); S. Linn, *The Case for Make Believe: Saving Play in a Commercial-ized World* (New York: New Press/W. W. Norton, 2008); M. Levine, *The Price of Privilege: How Parental Pressure and Material Advantage Are Creating a Generation of Disconnected and Unhappy Kids* (New York: HarperCollins, 2006); S. Smith and A. Granados, "Gender and the Media," *National PTA Magazine*, December/January 2009–2010, http://www.pta.org/3736.htm#14a; N. Singer, "Do Not Track? Adver-tisers Say 'Don't Tread on Us," *New York Times*, 13 October 2012, http://www.ny times.com/2012/10/14/technology/do-not-track-movement-is-drawing-advertisers-fire.html; K. O'Brien, "Privacy Advocates at Odds over Web Tracking," *New York Times*, 5 October 2012, http://www.nytimes.com/2012/10/05/technology/privacy-advocates-and-advertisers-at-odds-over-web-tracking.html?gwh=25CC86CF71D69 58540510F089155F000.

112 But those characters and plotlines: Smith and Granados, "Gender and the Media"; N. Kristof and S. WuDunn, *Half the Sky: Turning Oppression into Opportunity for Women Worldwide* (New York: Knopf, 2009); J. Herrett-Skjellum and M. Allen, "Television Programming and Sex Stereotyping: A Meta-Analysis," in *Communica-tion Yearbook* 19, edited by B. R. Burleson (Thousand Oaks, CA: Sage, 1995), 157–85;

Campaign for a Commercial-Free Childhood, http://commercialfreechildhood.org/; G. Fioravanti, D. Dèttore, and S. Casale, "Adolescent Internet Addiction: Testing the Association between Self-Esteem, the Perception of Internet Attributes, and Preference for Online Social Interactions," *Cyberpsychology, Behavior, and Social Networking* 15, no. 6 (2012): 318–23; R. Joiner, J. Gavin, M. Brosnan, J. Cromby, H. Gregory, J. Guiller, P. Maras, and A. Moon, "Gender, Internet Experience, Internet Identification, and Internet Anxiety: A Ten-Year Follow Up," *Cyberpsychology, Behavior, and Social Networking* 15, no. 7 (July 2012): 370–72.

113   Research suggests that as little as nine minutes: V. J. Rideout, U. G. Foehr, and D. F. Roberts, *Generation M2: Media in the Lives of 8- to 18-Year-Olds* (Menlo Park, CA: Henry J. Kaiser Family Foundation, January 2010); L. Guernsey, "How 'Screen Time' Impacts Kids—What Do Scientists Really Know?" Hatch Innovation webinar, 15 November 2011.

113   The neural map is always open: B. E. Wexler, *Brain and Culture: Neurobiology, Ideology, and Social Change* (Cambridge, MA: MIT Press, 2006); D. W. Winnicott, *Playing and Reality* (New York: Basic Books, 1971).

122   A proper diagnosis of ADD/ADHD: E. Hallowell and J. Ratey, *Driven to Distraction: Recognizing and Coping with Attention Deficit Disorder from Childhood through Adulthood* (New York: Anchor Books, 1994); J. M. Swanson, G. R. Elliot, L. L. Greenhill, T. Wigal, L. E. Arnold, M. Vitiello, L. Hechtman, J. N. Epstein, W. E. Pelham, B. Abikoff, J. H. Newcorn, B. S. G. Molina, S. G. Hinshaw, E. L. Swing, D. A. Gentile, C. A. Anderson, and D. A. Walsh, "Television and Video Game Exposure and the Development of Attention Problems," *Pediatrics* 126 (2010): 214–21; K. Bailey, R. West, and C. Anderson, "The Influence of Video Games on Social, Cognitive, and Affective Information Processing," in *Handbook of Social Neuroscience*, edited by J. Decety and J. Cacioppo (New York: Oxford University Press, in press); C. Rowan, "Unplug—Don't Drug: A Critical Look at the Influence of Technology on Child Behavior with an Alternative Way of Responding Other Than Evaluation and Drugging," *Ethical Human Psychology and Psychiatry* 12, no. 1 (2010): 60–67; N. Schulman and E. Birnbaum, *Practical Wisdom for Parents: Raising Self-Confident Children in Preschool Years* (New York: Random House, 2007).

124   Tech makes us get sloppy: E. J. Paavonen, M. Pennonen, and M. Roine, "Passive Exposure to TV Linked to Sleep Problems in Children," *Journal of Sleep Research* 15 (2006): 154–61; Swanson et al., "Television and Video Game Exposure and the Development of Attention Problems"; Schulman and Birnbaum, *Practical Wisdom for Parents.*

124   Since there are fewer boundaries: H. Seligson, "When the Work-Life Scales Are Unequal," *New York Times*, 1 September 2012, http://www.nytimes.com/2012/09/02/business/straightening-out-the-work-life-balance.html?pagewanted=all; Leslie Perlow, "Why 'Work-Life Balance' Doesn't Work," *Washington Post*, 11 July 2012, http://www.washingtonpost.com/national/on-leadership/step-away-from-the-smartphone/2012/07/11/gJQA3AhDdW_print.html; "Work-Life Balance: Tips to Reclaim Control: When Your Work Life and Personal Life Are out of Balance, Your Stress Level Is Likely to Soar," Mayo Clinic, http://www.mayoclinic.com/health/work-life-balance/WL00056; S. Shellenbarger, "Single and Off the Fast Track: It's Not

Just Working Parents Who Step Back to Reclaim a Life," *Wall Street Journal*, 23 May 2012, http://online.wsj.com/article/SB1000142405270230479170457742013027894886 6.html#printMode; M. Richtel, "Attached to Technology and Paying the Price," *New York Times*, 6 June 2010, http://www.nytimes.com/2010/06/07/technology/07brain.html?pagewanted=all&_r=0.

126   When we expose children at too young an age: Henry J. Kaiser Family Foundation, *Zero to Six: Electronic Media in the Lives of Infants, Toddlers and Preschoolers* (Menlo Park, CA: Henry J. Kaiser Family Foundation, 2003); Rideout, Foehr, and Roberts, *Generation M2*; J. Almon and E. Miller, "Crisis in Early Education: A Research-Based Case for More Play and Less Pressure," Alliance for Childhood, November 2011.

126   When Selma Fraiberg in *The Magic Years*: S. Fraiberg, *The Magic Years: Understanding and Handling the Problems of Early Childhood* (New York: Fireside, 1996).

127   The page and the screen are two: C. Feinberg, "The Mediatrician: Former Hollywood Filmmaker Michael Rich of HMS Studies How Media Affect Youth," *Harvard Magazine*, November/December 2011, http://harvardmagazine.com/2011/11/the-mediatrician; M. Rich, "Boy, Mediated: Effects of Entertainment Media on Adolescent Male Health," *Adolescent Medicine State of the Art Reviews* 14 (2003): 691–713; M. Rich and M. Bar-on, "Child Health in the Information Age: Media Education of Pediatricians," *Pediatrics* 107, no. 1 (2001): 156–62; M. Rich, S. Lamola, J. Gordon, and R. Chalfen, "Video Intervention/Prevention Assessment: A Patient-Centered Methodology for Understanding the Adolescent Illness Experience," *Journal of Adolescent Health* 27 (2000): 155–65; M. Rich, E. R. Woods, E. Goodman, S. J. Emans, and R. H. DuRant, "Aggressors or Victims: Gender and Race in Music Video Violence," *Pediatrics* 101, no. 4, part 1 (1998): 669–74.

Chapter 4: Fast-Forward Childhood

134   Erik Erikson observed: E. Erikson, *Childhood and Society* (New York: W. W. Norton, 1993).

135   This is a critical time for moral development: C. Wood, *Yardsticks: Children in the Classroom Ages 4–14*, 3rd ed. (Turner Falls, MA: Northeast Foundation for Children, 2007); M. Borba, *Building Moral Intelligence* (San Francisco: Jossey-Bass, 2001); R. Coles, *The Moral Intelligence of Children* (New York: Random House Digital, 2011); R. Coles, *The Moral Life of Children* (New York: Atlantic, 2000); R. Coles, *The Spiritual Life of Children* (Boston: Houghton Mifflin, 1990); D. Goleman, *Emotional Intelligence: Why It Can Matter More Than IQ* (New York: Bantam Books, 2006); M. Seligman, *The Optimistic Child: Proven Program to Safeguard Children from Depression and Build Lifelong Resilience* (New York: Houghton Mifflin, 1995); P. Tough, *How Children Succeed: Grit, Curiosity, and the Hidden Power of Character* (New York: Houghton Mifflin Harcourt, 2012); P. Bloom, "The Moral Life of Babies," *New York Times*, May 5, 2010, http://www.nytimes.com/2010/05/09/magazine/09babies-t.html?pagewanted=all&_r=0; Education Resources Information Center, http://www.eric.ed.gov/ERICWebPortal/search/detailmini.jsp?_nfpb=true&_&ERICExtSearch_SearchValue_0=EJ329627&ERICExtSearch_SearchType_0=no&accno=EJ329627.

135 They develop the voice of the inner critic: Interview with Janice Toben, February 2011.

137 Long-established insights into children's learning: Coles, *The Spiritual Life of Children*; J. Deak with T. Barker, *Girls Will Be Girls: Raising Confident, Courageous Daughters* (New York: Hyperion, 2003); M. Thompson with T. Barker, *The Pressured Child: Helping Your Child Find Success in School and Life* (New York: Ballantine Books, 2004); Erikson, *Childhood and Society*; J. Deak, *Your Fantastic Elastic Brain* (Belvedere, CA: Little Pickle Press, 2011); N. Carlsson-Paige, *Taking Back Childhood: Helping Your Kids Thrive in a Fast-Paced, Media-Saturated, Violence-Filled World* (New York: Hudson Street Press, 2008).

139 Other findings also show media exposure: D. Gentile, *Media Violence and Children: A Complete Guide for Parents and Professionals*, Advances in Applied Developmental Psychology (New York: Praeger, 2003); J. B. He, C. J. Liu, Y. Y. Guo, and L. Zhao, "Deficits in Early-Stage Face Perception in Excessive Internet Users," *Cyberpsychology, Behavior, and Social Networking* 14, no. 5 (2009): 303–8; L. Bensley and J. Van Eenwyk, "Video Games and Real-Life Aggression," review of literature, *Journal of Adolescent Health* 29, no. 4 (October 2001): 244–57; M. Choliz and C. Marco, "Patterns of Video Game Use and Dependence in Children and Adolescents," *Anales de Psicologia* 27, no. 2 (2011): 418–26; American Academy of Pediatrics, "Joint Statement on the Impact of Entertainment Violence on Children: Congressional Public Health Summit," 26 July 2000; S. Byun, C. Ruffini, J. Mills, A. Douglas, M. Niang, S. Stepchenkova, S. K. Lee, J. Loutfi, J. K. Lee, M. Atallah, and M. Blanton, "Internet Addiction: Metasynthesis of 1996–2006 Quantitative Research," *Cyberpsychology, Behavior, and Social Networking* 12, no. 2 (2010): 203–7; "Tech Addiction Symptoms Rife among Students," CBC News, 6 April 2011, http://www.cbc.ca/news/technology/story/2011/04/06/technology-addiction-students.html; D. A. Gentile, H. Choo, A. Liau, T. Sim, D. Li, D. Fung, and A. Khoo, "Pathological Video Game Use among Youths: A Two-Year Longitudinal Study," *Pediatrics* 127, no. 2 (2011): E319–29; D. A. Gentile, P. Lynch, J. Linder, and D. Walsh, "The Effects of Violent Video Game Habits on Adolescent Hostility, Aggressive Behaviors, and School Performance," *Journal of Adolescence* 27 (2004): 5–22.

139 Emotional and social development: J. Grafman, "Brain Development in a Hyper-Tech World," Dana Foundation, August 2008; Henry J. Kaiser Family Foundation, Program for the Study of Media and Health, "Media Multitasking among American Youth: Prevalence, Predictors and Pairings," 12 December 2006, http://kff.org/entmedia/7592.cfm.

139 Michael Friedlander, head of neuroscience: Grafman, "Brain Development in a Hyper-Tech World."

141 Thin is still in, but for ever younger girls: C. Steiner-Adair and L. Sjostrom, *Full of Ourselves: A Wellness Program to Advance Girl Power, Health, and Leadership* (New York: Teacher's College Press, 2006); J. Siebel Newsom, *Miss Representation*, documentary, 2011, http://www.missrepresentation.org/the-film/.

141 She is clearly in the early stages: Steiner-Adair and Sjostrom, *Full of Ourselves*; C. Steiner-Adair, "When the Body Speaks: Girls, Eating Disorders, and Psychotherapy," *Journal of Women and Therapy* 11, nos. 3 and 4 (1991), reprinted in *Women, Girls,*

and Psychotherapy: Reframing Resistance, edited by C. Gilligan, A. Rogers, and D. Tolman (Binghamton, NY: Haworth Press, 1991); P. Orenstein, *Cinderella Ate My Daughter: Dispatches from the Front Lines of the New Girlie-Girl Culture* (New York: Harper Paperbacks; reprint ed., 2012); N. Piran, M. Levine, and C. Steiner-Adair, *Preventing Eating Disorders: A Handbook of Interventions and Special Challenges* (Philadelphia: Brunner/Mizel, 1999).

142   Boys, too, are under pressure: M. Thompson, *It's a Boy! Understanding Your Son's Development from Birth to Eighteen* (Boston: Ballantine Books, 2008); M. Thompson and D. Kindlon, *Raising Cain: Protecting the Emotional Life of Boys* (New York: Ballantine Books, 2000); M. Thompson, *Speaking of Boys: Answers to the Most-Asked Questions about Raising Sons* (Boston: Ballantine Books, 2000); J. Katz, *Tough Guise: Violence, Media and the Crisis in Masculinity*, documentary, 1999, http://www.jacksonkatz.com/video2.html; J. Katz, *The Macho Paradox: Why Some Men Hurt Women and How All Men Can Help*, Media Education Foundation, University of Massachusetts, Amherst, www.mediaed.org; W. Pollock, *Real Boys: Rescuing Our Sons from the Myths of Boyhood* (New York: Henry Holt, 1999); M. Thompson, "Raising Cain: Boys in Focus," PBS/Oregon Public Broadcasting and Powderhouse, 12 January 2006; S. Lamb, L. Brown, and M. Tappen, *Packaging Boyhood: Saving Our Sons from Superheroes, Slackers, and Other Media Stereotypes* (New York: St. Martin's Press, 2009); M. Gurian, *The Purpose of Boys: Helping Our Sons Find Meaning, Significance, and Direction in Their Lives* (San Francisco: Jossey-Bass, 2010); J. Siebel Newsom, *Miss Representation*.

142   However, that's not the message they get: S. Goldberg, "TV Can Boost Self-Esteem of White Boys, Study Says," CNN, 1 June 2012, http://www.cnn.com/2012/06/01/showbiz/tv/tv-kids-self-esteem/index.html; K. Harrison and N. Martins, "Racial and Gender Differences in the Relationship between Children's Television Use and Self-Esteem : A Longitudinal Panel Study," *Communication Research* 39 (2012): 338; Steiner-Adair and Sjostrom, *Full of Ourselves*.

143   Research shows that a majority of TV content: S. Smith and A. Granados, "Gender and the Media," *National PTA Magazine*, December/January 2009–2010, http://www.pta.org/3736.htm#14a; R., Joiner, J. Gavin, M. Brosnan, J. Cromby, H. Gregory, J. Guiller, P. Maras, and A. Moon, "Gender, Internet Experience, Internet Identification, and Internet Anxiety: A Ten-Year Follow Up," *Cyberpsychology, Behavior, and Social Networking* 15, no. 7 (2012): 370–72; D. Levin and J. Kilbourne, *So Sexy So Soon: The New Sexualized Childhood and What Parents Can Do to Protect Their Kids* (New York: Ballantine Books, 2008).

143   Researchers Karen Dill and Kathryn Thill: K. Dill and K. Thill, "Video Game Characters and the Socialization of Gender Roles: Young People's Perceptions Mirror Sexist Media," *Business Media* 57 (2007): 851–64; S. A. Denham and R. P. Weissberg, "Social-Emotional Learning in Early Childhood: What We Know and Where to Go from Here," *Collaborative for Academic, Social and Emotional Learning*, 2004, http://casel.org/publications/social-emotional-learning-in-early-childhood-what-we-know-and-where-to-go-from-here/.

144   My class was billed as a "media literacy workshop": The Institute for Social and Emotional Learning Workshop for K–12 Educators and Administrators. Janice Toben, Rush Saviston Frank, Nick Haisnan, and Elizabeth McCloud, with Catherine Steiner-Adair.

147 Before *Rugrats, Gossip Girl*: C. Hedges, *Empire of Illusion: The End of Literacy and the Triumph of Spectacle* (New York: Nation Books, 2009).

149 The startle factor is much bigger: M. Seligman, "Chronic Fear Produced by Unpredictable Electric Shock," *Journal of Comparative and Physiological Psychology* 66 (1968): 402–11.

150 Mimi Ito: M. Ito, a cultural anthropologist: H. Horst, M. Bittanti, D. Boyd, B. Stephenson, P. Lange, C. J. Pascoe, and L. Robinson, "Living and Learning with New Media: Summary of Findings from the Digital Youth Project," John D. and Catherine T. MacArthur Foundation Reports on Digital Media and Learning (Cambridge, MA: MIT Press, 2008).

151 Graphic screen violence and pornography: M. Robins and G. Wilson, "Porn-Induced Sexual Dysfunction Is a Growing Problem," *Psychology Today*, 11 July 2011, http://www.psychologytoday.com/blog/cupids-poisoned-arrow/201107/porn-induced-sexual-dysfunction-growing-problem; M. L. Ybarra and K. J. Mitchell, "Exposure to Internet Pornography among Children and Adolescents: A National Survey," *Cyberpsychology, Behavior, and Social Networking* 8, no. 5 (2005): 473–82; S. J. Lee and Y. J. Chae,"Balancing Participation and Risks in Children's Internet Use: The Role of Internet Literacy and Parental Mediation," *Cyberpsychology, Behavior, and Social Networking* 15, no. 5 (2011): 257–62; A. Lenhart, "Protecting Teens Online," Pew Internet: Pew Internet and American Life Project, 17 March 2005, http://www.pewinternet.org/Reports/2005/Protecting-Teens-Online.aspx.

153 The connection between virtual and real life: *Kids and Media at the New Millennium: A Kaiser Family Foundation Report* (Menlo Park, CA: Henry J. Kaiser Family Foundation, 1999); American Academy of Pediatrics, Committee on Public Education, "Media and Violence," *Pediatrics* 108, no. 5 (2001): 1222–26; American Academy of Pediatrics, "Joint Statement on the Impact of Entertainment Violence on Children: Congressional Public Health Summit," 26 July 2000.

154 Researcher and pediatrician Dimitri Christakis: Interview with Dimitri Christakis, February 2012.

155 This is particularly hard for families: E. A. Vandewater, J. H. Lee, and M. Shim, "Family Conflict and Violent Electronic Media Use in School-Aged Children," *Media Psychology* 7 (2005): 73–86; R. S. Weisskirch, "No Crossed Wires: Cell Phone Communication in Parent-Adolescent Relationships," *Cyberpsychology, Behavior, and Social Networking* 14, nos. 7–8 (2005): 447–51; E. J. Paavonen, M. Pennonen, and M. Roine, "Passive Exposure to TV Linked to Sleep Problems in Children," *Journal of Sleep Research* 15 (2006): 154–61.

157 Craig Anderson, an Iowa State University professor: C. A. Anderson, L. Berkowitz, E. Donnerstein, L. R. Huesmann, J. D. Johnson, D. Linz, N. M. Malamuth, and E. Wartella, "The Influence of Media Violence on Youth," *Psychological Science in the Public Interest* 4, no. 3 (2003): 81–110, "Childhood Exposure to Media Violence Predicts Young Adult Aggressive Behavior," American Psychological Association, 9 March 2003; "Violent Video Games: Myths, Facts, and Unanswered Questions," PsychNet, 2004, APA Online.

157 South Korea and many other countries: Youkyung Lee, "South Korea: 160,000 Kids Between Age 5 and 9 Are Internet-Addicted," *Huffington Post*, 28 November 2012, http://

www.huffingtonpost.com/2012/11/28/south-korea-internet-addicted_n_2202371
.html.

158   "Neuroimaging suggests": Interview with Tina Payne Bryson, October 17, 2012.

158   "And at the same time the amygdala": Y. Wang, V. P. Mathews, A. J. Kalnin, K. M.
      Mosier, D. W. Dunn, A. J. Saykin, and W. G. Kronenberger, "Short-Term Exposure
      to a Violent Video Game Induces Changes in Frontolimbic Circuitry in Adolescents,"
      Brain Imaging and Behavior 3, no. 1 (2009): 38–50.

159   My colleague JoAnn Deak: Deak, Girls Will Be Girls.

Chapter 5: Going, Going, Gone

164   One superintendent called it "the Bermuda Triangle": Former Louisiana superinten-
      dent Cecil Picard, quoted by Peter Meyer in "The Middle School Mess," Education-
      next (Winter 2011), http://educationnext.org/the-middle-school-mess/.

165   We call them "tweens": C. Wood, Yardsticks: Children in the Classroom Ages 4–14,
      3rd ed. (Turner Falls, MA: Northeast Foundation for Children, 2007); P. Aftab,
      The Parents' Guide to Protecting Your Children in Cyberspace (New York: McGraw-
      Hill, 2000); S. Lamb, L. Brown, and M. Tappen, Packaging Boyhood: Saving
      Our Sons from Superheroes, Slackers, and Other Media Stereotypes (New York: St.
      Martin's Press, 2009); S. Lamb and L. M. Brown, Packaging Girlhood: Rescuing
      Our Daughters from Marketers' Schemes (New York: St. Martin's Press, 2006); N.
      Carlsson-Paige, Taking Back Childhood: Helping Your Kids Thrive in a Fast-Paced,
      Media-Saturated, Violence-Filled World (New York: Hudson Street Press, 2008);
      T. B. Brazelton, Touchpoints: The Essential Reference: Your Child's Emotional and
      Behavioral Development (Reading, MA: Perseus Books, 1992); American Medical
      Association Complete Guide to Your Children's Health (New York: Random House,
      1999); S. A. Denham and R. P. Weissberg, "Social-Emotional Learning in Early
      Childhood: What We Know and Where to Go from Here," in A Blueprint for the
      Promotion of Prosocial Behavior in Early Childhood, edited by E. Chesebrough, P.
      King, T. P. Gullotta, and M. Bloom (New York: Kluwer Academic/Plenum, 2004),
      13–50.

165   At a time when developmentally they need: C. Baker, "The Creator," Wired, August
      2012, 66; D. Goleman, Emotional Intelligence: Why It Can Matter More Than IQ
      (New York: Bantam Books, 2006); J. Gottman and J. Declaire, Raising an Emotion-
      ally Intelligent Child: The Heart of Parenting (New York: Simon and Schuster, 1998);
      Amanda Lanhart, "Teens, Cell Phones, and Texting: Text Messaging Becomes
      Centerpiece Communication," Pew Internet and American Life Project, April 2010,
      http://pewresearch.org/pubs/1572/teens-cell-phones-text-messages; J. Steyer, Talk-
      ing Back to Facebook: The Common Sense Guide to Raising Kids in the Digital Age
      (New York: Scribner, 2012); S. Turkle, Alone Together: Why We Expect More from
      Technology and Less from Each Other (New York: Basic Books, 2011); L. Rosen, iDis-
      order: Understanding Our Obsession with Technology and Overcoming Its Hold on Us
      (New York: Palgrave Macmillan, 2012).

165   Facebook alone is the after-school study group: "A Window onto Family Facebook
      Use: TRUSTe Study," Net Family News.org, 18 October 2012, http://www.netfami

lynews.org/a-window-onto-family-facebook-use-truste-study; "TRUSTe Releases Survey Results of Parents and Teenagers on Social Networking Behaviors: National Poll Conducted in Partnership with Lightspeed Research Reveals Alignment between Parents and Teens in Desire for Privacy," TRUSTe, 18 October 2010, http://www.truste.com/about-TRUSTe/press-room/news_truste_2010_survey_snsprivacy.

166   At times, even when they know: E. Bazelon, "Amanda Todd Was Stalked before She Was Bullied," *Slate*, 18 October 2012, www.slate.com/blogs/xx_factor/2012/10/18/suicide_victim_amanda_todd_stalked_before_she_was_bullied.html; C. Maag, "When the Bullies Turned Faceless," *New York Times*, 16 December 2007, http://www.nytimes.com/2007/12/16/fashion/16meangirls.html?ref=meganmeier&pagewanted=print; M. Schwartz, "The Trolls among Us," *New York Times*, 3 August 2008, http://www.nytimes.com/2008/08/03/magazine/03trolls-t.html?ref=meganmeier&pagewanted=print.

167   If you're not on top of this: J. Anderson, "Millennials Will Benefit and Suffer Due to Their Hyperconnected Lives," Pew Internet and American Life Project, 29 February 2012, http://www.pewinternet.org/Reports/2012/Hyperconnected-lives.aspx.

171   "A large part of this generation's social": J. Brown, ed., *Managing the Media Monster: The Influence of Media (from Television to Text Messages) on Teen Sexual Behavior and Attitudes* (Washington, DC: National Campaign to Prevent Teen and Unplanned Pregnancy, 2008); V. Carson, W. Pickett, and I. Janssen, "Screen Time and Risk Behaviors in 10- to 16-Year-Old Canadian Youth," *Preventive Medicine* 52, no. 2 (2011): 97–98; L. Guernsey, G. Troseth, E. Hartley-Brewer, M. Rich, and L. Rosen, "Wired Kids, Negligent Parents?" *New York Times*, 28 January 2010, http://roomfordebate.blogs.nytimes.com/2010/01/28/wired-kids-negligent-parents/; T. Lewin, "If Your Kids Are Awake, They're Probably Online," *New York Times*, 20 January 2010, http://www.nytimes.com/2010/01/20/education/20wired.html?_r=1&; D. S. Bickham and M. Rich, "Is Television Viewing Associated with Social Isolation?" *Archives of Pediatrics and Adolescent Medicine* 160, no. 4 (2006): 387–92; V. J. Rideout, U. G. Foehr, and D. F. Roberts, *Generation M2: Media in the Lives of 8- to 18-Year-Olds* (Menlo Park, CA: Henry J. Kaiser Family Foundation, January 2010); J. B. He, C. J. Liu, Y. Y. Guo, and L. Zhao, "Deficits in Early-Stage Face Perception in Excessive Internet Users," *Cyberpsychology, Behavior, and Social Networking* 14, no. 5 (2009): 303–308; L. Takeuchi, "Families Matter: Designing Media for a Digital Age," Joan Ganz Cooney at Sesame Workshop, June 2011; D. J. Siegel, *The Mindful Brain: Reflection and Attunement in the Cultivation of Well-Being* (New York: W. W. Norton, 2007).

171   And research confirms: For comprehensive information about SEL and research, see CASEL.org (Collaborative for Academic, Social and Emotional Learning).

172   A 2011 study of teens, kindness, and cruelty: J. Brenner, "Pew Internet: Teens," Pew Internet and American Life Project, 27 April 2012, http://pewinternet.org/Commentary/2012/April/Pew-Internet-Teens.aspx.

173   One in three younger teen girls: C. Lee, "Media Gender Divide: Boys v. Girls (Info-Graphic)," NMR, 5 July 2012, http://newmediarockstars.com/2012/07/infographic-social-media-gender-divide-boys-v-girls/; G. McMillan, "Study: Women Better at Using Social Media to Keep in Touch," *Techland*, 30 September 2011, http://tech

land.time.com/2011/09/30/study-women-better-at-using-social-media-to-keep-in-touch.

173   I like Mimi Ito and her colleagues' description: M. Ito et al., *Hanging Out, Messing Around, and Geeking Out: Kids Living and Learning with New Media* (Cambridge, MA: MIT Press, 2010); M. Ito, H. Horst, M. Bittanti, D. Boyd, B. Stephenson, P. Lange, C. J. Pascoe, and L. Laura Robinson, "Living and Learning with New Media: Summary of Findings from the Digital Youth Project," John D. and Catherine T. MacArthur Foundation Reports on Digital Media and Learning, November 2008.

177   We are now seeing them at younger ages: "8–18 Year Olds Pathologically Addicted to Games," http://www.sciencedaily.com/releases/2009/04/090420103547.htm#.UI MQbWnq5WQ.email; "Secret of Facebook's Success: Sharing Gossip with Friends Is Addictive and Arousing," Womenist.net, http://www.womenist.net/en/p5704/technology/secret_of_facebooks_success.html; "Tech Addiction Symptoms Rife among Students," CBC News, 6 April 2011, http://www.cbc.ca/news/technology/story/2011/04/06/technology-addiction-students.html; R. Kittinger, C. J. Correia, and J. G. Irons, "Relationship between Facebook Use and Problematic Internet Use among College Students," *Cyberpsychology, Behavior, and Social Networking* 15, no. 6 (2012): 324–27; G. Fioravanti, D. Dèttore, and S. Casale, "Adolescent Internet Addiction: Testing the Association between Self-Esteem, the Perception of Internet Attributes, and Preference for Online Social Interactions," *Cyberpsychology, Behavior, and Social Networking* 15, no. 6 (2012): 318–23.

183   As wonderful as the Internet is: Interview with Michael Thompson, June 2012; Interview with Jackson Katz, November 2011; K. Kaye et al., *The Fog Zone: How Misperceptions, Magical Thinking, and Ambivalence Put Young Adults at Risk for Unplanned Pregnancy* (Washington, DC: National Campaign to Prevent Teen and Unplanned Pregnancy, 2009); G. Martinez et al., "Teenagers in the United States: Sexual Activity, Contraceptive Use, and Childbearing, 2006–2010 National Survey of Family Growth," *Vital and Health Statistics*, ser. 23, no. 31, 2011.

184   Kids today know far more about sex: RTI International and Blue, "Prevention in Middle School Matters: A Summary of Findings of Teen Dating Violence Behaviors and Associated Risk Factors Among 7th-Grade Students," Robert Wood Johnson Foundation, 1 January 2011, www.rwjf.org/goto/middleschoolmatters; D. Weiss, "The New Normal? Youth Exposure to Online Pornography," *Rock*, 6 April 2011, http://www.myrocktoday.org/default.asp?q_areaprimaryid=7&q_areasecondaryid=74&q_areatertiaryid=0&q_articleid=861; Brown, ed., *Managing the Media Monster*; E. R. Buhi et al., "Quality and Accuracy of Sexual Health Information Web Sites Visited by Young People," *Journal of Adolescent Health* 47, no. 2 (2010): 206–208; A. Chandra et al., "Does Watching Sex on Television Predict Teen Pregnancy? Findings from a National Longitudinal Survey of Youth," *Pediatrics* 122, no. 5 (2008): 1047–54; G. Martinez, J. Abma, and C. Casey, "Educating Teenagers about Sex in the United States," *NCHS Data Brief*, no. 44 (2010).

184   When the psychologist Michael Thompson: M. Thompson, "Raising Cain: Boys in Focus," PBS/Oregon Public Broadcasting and Powderhouse, 12 January 2006.

184   "The average American child sees pornography": Interview with Michael Thompson, June 2012; *Social and Demographic Trends: Millennials: A Portrait of Generation*

Next, Pew Research Center, February 2010; "Parents Taking New Steps to Protect Their Children from New Porn," www.netnanny.com; J. Bryant and D. Brown, "Use of Pornography," in *Pornography: Research Advances and Policy Considerations*, edited by D. Zillmann and J. Bryant (Hillsdale, NJ: Erlbaum, 1989), 25–55.

184   With Internet porn so accessible: Interview with Michael Thompson, June 2011.

186   That peer culture now accepts slut-chic: D. Levin and J. Kilbourne, *So Sexy So Soon: The New Sexualized Childhood and What Parents Can Do to Protect Their Kids* (New York: Ballantine Books, 2008); Interview with Jackson Katz, November 2011; S. Smith and A. Granados, "Gender and the Media," *National PTA Magazine*, December/January 2009–2010, http://www.pta.org/3736.htm#14a; M. Flood and C. Hamilton, "Youth and Pornography in Australia: Evidence on the Extent of Exposure and Likely Effects," Australia Institute, Discussion Paper 52, February 2003; Mandy Marlette, cartoon, http://thesocietypages.org/socimages/files/2009/10/1-499x387.gif.

186   A study of 1,430 seventh-grade students: RTI International and Blue, "Prevention in Middle School Matters."

186   While the hookup culture may seem hip: C. Bui, "Hookup Culture," *Tufts Daily* (Tufts University), 24 January 2011, http://www.tuftsdaily.com/op-ed/hookup-culture-1.2445270; C. Bui, "Teaching Good Sex," *New York Times*, 16 November 2011, http://www.nytimes.com/2011/11/20/magazine/teaching-good-sex.html?pagewanted=print; Brown, ed., *Managing the Media Monster*.

186   Jackson Katz says media has become: Interview with Jackson Katz, November 2011; J. Wolak, K. Mitchell, and D. Finkelhor, "Unwanted and Wanted Exposure to Online Pornography in a National Sample of Youth Internet Users," *Pediatrics* 119, no. 2 (2007): 247–57, http://pediatrics.aappublications.org/cgi/reprint/119/2/247; C. Sabina, J. Wolak, and D. Finkelhor, "The Nature and Dynamics of Internet Pornography Exposure for Youth," *Cyberpsychology, Behavior, and Social Networking* 11, no. 6 (2008): 691–93.

191   *This is not your computer*: For good examples of family contracts for technology use, see Common Sense Media, cyberbullying.com, safekid.com, and netfamily.com.

Chapter 6: Teens, Tech, Temptation, and Trouble

195   Instead of opening a way into deeper experience: E. Aboujaoude, *Virtually You: The Dangerous Powers of the E-Personality* (New York: W. W. Norton, 2011); E. Hallowell, *Connect: 12 Vital Ties That Open Your Heart, Lengthen Your Life, and Deepen Your Soul* (New York: Pocket Books, 1999); J. Brown, ed., *Managing the Media Monster: The Influence of Media (from Television to Text Messages) on Teen Sexual Behavior and Attitudes* (Washington, DC: National Campaign to Prevent Teen and Unplanned Pregnancy, 2008); R. Coles, *The Moral Intelligence of Children* (New York: Random House Digital, 2011); D. Goleman, *Emotional Intelligence: Why It Can Matter More Than IQ* (New York: Bantam Books, 2006); N. Carr, *The Shallows: What the Internet Is Doing to Our Brains* (New York: W. W. Norton, 2010); G. Fioravanti, D. Dèttore, and S. Casale, "Adolescent Internet Addiction: Testing the Association between Self-Esteem, the Perception of Internet Attributes, and Preference for Online Social

Interactions," *Cyberpsychology, Behavior, and Social Networking* 15, no. 6 (2012): 318–23; S. A. Denham and R. P. Weissberg, "Social-Emotional Learning in Early Childhood: What We Know and Where to Go from Here," Collaborative for Academic, Social, and Emotional Learning, 2004, http://casel.org/publications/social-emotional-learning-in-early-childhood-what-we-know-and-where-to-go-from-here/; C. Steiner-Adair, "Raising Sustainable Girls," talk given at Innovation in Education Summit, 2011; C. Steiner-Adair, "Cultural Literacy: Teaching Smart Kids to Outsmart Unhealthy Stereotypes."

196 Adolescence has always been characterized: M. McConville, *Adolescence: Psychotherapy and the Emergent Self* (San Francisco: Jossey-Bass, 1995); T. B. Brazelton, *Touchpoints: The Essential Reference: Your Child's Emotional and Behavioral Development* (Reading, MA: Perseus Books, 1992); L. Kessler, *My Teenage Werewolf: A Mother, a Daughter, a Journey through the Thicket of Adolescence* (New York, Penguin, 2010).

203 Tech often amplifies gender differences: Kelly Schryver, "Keeping Up Appearances," honors thesis, Brown University, 2011; *Girls, Boys, and Media Messages: A Gender and Digital Life Toolkit*," Common Sense Media online, http://www.commonsensemedia.org/educators/gender.

206 "Kids sometimes post comments without thinking": Schryver, "Keeping Up Appearances."

207 In fact, for teens with eating or body image disorders: R. Joiner, J. Gavin, M. Brosnan, J. Cromby, H. Gregory, J. Guiller, P. Maras, and A. Moon, "Gender, Internet Experience, Internet Identification, and Internet Anxiety: A Ten-Year Follow Up," *Cyberpsychology, Behavior, and Social Networking* 15, no. 7 (2012): 370–72; P. Cramer and T. Steinwert, "Thin Is Good, Fat Is Bad: How Early Does It Begin?" *Journal of Applied Developmental Psychology* 19 (1998): 429–51; S. Shapiro, M. Newcomb, and T. B. Loeb, "Fear of Fat, Disregulated-Restrained Eating, and Body-Esteem: Prevalence and Gender Differences among Eight- to Ten-Year-Old Children," *Journal of Clinical Child Psychology* 26 (1997): 358–65; C. Steiner-Adair and L. Sjostrom, *Full of Ourselves: A Wellness Program to Advance Girl Power, Health, and Leadership* (New York: Teacher's College Press, 2006).

210 In a study of sexting among about six hundred students: D. S. Strassberg, R. K. McKinnon, M. A. Sustaita, and J. Rullo, "Sexting by High School Students: An Exploratory and Descriptive Study," *Archives of Sexual Behaviour* 42, no. 1 (2012): 15–21.

214 Teens who are emotionally neglected: S. S. Luthar, "The Culture of Affluence: Psychological Costs of Material Wealth," *Child Development* 74 (2003): 1581–93; S. S. Luthar and B. E. Becker, "Privileged but Pressured? A Study of Affluent Youth," *Child Development* 73 (2002): 1593–1610; M. Levine, *The Price of Privilege: How Parental Pressure and Material Advantage Are Creating a Generation of Disconnected and Unhappy Kids* (New York: HarperCollins, 2006); L. Ponton, *The Romance of Risk: Why Teenagers Do the Things They Do* (New York: Basic Books, 1997), C. Steiner-Adair, *Stress for Success: Overinvesting in a Child's Performance Presents a High-Risk Exercise for Parents, Worth* magazine, April 2007.

214 Binge drinking and other substance abuse: Levine, *The Price of Privilege*; D. Pope, *Doing School: How We Are Creating a Generation of Stressed-Out, Materialistic, and Miseducated Students* (New Haven, CT: Yale University Press, 2003); Luthar, "The Culture of Affluence"; Luthar and Becker, "Privileged but Pressured?"; T. L. Lam

and Z.-W. Peng, "Effect of Pathological Use of the Internet on Adolescent Mental Health: A Prospective Study," *Archives of Pediatrics and Adolescent Medicine* 164, no. 10 (2010): 901–906; R. A. Mentzoni, G. S. Brunborg, H. Molde, H. Myrseth, K. J. Mar Skouveroe, J. Hetland, and S. Pallesen, "Problematic Video Game Use: Estimated Prevalence and Associations with Mental and Physical Health," *Cyberpsychology, Behavior, and Social Networking* 14, no. 10 (2011): 591–96; S. Shellenbarger, "Understanding the Zombie Teen's Body Clock," *Wall Street Journal*, updated 16 October 2012, http://online.wsj.com/article/work_and_family.html; S. Shellenbarger, "Tips for Improving Your Teens' Sleep Schedule," The Juggle: *Wall Street Journal* Blogs, 17 October 2012, http://blogs.wsj.com/juggle/2012/10/17/tips-for-improving-your-teens-sleep-schedule/.

220 In the adolescent flux of identity and sexuality: Interview with Jackson Katz, November 2011.

221 Fame is the do-it-yourself dream: Y. T. Uhls and P. Greenfield, "Rise of Fame: A Historical Content Analysis," *Journal of Psychosocial Research on Cyberspace* 5, no. 1 (2011): article 1, http://www.cyberpsychology.eu/view.php?cisloclanku=2011061601; A. Hawgood, "How Teenagers Handle the Web's Instant Fame: No Stardom Until after Homework," *New York Times*, 15 July 2011, http://www.nytimes.com/2011/07/17/fashion/how-teenagers-handle-the-webs-instant-fame.html?pagewanted=all.

225 Teenagers have always been drawn: Ponton, *The Romance of Risk*.

Chapter 7: Scary, Crazy, and Clueless

230 This is a difficult, often unsettling time: M. Levine, *Teach Your Children Well: Parenting for Authentic Success* (New York: HarperCollins, 2012); P. Bronson and A. Merryman, *Nurtureshock: New Thinking about Children* (New York: Hachette Book Group, 2009); M. Thompson, with T. Barker, *The Pressured Child: Helping Your Child Find Success in School and Life* (New York: Ballantine Books, 2004); M. Thompson, *Homesick and Happy: How Time Away from Parents Can Help a Child Grow* (Boston: Ballantine Books, 2012); E. H. Marano, *A Nation of Wimps: The High Cost of Invasive Parenting* (New York: Broadway Books, 2008); S. Shellenbarger, "When Curious Parents See Math Grades in Real Time," *Wall Street Journal*, 2 October 2012, http://online.wsj.com/article/SB10000872396390444592404578032360255233782.html; A. Shrivastava, "Is It Okay to Be a 'Helicopter Parent?'" The Juggle: *Wall Street Journal* Blogs, 12 October 2012, http://responsibility-project.libertymutual.com/articles/the-case-for-the-relaxed-parent#fbid=k_VjLUEe6bn; S. O'Keeffe and K. Clarke-Pearson, Council on Communications and Media, "Clinical Report: The Impact of Social Media on Children, Adolescents, and Families," *Pediatrics* 127 (2011): 800–804, posted online 28 March 2011, http://pediatrics.aappublications.org/content/127/4/800.full.pdf; A. Chua, *Battle Hymn of the Tiger Mother* (New York: Penguin, 2011); L. Kessler, *My Teenage Werewolf: A Mother, a Daughter, a Journey Through the Thicket of Adolescence* (New York: Penguin, 2010).

243 She acknowledged that she was not eating: C. Steiner-Adair, "When the Body Speaks: Girls, Eating Disorders, and Psychotherapy," *Journal of Women and Therapy* 11, nos. 3 and 4 (Fall 1991), reprinted in *Women, Girls, and Psychotherapy: Reframing Resis-*

tance, edited by C. Gilligan, A. Rogers, and D. Tolman (Binghamton, NY: Haworth Press, 1991).

249 Sometimes adults overreact in efforts: M. Thomas and O. G. Catherine with L. Cohen, *Best Friends, Worst Enemies: Understanding the Social Lives of Children* (New York: Ballantine, 2001); M. Thompson and L. Cohen with C. O'Neill Grace, *Mom, They're Teasing Me: Helping Your Child Solve Social Problems* (Boston: Ballantine, 2002); R. Wiseman, *Queen Bees and Wannabes: Helping Your Daughter Survive Cliques, Gossip, Boyfriends, and Other Realities of Adolescence* (New York: Crown, 2001).

249 As Paul Tough writes in *How Children Succeed*: P. Tough, *How Children Succeed: Grit, Curiosity, and the Hidden Power of Character* (New York: Houghton Mifflin Harcourt, 2012); C. Steiner-Adair, "Got Grit? The Call to Educate Smart, Savvy, and Socially Intelligent Students," *Independent Schools*, Winter 2013; T. Wagner, *Creating Innovators: The Making of Young People Who Will Change the World* (New York: Scribner, 2012); W. Mogel, *The Blessing of a Skinned Knee: Using Jewish Teachings to Raise Self-Reliant Children* (New York: Scribner, 2008); G. Steffgen, M. S, Konig, J. Pfetsch, and A. Melzer, "Are Cyberbullies Less Empathic? Adolescents' Cyberbullying Behavior and Empathic Responsiveness," *Cyberpsychology, Behavior, and Social Networking* 14, no. 11 (2011): 643–48, posted online 9 May 2011, http://www.liebert online.com/doi/abs/10.1089/cyber.2010.0445.

252 Michael Thompson described: Interview with Michael Thompson, June 2012.

257 Effective sex education involves: C. Bui, "Teaching Good Sex," *New York Times*, 16 November 2011, http://www.nytimes.com/2011/11/20/magazine/teaching-good-sex.html?pagewanted=print; C. Steiner-Adair, talk, "Educating Smart Girls to Outsmart the Media," presented at the National Coalition of Girls Schools Conference, June 2012.

Chapter 8: The Sustainable Family

261 Sustainability has been described: B. Norton, *Sustainability: A Philosophy of Adaptive Ecosystem Management* (Chicago: University of Chicago Press, 2005); C. Steiner-Adair, "Raising Sustainable Girls," presented at Education Innovation Summit, Hawthorne Brown School, Cleveland, OH, October 2011.

263 Neil Postman concluded his 1982 book: N. Postman, *The Disappearance of Childhood* (New York: Delacorte Press, 1982; New York: Vintage Books, 1994).

264 But not all life's lessons: T. Wagner, *The Global Achievement Gap: Why Even Our Best Schools Don't Teach the New Survival Skills Our Children Need—And What We Can Do about It* (New York: Basic Books, 2008).

264 Martin Seligman, a psychologist: M. Seligman, *Flourish: A Visionary New Understanding of Happiness and Well-Being* (New York: Free Press, 2011).

270 In addition, each family needs: J. Steyer, *Talking Back to Facebook: The Common Sense Guide to Raising Kids in the Digital Age* (New York: Scribner, 2012); D. Christakis and F. Zimmerman, *The Elephant in the Living Room: Make Television Work for Your Kid* (Emmaus, PA: Rodale Books, 2006); J. Farley-Gillispie and J. Gackenbach, *cyber.rules: Negotiating Healthy Internet Use* (New York: W. W. Norton, 2007);

P. Aftab, *The Parents' Guide to Protecting Your Children in Cyberspace* (New York: McGraw-Hill, 2000); Common Sense Media, http://www.commonsensemedia.org/; Dr. Michael Rich, Center on Media and Child Health, Children's Hospital, Boston, http://www.cmch.tv.

273 In this "crazy busy" world: E. Hallowell, *Crazy Busy: Overstretched, Overbooked, and About to Snap!* (New York: Ballantine, 2007). S. Newman, *Little Things Mean a Lot: Creating Happy Memories with Your Grandchildren* (New York: Crown, 1996).

275 Play is where children discover: M. Merrill, *Play Again: What Are the Consequences of a Childhood Removed from Nature?* documentary, playagainfilm.com.; Wagner, *Creating Innovators.*

279 These conversations are a place: K. Robinson, *Out of Our Minds: Learning to be Creative* (Oxford: Capstone, 2011; New York: Wiley, 2011).

279 Dinner first: S. Dominus, "Table Talk: The New Family Dinner," *New York Times*, 29 April 2012; M. Pipher, *Shelter of Our Families: Rebuilding Our Families* (New York: Ballantine, 1996).

281 Finally, we can even fine-tune: D. J. Siegel, *The Whole-Brain Child: 12 Revolutionary Strategies to Nurture Your Child's Developing Mind, Survive Everyday Parenting Struggles, and Help Your Family Thrive* (New York: Delacorte Press, 2011); Tough, *How Children Succeed.*

291 Nature, in particular, also reminds us: H. L. Burdette and R. C. Whitaker, "A National Study of Neighborhood Safety, Outdoor Play, Television Viewing, and Obesity in Preschool Children," *Pediatrics* 116, no. 3 (2005): 657–62; A. L. Philips, "A Walk in the Woods—Evidence Builds That Time Spent in the Natural World Benefits Human Health," *American Scientist* 99, no. 4 (2001): 301.

292 TV shows and documentaries: S. Kaplan, "The Restorative Benefits of Nature: Toward an Integrative Framework," *Journal of Environmental Psychology* 15 (1995): 169–82; Philips, "A Walk in the Woods"; K. Muldoon, "Animal Nature: New Map Brims with Great Spots to Watch Oregon Wildlife; Party Time at Bandon March," *The Oregonian*, 29 September 2011.

293 *New York Times* columnist Nicholas Kristof: N. Kristof, "Blissfully Lost in the Woods," *New York Times*, 28 July 2012, http://www.nytimes.com/2012/07/29/opinion/sunday/kristof-blissfully-lost-in-the-woods.html?_r=0.

Bibliography

Books

Aboujaoude, E. *Virtually You: The Dangerous Powers of the E-Personality*. New York: W. W. Norton, 2011.

Aftab, P. *The Parents' Guide to Protecting Your Children in Cyberspace*. New York: McGraw-Hill, 2000.

American Medical Association. *American Medical Association Complete Guide to Your Children's Health*. New York: Random House, 1999.

Ames, L. B. *Is Your Child in the Wrong Grade?* New Haven, CT: Gesell Institute, 1966.

Ames, L. B., and F. Ilg. *Your Five-Year-Old: Sunny and Serene*. New York: Dell, 1982.

———. *Your Four-Year-Old: Wild and Wonderful*. New York: Dell, 1982.

———. *Your Three-Year-Old: Friend or Enemy*. New York: Dell Trade Paperback, 1982.

Bauman, S. *Cyberbullying: What Counselors Need to Know*. Alexandria, VA: American Counseling Association, 2011.

Borba, M. *Building Moral Intelligence: The Seven Essential Virtues That Teach Kids to Do the Right Thing*. San Francisco: Jossey-Bass, 2001.

Bowlby, J. *The Making and Breaking of Affectional Bonds*. London: Tavistock, 1979.

Brazelton, T. B. *Touchpoints: The Essential Reference: Your Child's Emotional and Behavioral Development*. Reading, MA: Perseus Books, 1992.

Bronson, P., and A. Merryman. *Nurtureshock: New Thinking about Children*. New York: Hachette Book Group, 2009.

Brown J., ed., *Managing the Media Monster: The Influence of Media (from Television to Text Messages) on Teen Sexual Behavior and Attitudes*. Washington, DC: National Campaign to Prevent Teen and Unplanned Pregnancy, 2008.

Carlsson-Paige, N. *Taking Back Childhood: Helping Your Kids Thrive in a Fast-Paced, Media-Saturated, Violence-Filled World*. New York: Hudson Street Press, 2008.

Carr, N. *The Shallows: What the Internet Is Doing to Our Brains*. New York: W. W. Norton, 2010.

Christakis, D. A., and F. Zimmerman. *The Elephant in the Living Room: Make Television Work for Your Kid*. Emmaus, PA: Rodale Books, 2006.

Christakis, N. A., and J. Fowler. *Connected: How Your Friends' Friends' Friends Affect Everything You Feel, Think, and Do*. New York: Back Bay Books/Little Brown, 2009.

Chua, A. *Battle Hymn of the Tiger Mother*. New York: Penguin, 2011.

Chudacoff, H. *Children at Play: An American History*. New York: New York University Press, 2007.

Cohen, G. D. *The Creative Age: Awakening Potential in the Second Half of Life*. New York: HarperCollins, 2001.

———. *The Mature Mind: The Positive Power of the Aging Brain*. New York: Basic Books, 2006.

Coles, R. *The Moral Intelligence of Children*. New York: Random House Digital, 2011.

———. *The Moral Life of Children*. New York: Atlantic Monthly Press, 2000.

———. *The Spiritual Life of Children*. Boston: Houghton Mifflin, 1990.

Connor, B. *Unplugged Play: No Batteries. No Plugs. Pure Fun*. New York: Workman Publishing Company, 2007.

Crittenden, P. M. *Raising Parents: Attachment, Parenting and Child Safety*. Milton Park, Oxfordshire, UK: Willan Publishing, 2008.

Davidson, C. *Now You See It: How the Brain Science of Attention Will Transform the Way We Live, Work, and Learn*. New York: Viking Press, 2011.

Deak, J. *Girls Will Be Girls: Raising Confident, Courageous Daughters*. New York: Hyperion, 2003.

———. *Your Fantastic Elastic Brain*. Belvedere, CA: Little Pickle Press, 2011.

Diamond, M., and J. Hopson. *Magic Trees of the Mind: How to Nurture Your Child's Intelligence, Creativity, and Healthy Emotions from Birth through Adolescence*. New York: Plume, 1999.

Dweck, C. *Mindset: The New Psychology of Success*. New York: Random House, 2006.

Elkind, D. *All Grown Up and No Place to Go: Teenagers in Crisis*. Reading, MA: Addison-Wesley, 1984.

———. *The Power of Play: Learning What Comes Naturally*. Philadelphia: Da Capo Press, 2007.

Erikson, E. *Childhood and Society*. New York: Penguin, 1973.

Evans, F. B., III. *Harry Stack Sullivan*. London and New York: Routledge, 1996.

Farley-Gillispie, J., and J. Gackenbach. *cyber.rules: Negotiating Healthy Internet Use*. New York: W. W. Norton, 2007.

Farrell, W. *Father and Child Reunion: How to Bring the Dads We Need to the Children We Love*. New York: Penguin Putnam, 2001.

Flores, P. *Addiction as an Attachment Disorder*. Oxford, UK: Rowman and Littlefield, 2004.

Fogg Phillips, L., and B. Fogg. *Facebook for Parents: Answers to the Top 25 Questions*. New York: Wiley, 2012.

Fraiberg, S. *The Magic Years: Understanding and Handling the Problems of Early Childhood*. New York: Fireside, 1996.

Freidman, T., and M. Mandelbaum. *That Used to Be Us: How America Fell Behind in the World It Invented and How We Can Come Back*. New York: Farrar, Straus and Giroux, 2011.

Gallagher, W. *Rapt: Attention and the Focused Life*. New York: Penguin, 2009.

Gentile, D. *Media Violence and Children: A Complete Guide for Parents and Professionals*. Advances in Applied Developmental Psychology. New York: Praeger, 2003.

Goleman, D. *Emotional Intelligence: Why It Can Matter More Than IQ*. New York: Bantam Books, 2006.

Gottman, J., and J. Declaire. *Raising an Emotionally Intelligent Child: The Heart of Parenting*. New York: Simon and Schuster, 1998.

Greenspan, S., and S. Shakner. *The First Idea: How Symbols, Language, and Intelligence Evolved from Our Primate Ancestors to Modern Humans*. Cambridge, MA: Da Capo Press, 2004.

Guernsey, L. *Screen Time: How Electronic Media—From Baby Videos to Educational Software—Affects Your Young Child*. New York: Basic Books, 2007.

Gurian, M. *The Purpose of Boys: Helping Our Sons Find Meaning, Significance, and Direction in Their Lives*. San Francisco: Jossey-Bass, 2010.

Hallowell E. *The Childhood Roots of Adult Happiness: Five Steps to Help Kids Create and Sustain Lifelong Joy*. New York: Ballantine Books, 2002.

———. *Connect: 12 Vital Ties That Open Your Heart, Lengthen Your Life, and Deepen Your Soul*. New York: Pocket Books, 1999.

Hallowell, E. *Crazy Busy: Overstretched, Overbooked, and About to Snap! Strategies for Handling Your Fast-Paced Life*. New York: Ballantine, 2006.

Hallowell E., and J. Ratey. *Driven to Distraction: Recognizing and Coping with Attention Deficit Disorder from Childhood through Adulthood*. New York: Anchor Books, 1994.

Harris, R., and M. Emberley. *It's Not the Stork: A Book about Girls, Boys, Babies, Bodies, Family, and Friends*. Somerville, MA: Candlewick Press, 2008.

———. *It's Perfectly Normal: Changing Bodies, Growing Up, Sex and Sexual Health*. Somerville, MA: Candlewick Press, 2004.

————. *It's So Amazing!: A Book about Eggs, Sperm, Birth, Babies, and Families*. Somerville, MA: Candlewick Press, 2004.

Harris, R., and N. Westcott. *Who Has What? All About Girls' Bodies, Boys' Bodies (Let's Talk About You and Me)*. Somerville, MA: Candlewick Press, 2011.

Hedges, C. *Empire of Illusion: The End of Literacy and the Triumph of Spectacle*. New York: Nation Books, 2009.

Hochschild, A. R. *The Time Bind: When Work Becomes Home and Home Becomes Work*. New York: Henry Holt, 1997.

Ito, M., et al. *Hanging Out, Messing Around, and Geeking Out: Kids Living and Learning with New Media*. Cambridge, MA: MIT Press, 2010.

Jenkins, H. *Convergence Culture: Where Old and New Media Collide*. New York: New York University Press, 2006.

Katz, J. *The Macho Paradox: Why Some Men Hurt Women and How All Men Can Help*. Naperville, IL: Sourcebooks, 2006.

Kaye, K., et al., *The Fog Zone: How Misperceptions, Magical Thinking, and Ambivalence Put Young Adults at Risk for Unplanned Pregnancy*. Washington, DC: National Campaign to Prevent Teen and Unplanned Pregnancy, 2009.

Kelly, K. *What Technology Wants*. New York: Viking, 2010.

Kessler, L. *My Teenage Werewolf: A Mother, a Daughter, a Journey through the Thicket of Adolescence*. New York, Penguin, 2010.

Kindlon, D., and M. Thompson, with T. Barker. *Raising Cain: Protecting the Emotional Life of Boys*. New York: Ballantine Books, 2000.

Kowalski, R., S. Limber, and P. Agatston. *Cyber Bullying: Bullying in the Digital Age*. Malden, MA: Blackwell, 2008.

Kristof, N., and S. WuDunn. *Half the Sky: Turning Oppression into Opportunity for Women Worldwide*. New York: Alfred A. Knopf, 2009.

Kutner, L., and C. Olson. *Grand Theft Childhood: The Surprising Truth about Violent Video Games and What Parents Can Do*. New York: Simon and Schuster, 2008.

Lamb, S., and L. M. Brown. *Packaging Girlhood: Rescuing Our Daughters from Marketers' Schemes*. New York: St. Martin's Press, 2006.

Lamb, S., L. M. Brown, and M. Tappen. *Packaging Boyhood: Saving Our Sons from Superheroes, Slackers, and Other Media Stereotypes*. New York: St. Martin's Press, 2009.

Lasch, C. *The Culture of Narcissism: American Life in an Age of Diminishing Expectations*. Revised edition. New York: W. W. Norton, 1991.

Leach, P. *Your Baby and Child from Birth to Age 5*. New York: Alfred A. Knopf, 2003.

Levin, D., and J. Kilbourne. *So Sexy So Soon: The New Sexualized Childhood and What Parents Can Do to Protect Their Kids*. New York: Ballantine Books, 2008.

Levine, M. *The Price of Privilege: How Parental Pressure and Material Advantage Are Creating a Generation of Disconnected and Unhappy Kids.* New York: HarperCollins, 2006.

———. *Teach Your Children Well: Parenting for Authentic Success.* New York: Harper-Collins, 2012.

Linn, S. *The Case for Make Believe: Saving Play in a Commercialized World.* New York: The New Press/W. W. Norton, 2008.

———. *Consuming Kids: Protecting Our Children from the Onslaught of Marketing and Advertising.* New York: Anchor Books/Random House, 2004.

Louv, R. *Last Child in the Woods: Saving Our Children from Nature Deficit Disorder.* New York: Algonquin Books, 2008.

Marano, E. H. *A Nation of Wimps: The High Cost of Invasive Parenting.* New York: Broadway Books, 2008.

McConville, M. *Adolescence: Psychotherapy and the Emergent Self.* San Francisco: Jossey-Bass, 1995.

Mogel, W. *The Blessing of a Skinned Knee: Using Jewish Teachings to Raise Self-Reliant Children.* New York: Scribner, 2008.

Murphy, C. *Are We Rome? The Fall of an Empire and the Fate of America.* New York: Houghton Mifflin, 2007.

Nass, C., and C. Yen. *The Man Who Lied to His Laptop: What Machines Teach Us about Human Relationships.* New York: Penguin, 2010.

Newman, S. *Little Things Mean a Lot: Creating Happy Memories with Your Grandchildren.* New York: Crown, 1996.

Norton, B. G. *Sustainability: A Philosophy of Adaptive Ecosystem Management.* Chicago: University of Chicago Press, 2005.

Orenstein, P. *Cinderella Ate My Daughter: Dispatches from the Front Lines of the New Girlie-Girl Culture.* New York: HarperCollins. 2011.

Paley, V. G. *You Can't Say You Can't Play.* Cambridge, MA: Harvard University Press, 1992.

Palfrey, J., and U. Gasser. *Born Digital: Understanding the First Generation of Digital Natives.* New York: Basic Books, 2008.

Pipher, M. *Shelter of Our Families: Rebuilding Our Families.* New York: Ballantine Books, 1996.

Piran, N., M. Levine, and C. Steiner-Adair. *Preventing Eating Disorders: A Handbook of Interventions and Special Challenges.* Philadelphia: Brunner/Mazel, 1999.

Pollock, W. *Real Boys: Rescuing Our Sons from the Myths of Boyhood.* New York: Henry Holt, 1999.

Ponton, L. *The Romance of Risk: Why Teenagers Do the Things They Do.* New York: Basic Books, 1997.

Pope, D. *Doing School: How We Are Creating a Generation of Stressed-Out, Materialistic, and Miseducated Students*. New Haven, CT: Yale University Press, 2003.

Postman, N. *The Disappearance of Childhood*. New York: Delacorte Press, 1982.

———. *The Disappearance of Childhood*. New York: Vintage Books, 1994.

Powers, W. *Hamlet's Blackberry: A Practical Philosophy for Building a Good Life in the Digital Age*. New York: HarperCollins, 2010.

Rich, H., and T. Barker. *In the Moment: Celebrating the Everyday*. New York: HarperCollins, 2002.

Robinson, K. *Out of Our Minds: Learning to Be Creative*. Oxford, UK: Capstone, 2011; New York: Wiley, 2011.

Rosen, L. *iDisorder: Understanding Our Obsession with Technology and Overcoming Its Hold on Us*. New York: Palgrave Macmillan, 2012.

Rothman, D. *Sex and Sensibility: The Thinking Parent's Guide to Talking Sense About Sex*. Cambridge, MA: Da Capo Press, 2001.

Schor, J. *Born to Buy: The Commercialized Child and the New Consumer Culture*. New York: Scribner, 2004.

Schulman, N., and E. Birnbaum. *Practical Wisdom for Parents: Demystifying the Preschool Years*. New York: Alfred A. Knopf, 2007.

———. E. *Practical Wisdom for Parents: Raising Self-Confident Children in Preschool Years*. New York: Random House, 2007.

Seligman, M. *Flourish: A Visionary New Understanding of Happiness and Well-Being*. New York: Free Press, 2011.

———. *The Optimistic Child: Proven Program to Safeguard Children from Depression and Build Lifelong Resilience*. New York: Houghton Mifflin, 1995.

Siegel, D. J. *The Developing Mind: How Relationships and the Brain Interact to Shape Who We Are*. 2nd ed. New York: Guilford Press, 2012.

———. *The Mindful Brain: Reflection and Attunement in the Cultivation of Well-Being*. New York: W. W. Norton, 2007.

———. *Mindsight: The New Science of Personal Transformation*. New York: Bantam Books, 2011.

Siegel, D. J., and T. P. Bryson, *The Whole-Brain Child: 12 Revolutionary Strategies to Nurture Your Child's Developing Mind, Survive Everyday Parenting Struggles, and Help Your Family Thrive*. New York: Delacorte Press, 2011.

Siegel, D. J., and M. Hartzell. *Parenting from the Inside Out: How a Deeper Self-Understanding Can Help You Raise Children Who Thrive*. New York: Jeremy P. Tarcher/Penguin, 2004.

Steiner-Adair, C., and L. Sjostrom. *Full of Ourselves: A Wellness Program to Advance Girl Power, Health, and Leadership*. New York: Teacher's College Press, 2006.

Stern, D. *The Interpersonal World of the Infant*. New York: Basic Books, 1987.

Steyer, J. *The Other Parent: The Inside Story of the Media's Effect on Our Children.* New York: Atria, 2002.

————. *Talking Back to Facebook: The Common Sense Guide to Raising Kids in the Digital Age.* New York: Scribner, 2012.

Taffel, R. *Parenting by Heart: How to Stay Connected to Your Child in a Disconnected World.* Cambridge, MA: Perseus Publishing, 2011.

Thomas, M., and O. G. Catherine, with L. J. Cohen. *Best Friends, Worst Enemies: Understanding the Social Lives of Children.* New York: Ballantine Books, 2001.

Thompson, M. *Homesick and Happy: How Time Away from Parents Can Help a Child Grow.* Boston: Ballantine Books, 2012.

Thompson, M., and T. Barker. *It's a Boy! Understanding Your Son's Development from Birth to Eighteen.* Boston: Ballantine Books, 2008.

————. *Speaking of Boys: Answers to the Most-Asked Questions about Raising Sons.* Boston: Ballantine Books, 2000.

————. *The Pressured Child: Helping Your Child Find Success in School and Life.* New York: Ballantine Books, 2004.

Thompson, M., and L. Cohen, with C. O'Neill Grace. *Mom, They're Teasing Me: Helping Your Child Solve Social Problems.* Boston: Ballantine Books, 2002.

Tough, P. *How Children Succeed: Grit, Curiosity, and the Hidden Power of Character.* New York: Houghton Mifflin Harcourt, 2012.

Turkle, S. *Alone Together: Why We Expect More from Technology and Less from Each Other.* New York: Basic Books, 2011.

Wagner, T. *Creating Innovators: The Making of Young People Who Will Change the World.* New York: Scribner, 2012.

————. *The Global Achievement Gap: Why Even Our Best Schools Don't Teach the New Survival Skills Our Children Need—And What We Can Do about It.* New York: Basic Books, 2008.

Walsh, D. *Why Do They Act That Way? A Survival Guide to the Adolescent Brain for You and Your Teen.* New York: Free Press, 2004.

Wexler, B. E. *Brain and Culture: Neurobiology, Ideology, and Social Change.* Cambridge, MA: MIT Press, 2006.

Winnicott, D. W. *Playing and Reality.* New York: Basic Books, 1971.

Wiseman, R. *Queen Bees and Wannabes: Helping Your Daughter Survive Cliques, Gossip, Boyfriends, and Other Realities of Adolescence.* New York: Crown, 2001.

Wolf, M. *Proust and the Squid: The Story and Science of the Reading Brain.* New York: HarperCollins, 2007.

Wood, C. *Yardsticks: Children in the Classroom Ages 4–14.* 3rd ed. Turner Falls, MA: Northeast Foundation for Children, 2007.

Journal Articles and Contributed Articles

American Academy of Pediatrics, Committee on Communications. "Children, Adolescents and Advertising." *Pediatrics* 118, no. 6 (2006): 2562–69.

American Academy of Pediatrics, Committee on Public Education. "Children, Adolescents and Television." *Pediatrics* 107, no. 2 (2001): 423–26.

———. "Media and Violence." *Pediatrics* 108, no. 5 (2001): 1222–26.

Anderson, C. A., L. Berkowitz, E. Donnerstein, L. R. Huesmann, J. D. Johnson, D. Linz, N. M. Malamuth, and E. Wartella. "The Influence of Media Violence on Youth." *Psychological Science in the Public Interest* 4, no. 3 (2003): 81–110.

Bailey, K., R. West, and C. Anderson. "The Influence of Video Games on Social, Cognitive, and Affective Information Processing." In *Handbook of Social Neuroscience*, edited by J. Decety and J. Cacioppo (New York: Oxford University Press, in press).

Baranowski, T., D. Abdelsamad, J. Baranowski, M. T. O'Connor, D. Thompson, A. Barnett, E. Cerin, and T. A. Chen. "Impact of an Active Video Game on Healthy Children's Physical Activity," *Pediatrics* 129 (2012): e636–42.

Bensley, L., and J. Van Eenwyk. "Video Games and Real-Life Aggression," review of literature. *Journal of Adolescent Health* 29, no. 4 (2001): 244–57.

Bickham, D. S., and M. Rich. "Is Television Viewing Associated with Social Isolation?" *Archives of Pediatrics and Adolescent Medicine* 160, no. 4 (2006): 387–92.

Biegler, S., and D. Boyd. "Risky Behaviors and Online Safety: A 2010 Literature Review." Berkman Center for Internet and Society, Harvard University, 4 November 2010, http://cyber.law.harvard.edu/research/youthandmedia/digitalnatives.

Brasel, A., and J. Gips. "Media Multitasking Behavior: Concurrent Television and Computer Usage." *Cyberpsychology, Behavior, and Social Networking*, 15 March 2011, http://www.liebertonline.com/doi/pdfplus/10.1089/cyber.2010.0350.

Bryant, J., and D. Brown. "Use of Pornography." In *Pornography: Research Advances and Policy Considerations*, edited by D. Zillmann and J. Bryant (Hillsdale, NJ: Erlbaum, 1989), 25–55.

Buhi, E. R., et al. "Quality and Accuracy of Sexual Health Information Web Sites Visited by Young People." *Journal of Adolescent Health* 47, no. 2 (2010): 206–8.

Burdette, H. L., and R. C. Whitaker. "A National Study of Neighborhood Safety, Outdoor Play, Television Viewing, and Obesity in Preschool Children." *Pediatrics* 116, no. 3 (2005): 657–62.

Byun, S., C. Ruffini, J. Mills, A. Douglas, M. Niang, S. Stepchenkova, S. K. Lee, J. Loutfi, J. K. Lee, M. Atallah, and M. Blanton. "Internet Addiction: Metasynthesis of 1996–2006 Quantitative Research." *Cyberpsychology, Behavior, and Social Networking* 12, no. 2 (2010): 203–7.

Carson, V., W. Pickett, and I. Janssen. "Screen Time and Risk Behaviors in 10- to 16-Year-Old Canadian Youth." *Preventive Medicine* 52, no. 2 (2011): 97–98.

Chandra, A., et al. "Does Watching Sex on Television Predict Teen Pregnancy? Findings from a National Longitudinal Survey of Youth." *Pediatrics* 122, no. 5 (2008): 1047–54.

Choliz, M., and C. Marco. "Patterns of Video Game Use and Dependence in Children and Adolescents." *Anales de Psicologia* 27, no. 2 (2011): 418–26.

Christakis, D. A. "Effect of Block Play on Language Acquisition and Attention in Toddlers: A Pilot Randomized Controlled Trial." *Archives of Pediatric and Adolescent Medicine* 161, no. 10 (2007): 967–71.

———. "The Effects of Fast-Paced Cartoons." *Pediatrics* 128, no. 4 (2011): 772–74.

Christakis, D. A., and F. J. Zimmerman. "Violent Television during Preschool Is Associated with Antisocial Behavior during School Age." *Pediatrics* 120, no. 5 (2007): 993–99.

Cohn, J. F., R. Matias, E. Z. Tronick, D. Connell, and K. Lyons-Ruth. "Face-to-Face Interactions of Depressed Mothers and Their Infants." *New Directions for Child and Adolescent Development* 1986, no. 34 (2006): 31–45.

Conners-Burrow, N. A., L. M. McKelvey, and J. J. Fussell. "Social Outcomes Associated with Media Viewing Habits of Low-Income Preschool Children." *Early Education and Development* 22, no. 2 (2011): 256–73.

Cook, C., N. Goodman, and L. E. Schulz. "Where Science Starts: Spontaneous Experiments in Preschoolers' Exploratory Play." *Cognition* 120, no. 3 (2011): 341–49.

Council on Communications and Media and A. Brown. "Media Use by Children Younger Than 2 Years." *Pediatrics* 128, no. 5 (2011): 1040–45.

Coyne, S. M., L. A. Stockdale, D. A. Nelson, and A. Fraser. "Profanity in Media Associated with Attitudes and Behavior Regarding Profanity Use and Aggression." *Pediatrics* 128, no. 5 (2011): 867–72.

Cramer, P., and T. Steinwert. "Thin Is Good, Fat Is Bad: How Early Does It Begin?" *Journal of Applied Developmental Psychology* 19 (1998): 429–51.

Denham, S. A., and R. P. Weissberg. "Social-Emotional Learning in Early Childhood: What We Know and Where to Go from Here." Collaborative for Academic, Social and Emotional Learning, 2004, http://casel.org/publications/social-emotional-learning-in-early-childhood-what-we-know-and-where-to-go-from-here/.

———. "Social-Emotional Learning in Early Childhood: What We Know and Where to Go from Here?" In *A Blueprint for the Promotion of Prosocial Behavior in Early Childhood*, edited by E. Chesebrough, P. King, T. P. Gullotta, and M. Bloom (New York: Kluwer Academic/Plenum, 2004), 13–50.

Dill, K., and K. Thill. "Video Game Characters and the Socialization of Gender Roles: Young People's Perceptions Mirror Sexist Media." *Business Media* 57 (2007): 851–64.

Din, F., and J. Calao. "The Effects of Playing Educational Video Games on Kindergarten Achievement." *Child Study Journal* 31, no. 2 (2001): 95–102.

Drews, F. M., M. Pasupathi, and D. L. Strayer. "Passenger and Cell Phone Conversations in Simulated Driving." *Journal of Experimental Psychology* 14, no. 2 (2008): 392–400.

Feng, D., D. B. Reed, M. C. Esperat, and M. Uchida. "Effects of TV in the Bedroom on Young Hispanic Children." *American Journal of Health Promotion* 25, no. 5 (2011): 310–18.

Field, T., D. Sandberg, R. Garcia, N. Vega-Lahr, S. Goldstein, and L. Guy. "Pregnancy Problems, Postpartum Depression, and Early Mother-Infant Interactions." *Developmental Psychology* 21, no. 6 (1985): 1152–56.

Fioravanti, G., D. Dèttore, and S. Casale. "Adolescent Internet Addiction: Testing the Association between Self-Esteem, the Perception of Internet Attributes, and Preference for Online Social Interactions." *Cyberpsychology, Behavior, and Social Networking* 15, no. 6 (2012): 318–23.

Gandhi, O. P., L. L. Morgan, A. A. de Salles, Y. Y. R. B. Herberman, and D. L. Davis. "Exposure Limits: The Underestimation of Absorbed Cell Phone Radiation, Especially in Children." *Electromagnetic Biology and Medicine* 31 (2011): 1–18.

Gentile, D. A., H. Choo, A. Liau, T. Sim, D. Li, D. Fung, and A. Khoo. "Pathological Video Game Use among Youths: A Two-Year Longitudinal Study." *Pediatrics* 127, no. 2 (2011): e319–29.

Gentile, D. A., P. Lynch, J. Linder, and D. Walsh. "The Effects of Violent Video Game Habits on Adolescent Hostility, Aggressive Behaviors, and School Performance." *Journal of Adolescence* 27 (2004): 5–22.

Hancox, R. J., B. J. Milne, and R. Poulton. "Association of Television during Childhood with Poor Educational Achievement." *Archives of Pediatric and Adolescent Medicine* 159, no. 7 (2005): 614–18.

Harrison, K., and N. Martins. "Racial and Gender Differences in the Relationship between Children's Television Use and Self-Esteem: A Longitudinal Panel Study." *Communication Research* 39 (2012): 338.

Hartmann, W., and G. Brougere. "Toy Culture in Preschool Education and Children's Toy Preferences." In *Toys, Games and Media,* edited by J. Goldstein, D. Buckingham, and G. Brougere (Mahwah, NJ: Erlbaum, 2004), 37–53.

He, J. B., C. J. Liu, Y. Y. Guo, and L. Zhao. "Deficits in Early-Stage Face Perception in Excessive Internet Users." *Cyberpsychology, Behavior, and Social Networking* 14, no. 5 (2009): 303–308.

Herrett-Skjellum, J., and M. Allen. "Television Programming and Sex Stereotyping: A Meta-Analysis." In *Communication Yearbook* 19, edited by B. R. Burleson (Thousand Oaks, CA: Sage, 1995), 157–85.

Insel, T. R., and L. J. Young. "The Neurobiology of Attachment." *Nature Reviews Neuroscience* 2 (2001): 129–36.

Joiner, R., J. Gavin, M. Brosnan, J. Cromby, H. Gregory, J. Guiller, P. Maras, and A. Moon. "Gender, Internet Experience, Internet Identification, and Internet Anxiety: A Ten-Year

Follow Up." *Cyberpsychology, Behavior, and Social Networking* 15, no. 7 (2012): 370–72.

Jostmann, N., D. Lakens, and T. Schubert. "Weight as an Embodiment of Importance." *Psychological Science* 20, no. 9 (2009): 1169–74.

Kaplan, S. "The Restorative Benefits of Nature: Toward an Integrative Framework." *Journal of Environmental Psychology* 15 (1995): 169–82.

Keung, M. H. "Internet Addiction and Antisocial Internet Behavior of Adolescents." *Scientific World Journal* 3 (2011): 2187–96.

Kittinger, R., C. J. Correia, and J. G. Irons. "Relationship between Facebook Use and Problematic Internet Use Among College Students." *Cyberpsychology, Behavior, and Social Networking* 15, no. 6 (2012): 324–27.

Konrath, S., E. O'Brian, and C. Hsing. "Changes in Dispositional Empathy in American College Students over Time: A Meta-Analysis." *Personality and Social Psychology Review* 15, no. 2 (2011): 180–98.

Lam, T. L., and Z.-W. Peng. "Effect of Pathological Use of the Internet on Adolescent Mental Health: A Prospective Study." *Archives of Pediatrics and Adolescent Medicine* 164, no. 10 (2010): 901–906.

Lane, S. J., and R. C. Schaaf. "Examining the Neuroscience Evidence for Sensory-Driven Neuroplasticity: Implications for Sensory-Based Occupational Therapy for Children and Adolescents." *American Journal of Occupational Therapy* 64, no. 3 (2010): 375–90.

Lee, H. W., J. S. Choi, Y. C. Shin, J. Y. Lee, H. Y. Jung, and J. S. Kwon. "Impulsivity in Internet Addiction: A Comparison with Pathological Gambling." *Cyberpsychology, Behavior, and Social Networking* 15, no. 7 (2012): 373–77.

Lee, S. J., and Y. J. Chae. "Balancing Participation and Risks in Children's Internet Use: The Role of Internet Literacy and Parental Mediation." *Cyberpsychology, Behavior, and Social Networking* 15, no. 5 (2011): 257–62.

Luthar, S. S. "The Culture of Affluence: Psychological Costs of Material Wealth." *Child Development* 74 (2003): 1581–93.

Luthar, S. S., and B. E. Becker. "Privileged but Pressured? A Study of Affluent Youth." *Child Development* 73, no. 5 (2002): 1593–1610.

Maguire, E. A., R. Frackowiak, and C. Frith. "Recalling Routes around London: Activation of the Right Hippocampus in Taxi Drivers." *Journal of Neuroscience* 17, no. 18 (1997): 7103–10.

McNally, M. A., D. Crocetti, M. E. Mahone, M. B. Denckla, S. J. Suskauer, and S. H. Mostofsky. "Corpus Callosum Segment Circumference Is Associated with Response Control in Children with Attention-Deficit Hyperactivity Disorder (ADHD)." *Journal of Child Neurology* 25, no. 4 (2010): 453–62.

Mentzoni, R. A., G. S. Brunborg, H. Molde, H. Myrseth, K. J. Mar Skouveroe, J. Hetland, and S. Pallesen. "Problematic Video Game Use: Estimated Prevalence and Associations with

Mental and Physical Health." *Cyberpsychology, Behavior, and Social Networking* 14, no. 10 (2011): 591–96.

Miles, L., L. Nink, and C. N. Macrae. "Moving through Time." *Psychological Science* 21, no. 2 (2012): 222–23.

Mistry, K., C. Minkovitz, D. Strobino, and D. Borzekowski. "Children's Television Exposure and Behavioral and Social Outcomes at 5.5 Years: Does Timing of Exposure Matter?" *Pediatrics* 120, no. 4 (2007): 762–69.

Mossle, T., M. Kleimann, F. Rehbein, and C. Pfeiffer. "Media Use and School Achievement—Boys at Risk?" *British Journal of Developmental Psychology* 28, no. 3 (2010): 699–725.

O'Keeffe, S., and K. Clarke-Pearson. "Clinical Report—the Impact of Social Media on Children, Adolescents, and Families." *Pediatrics* 127 (2011): 800–804, posted online 28 March 2011, http://pediatrics.aappublications.org/content/127/4/800.full.pdf.

Paavonen, E. J., M. Pennonen, and M. Roine. "Passive Exposure to TV Linked to Sleep Problems in Children." *Journal of Sleep Research* 15 (2006): 154–61.

Pagani, L. S., M. A. Fitzpatrick, T. A. Barnett, and E. Dubow. "Prospective Associations between Early Childhood Television Exposure and Academic, Psychosocial, and Physical Well-Being by Middle Childhood." *Archives of Pediatric and Adolescent Medicine* 164, no. 5 (2010): 425–31.

Pellegrini, A. D., and C. M. Bohn. "The Role of Recess in Children's Cognitive Performance and School Adjustment." *Educational Researcher* 34, no. 1 (2005): 13–19.

Philips, A. L. "A Walk in the Woods—Evidence Builds That Time Spent in the Natural World Benefits Human Health." *American Scientist* 99, no. 4 (2001): 301.

Rich, M. "Boy, Mediated: Effects of Entertainment Media on Adolescent Male Health." *Adolescent Medicine State of the Art Reviews* 14 (2003): 691–713.

———. "Health Literacy via Media Literacy: Video Intervention/Prevention Assessment." *American Behavioral Scientist* 48 (2004): 165–88.

Rich, M., and M. Bar-on. "Child Health in the Information Age: Media Education of Pediatricians." *Pediatrics* 107, no. 1 (2001): 156–62.

Rich, M., S. Lamola, J. Gordon, and R. Chalfen. "Video Intervention/Prevention Assessment: A Patient-Centered Methodology for Understanding the Adolescent Illness Experience." *Journal of Adolescent Health* 27 (2000): 155–65.

Rich, M., E. R. Woods, E. Goodman, S. J. Emans, and R. H. DuRant. "Aggressors or Victims: Gender and Race in Music Video Violence." *Pediatrics* 101, no. 4 (1998): 669–74.

Robinson, J. P., and S. Martin. "What Do Happy People Do?" *Journal of Social Indicators Research* 89 (2008): 565–71.

Robinson, T. "Reducing Children's Television Viewing to Prevent Obesity." *JAMA* 282, no. 16 (1999): 1561–67.

Rowan, C. "Unplug—Don't Drug: A Critical Look at the Influence of Technology on Child

Behavior with an Alternative Way of Responding Other Than Evaluation and Drugging." *Ethical Human Psychology and Psychiatry* 12, no. 1 (2010): 60–67.

Sabina, C., J. Wolak, and D. Finkelhor. "The Nature and Dynamics of Internet Pornography Exposure for Youth." *Cyberpsychology, Behavior, and Social Networking* 11, no. 6 (2008): 691–93.

Schlaug, G., M. Forgeard, L. Zhu, A. Norton, and E. Winner. "Training-Induced Neuroplasticity in Young Children." *Annals of the New York Academy of Sciences* 1169 (2009): 205–8.

Schmidt, M. E., M. Rich, S. L. Rifas-Shiman, E. Oken, and E. L. Taveras. "Television Viewing in Infancy and Child Cognition at 3 Years of Age in a US Cohort." *Pediatrics* 123, no. 3 (2009): e370–75.

Seligman, M. "Chronic Fear Produced by Unpredictable Electric Shock." *Journal of Comparative and Physiological Psychology* 66 (1968): 402–11.

Shapiro, S., M. Newcomb, and T. B. Loeb. "Fear of Fat, Disregulated-Restrained Eating, and Body-Esteem: Prevalence and Gender Differences among Eight- to Ten-Year-Old Children." *Journal of Clinical Child Psychology* 26 (1997): 358–65.

Siegel, D. J. "An Interpersonal Neurobiology Approach to Psychotherapy: How Awareness, Mirror Neurons and Neural Plasticity Contribute to the Development of Well-Being." *Psychiatric Annals* 36, no. 4 (2006): 248–58.

Soifer, L. "Development of Oral Language and Its Relationship to Literacy." In *Multi-Sensory Teaching of Basic Language Skills*, edited by Judith Birsh (Baltimore: Paul H. Brooks, 1999).

Steffgen, G., M. S. Konig, J. Pfetsch, and A. Melzer. "Are Cyberbullies Less Empathic? Adolescents' Cyberbullying Behavior and Empathic Responsiveness." *Cyberpsychology, Behavior, and Social Networking* 14, no. 11 (2011): 643–48, posted online 9 May 2011, http://www.liebertonline.com/doi/abs/10.1089/cyber.2010.0445.

Steiner-Adair, C. "Got Grit? The Call to Educate Smart, Savvy and Socially Intelligent Kids." *Independent School* (Winter 2013).

———. "When the Body Speaks: Girls, Eating Disorders, and Psychotherapy." In *Women, Girls and Psychotherapy: Reframing Resistance*, edited by C. Gilligan, A. Rogers, and D. Tolman (Binghamton, NY: Haworth Press, 1991). First published in *Journal of Women and Therapy* 11, nos. 3 and 4 (1991).

Strassberg, D. S., R. K. McKinnon, M. A. Sustaita, and J. Rullo. "Sexting by High School Students: An Exploratory and Descriptive Study." *Archives of Sexual Behaviour* 42, no. 1 (2012): 15–21.

Swanson, J. M., G. R. Elliot, L. L. Greenhill, T. Wigal, L. E. Arnold, M. Vitiello, L. Hechtman, J. N. Epstein, W. E. Pelham, B. Abikoff, J. H. Newcorn, B. S. G. Molina, S. G. Hinshaw, E. L. Swing, D. A. Gentile, C. A. Anderson, and D. A. Walsh. "Television and Video Game Exposure and the Development of Attention Problems." *Pediatrics* 126 (2010): 214–21.

Uhls, Y. T., and P. Greenfield. "Rise of Fame: A Historical Content Analysis," *Journal of Psychosocial Research on Cyberspace* 5, no. 1 (2011): article 1, http://www.cyberpsychology.eu/view.php?cisloclanku=2011061601.

Vandewater , E. A., et al. "Digital Childhood: Electronic Media and Technology Use among Infants, Toddlers, and Preschoolers." *Pediatrics* 119, no. 5 (2007): e1006–15.

Vandewater, E. A., D. S. Bickham, J. H. Lee, H. M. Cummings, E. A. Wartella, and V. J. Rideout. "When the Television Is Always On: Heavy Television Exposure and Young Children's Development." *American Behavioral Scientist* 48 (2005): 562–77.

Vandewater, E. A., J. H. Lee, and M. Shim. "Family Conflict and Violent Electronic Media Use in School-Aged Children." *Media Psychology* 7 (2005): 73–86.

Weisskirch, R. S. "No Crossed Wires: Cell Phone Communication in Parent-Adolescent Relationships." *Cyberpsychology, Behavior, and Social Networking* 14, nos. 7–8 (2005): 447–51.

Williams, L. E. "Experiencing Physical Warmth Promotes Interpersonal Warmth." *Science* 322 (2008): 606–607.

Wilson, B. "Media and Children's Aggression, Fear and Altruism." *The Future of Children* 18, no. 1 (2008), http://futureofchildren.org/publications/journals/article/index.xml?journalid=32&articleid=58&sectionid=270.

Winerman, L. "Playtime in Peril." *American Psychological Association* 40, no. 8 (2009): 50.

Wolak, J., K. Mitchell, and D. Finkelhor. "Unwanted and Wanted Exposure to Online Pornography in a National Sample of Youth Internet Users." *Pediatrics* 119, no. 2 (2007): 247–57, http://pediatrics.aappublications.org/cgi/reprint/119/2/247.

Xiuqin, H., Z. Huimin, L. Mengchen, W. Jinan, Z. Ying, and T. Ran. "Mental Health, Personality, and Parental Rearing Styles of Adolescents with Internet Addiction Disorder." *Cyberpsychology, Behavior, and Social Networking* 13, no. 4 (2010): 401–6.

Ybarra, M. L., and K. J. Mitchell. "Exposure to Internet Pornography among Children and Adolescents: A National Survey." *Cyberpsychology, Behavior, and Social Networking* 8, no. 5 (2005): 473–82.

Yen, J. Y., C. F. Yen, C. S. Chen, T. C. Tang, and C. H. Ko. "The Association between Adult ADHD Symptoms and Internet Addiction among College Students: The Gender Difference." *Cyberpsychology, Behavior, and Social Networking* 12, no. 2 (2009): 187–91.

Zimmerman, F., and D. Christakis. "Associations between Content Types of Early Media Exposure and Subsequent Attentional Problems." *Pediatrics* 120, no. 5 (2007): 986–92.

Reports

Almon, J., and Edward Miller. "Crisis in Early Education: A Research-Based Case for More Play and Less Pressure." Alliance for Childhood, November 2011.

American Academy of Pediatrics. "Joint Statement on the Impact of Entertainment Violence on Children: Congressional Public Health Summit." 26 July 2000.

Anderson, J. "Millennials Will Benefit and Suffer Due to Their Hyperconnect Lives." Pew

Internet, Pew Internet and American Life Project, 29 February 2012, http://www.pewinter net.org/Reports/2012/Hyperconnected-lives.aspx.

Borse, N. N., J. Gilchrist, A. M. Dellinger, R. A. Rudd, M. F. Ballesteros, and D. A. Sleet. *Childhood Injury Report: Patterns of Unintentional Injuries among 0–19 Year Olds in the United States, 2000–2006.* Centers for Disease Control and Prevention, National Center for Injury Prevention and Control. Atlanta: Centers for Disease Control, 2008.

Brenner, J. "Pew Internet: Teens." Pew Internet and American Life Project, 27 April 2012, http://pewinternet.org/Commentary/2012/April/Pew-Internet-Teens.aspx.

Center on Media and Child Health, Children's Hospital Boston. "The Effects of Electronic Media on Children Ages Zero to Six: A History of Research." Henry J. Kaiser Family Foundation Issue Brief. Menlo Park, CA: Henry J. Kaiser Family Foundation, January 2005.

"Distracted Driving and Driver, Roadway, and Environmental Factors." National Highway Traffic Safety Administration, September 2010.

Flood, M., and C. Hamilton. "Youth and Pornography in Australia: Evidence on the Extent of Exposure and Likely Effects." Australia Institute, Discussion Paper 52, February 2003, http://thesocietypages.org/socimages/files/2009/10/1-499x387.gif.

Henry J. Kaiser Family Foundation. *Kids and Media at the New Millennium: A Kaiser Family Foundation Report.* Menlo Park, CA: Henry J. Kaiser Family Foundation, 1999.

———. *Zero to Six: Electronic Media in the Lives of Infants, Toddlers and Preschoolers.* Menlo Park, CA: Henry J. Kaiser Family Foundation, 2003.

Henry J. Kaiser Family Foundation, Program for the Study of Media and Health. "Media Multitasking among American Youth: Prevalence, Predictors, and Pairings." 12 December 2006, http://kff.org/entmedia/7592.cfm.

Lenhart, A. "Protecting Teens Online." Pew Internet: Pew Internet and American Life Project, 17 March 2005, http://www.pewinternet.org/Reports/2005/Protecting-Teens-Online .aspx.

———. "Teens, Cell Phones, and Texting: Text Messaging Becomes Centerpiece Communication." Pew Internet and American Life Project, April 2010, http://pewresearch.org/ pubs/1572/teens-cell-phones-text-messages.

Lenhart, A., R. Ling, S. Campbell, and K. Purcell. "Teens and Mobile Phones." Pew Internet, Pew Internet and American Life Project, 20 April, 2012, http://pewinternet.org/Re ports/2010/Teens-and-Mobile-Phones/Summary-of-findings/Findings.aspx.

Living and Learning with New Media: Summary of Findings from the Digital Youth Project. John D. and Catherine T. MacArthur Foundation Reports on Digital Media and Learning. Cambridge, MA: MIT Press, 2008.

Madden, M., and A. Lenhart. "Teens and Distracted Driving: Texting, Talking and Other Uses of the Cell Phone behind the Wheel." Pew Internet and American Life Project, 16 November 2009, http://pewinternet.org/Reports/2009/Teens-and-Distracted-Driving.aspx.

Martinez, G., J. Abma, and C. Casey. "Educating Teenagers about Sex in the United States." NCHS Data Brief no. 44, 2010.

Martinez, G., et al. "Teenagers in the United States: Sexual Activity, Contraceptive Use, and Childbearing, 2006–2010 National Survey of Family Growth." *Vital and Health Statistics,* ser. 23, no. 31, 2011.

National Safety Council. *Compilation of Research Comparing Handheld and Hands-Free Devices.* December 2009.

———. *Attributable Risk Estimate Model.* December 2009.

Pew Research Center. "Social and Demographic Trends: Millennials: A Portrait of Generation Next." February 2010.

Rideout, V. J., U. G. Foehr, and D. F. Roberts. *Generation M2: Media in the Lives of 8- to 18-Year-Olds.* Menlo Park, CA: Henry J. Kaiser Family Foundation, January 2010.

Rideout, V. J., and E. Hamel. *The Media Family: Electronic Media in the Lives of Infants, Toddlers, Preschoolers and Their Parents.* Menlo Park, CA: Henry J. Kaiser Family Foundation, 2006.

Schulz, L. "Intelligence, Commonsense, and Cognitive Development," MIT Department of Brain and Cognitive Sciences Early Cognition Lab, n.d.

Social and Demographic Trends: Millennials: A Portrait of Generation Next. Pew Research Center, February 2010.

Special Report: America at the Digital Turning Point, University of Southern California Annenberg School Center for the Digital Future, 2012, http://www.digitalcenter.org/.

"Will Hyperconnected Millennials Suffer Cognitive Consequences?" The Daily Circuit, Minnesota Public Radio (audio). Pew Internet: Pew Internet and American Life Project, 1 March 2012, http://www.pewinternet.org/Media-Mentions/2012/MPR-hyperconnected-millennials.aspx.

"Work-Life Balance: Tips to Reclaim Control: When Your Work Life and Personal Life Are out of Balance, Your Stress Level Is Likely to Soar." Mayo Clinic, http://www.mayoclinic.com/health/work-life-balance/WL00056.

Newspaper and Magazine Articles, Broadcasts, and Online Resources

"Ads Touting 'Your Baby Can Read' Were Deceptive, FTC Complaint Alleges Two of Three Defendants Settle Charges for Claims Made about Children's Learning Program; Order Bans Use of Product Name." Federal Trade Commission, 28 August 2012, http://ftc.gov/opa/2012/08/babyread.shtm.

"Almost All Americans Believe They Are Safe Drivers." Nationwide Insurance, May 2008, http://www.nationwide.com/newsroom/press-release-almost-all-americans-believe-they-are-safe-drivers-2008.jsp.

Angier, N. "Abstract Thoughts? The Body Takes Them Literally." *New York Times*, 2 February 2010, http://www.nytimes.com/2010/02/02/science/02angier.html?pagewanted=all&_r=0.

Baker, C. "The Creator." *Wired*, August 2012, 66.

Bazelon, E. "Amanda Todd Was Stalked before She Was Bullied." *Slate*, 18 October 2012, www.slate.com/blogs/xx_factor/2012/10/18/suicide_victim_amanda_todd_stalked_before_she_was_bullied.html.

"Best Websites for Kids." Common Sense Media, http://www.commonsensemedia.org/website-lists.

Bielski, Z. "Today's College Kids Are 40-per-cent Less Empathetic, Study Finds." *Globe and Mail* (Toronto), 1 June 2010, http://www.theglobeandmail.com/life/work/todays-college-kids-are–40-per-cent-less-empathetic-study-finds/article1587609/.

Blanchard, J., and T. Moore. "The Digital World of Young Children: Impact on Emergent Literacy." Pearson Foundation, 1 March 2010.

Bloom, P. "The Moral Life of Babies." *New York Times*, 5 May 2010, http://www.nytimes.com/2010/05/09/magazine/09babies-t.html?pagewanted=all&_r=0.

Buckleitner, W. "Tablets for Children, Including Apps." *New York Times*, 28 March 2012, http://www.nytimes.com/2012/03/29/technology/personaltech/tablets-for-children-including-apps.html.

Bui, C. "Hookup Culture." *Tufts Daily* (Tufts University), 24 January 2011, http://www.tuftsdaily.com/op-ed/hookup-culture–1.2445270.

———. "Teaching Good Sex." *New York Times*, 16 November 2011, http://www.nytimes.com/2011/11/20/magazine/teaching-good-sex.html?pagewanted=print.

Campaign for a Commercial Free Childhood, http://commercialfreechildhood.org/.

Carey, B. "Parents Urged Again to Limit TV for Youngest." *New York Times*, 18 October 2011, http://www.nytimes.com/2011/10/19/health/19babies.html.

———. "Shooting in the Dark," *New York Times*, 11 February 2013, http://www.nytimes.com/2013/02/12/science/studying-the-effects-of-playing-violent-video-games.html.

"Cellphones and Driving." Insurance Information Institute, October 2008.

"Cell Phone Use in Pregnancy May Cause Behavioral Disorders in Offspring, Mouse Study Suggests," *Science Daily*, March 15, 2012, http://www.sciencedaily.com/releases/2012/03/120315110138.htm#.UHriJP43q98.email.

"Cell Phone Use May Reduce Male Fertility, Austrian-Canadian Study Suggests." *Science Daily*, May 19, 2011, http://www.sciencedaily.com/releases/2011/05/110519113022.htm#.UHrhAoPFtaE.email.

Chapin, Harry. "Cat's in the Cradle." Elektra Records, 1974.

Child Development Institute Parenting Today. *Ages & Stages*, http://childdevelopmentinfo.com/ages-stages.shtml.

Clifford, S. "Teaching Teenagers about Harassment." *New York Times*, 26 January 2009.

Common Sense Media, http://www.commonsensemedia.org/.

"Cyberbullying: One in Two Victims Suffer from Distribution of Embarrassing Photos and Videos." *Science Daily*, 25 July 2012, www.sciencedaily.com/releases/2012/07/120725090048 .htm.

Dell'Antonia, K. J. "Why Books Are Better Than e-Books for Children." *New York Times*, 28 December 2011, http://parenting.blogs.nytimes.com/2011/12/28/why-books-are-better-than-e-books-for-children/?partner=rssnyt&emc=rss.

"Distractions Challenge Teen Drivers." *USA Today*, 26 January 2007.

Dominus, S. "A Facebook Movement against Mom and Dad." *New York Times*, 16 January 2010, A-14.

———. "Table Talk: The New Family Dinner." *New York Times*, 29 April 2012.

"Driver Electronic Device Use in 2010." *Traffic Safety Facts*, National Highway Traffic Safety Administration, December 2011.

Education Resources Information Center, http://www.eric.ed.gov/ERICWebPortal/ search/detailmini.jsp?_nfpb=true&_&ERICExtSearch_SearchValue_0=EJ329627&ERIC ExtSearch_SearchType_0=no&accno=EJ329627.

"8–18 Year Olds Pathologically Addicted to Games." *Science Daily*, http://www.science daily.com/releases/2009/04/090420103547.htm#.UIMQbWnq5WQ.email.

"Empathy: College Students Don't Have as Much as They Used to, Study Finds." *Science Daily*, 29 May 2012, http://www.sciencedaily.com/releases/2010/05/100528081434.htm.

"Exposure to Mobile Phones before and after Birth Linked to Kids' Behavioral Problems." *BMJ–British Medical Journal*, 7 December 2011, http://www.sciencedaily.com/releas es/2010/12/101206201242.htm.

Feinberg, C. "The Mediatrician: Former Hollywood Filmmaker Michael Rich of HMS Studies How Media Affect Youth." *Harvard Magazine*, November/December 2011, http:// harvardmagazine.com/2011/11/the-mediatrician.

"France Pulls Plug on TV Shows Aimed at Babies." CBC News, 20 August 2008, http:// www.cbc.ca/world/story/2008/08/20/french-baby.html.

Gaudin, S. "Families Spending More Time on Social Networks, Less Time Together." *Computerworld*, 16 June 2009.

Girls, Boys, and Media Messages: A Gender and Digital Life Toolkit. Common Sense Media online with C. Steiner-Adair, http://www.commonsensemedia.org/educators/gender.

Goldberg, S. "Parents Using Smartphone to Entertain Bored Kids." *CNN Technology*, 26 April 2010.

———. "TV Can Boost Self-Esteem of White Boys, Study Says." CNN, 1 June 2012, http:// www.cnn.com/2012/06/01/showbiz/tv/tv-kids-self-esteem/index.html.

Grafman, J. "Brain Development in a Hyper-Tech World." Dana Foundation, August 2008.

Guernsey, L. "How 'Screen Time' Impacts Kids—What Do Scientists Really Know?" Hatch Innovation webinar, 15 November 2011.

———. "Why eReading with Your Kid Can Impede Learning." *Time*, 20 December 2011, http://ideas.time.com/2011/12/20/why-ereading-with-your-kid-can-impede-learning/.

Guernsey, L., G. Troseth, E. Hartley-Brewer, M. Rich, and L. Rosen. "Wired Kids, Negligent Parents?" *New York Times*, 28 January 2010, http://roomfordebate.blogs.nytimes .com/2010/01/28/wired-kids-negligent-parents/.

Guernsey, L., B. Worthen, H. Kirkorian, and L. Perle. "Touch-Screen Devices and Very Young Children." *Diane Rehm Show*, 23 May 2012, http://thedianerehmshow.org/ shows/2012–05–23/touch-screen-devices-and-very-young-children?page=1.

Gutnick, A. L. "Always Connected: The New Digital Media Habits of Young Children." Joan Ganz Cooney Center at Sesame Workshop, March 2011, http://joanganzcooneycenter .org/Reports–28.html.

Halsey, A., III. "Teen Drivers Are Texting, Just Like Their Parents." *Washington Post*, 13 May 2012, http://www.washingtonpost.com/local/trafficandcommuting/teen-drivers-are-texting-just-like-their-parents/2012/05/13/gIQA8raQNU_story.html.

Haskins, W. "Do Educational DVDs Make Babies Blockheads?" *Technology News World: Science*, 8 August 2007, http://www.technewsworld.com/story/58735.html.

Hawgood, A. "How Teenagers Handle the Web's Instant Fame: No Stardom Until after Homework." *New York Times*, 15 July 2011, http://www.nytimes.com/2011/07/17/fashion/ how-teenagers-handle-the-webs-instant-fame.html?pagewanted=all.

Heather, K. "Survey: Most Americans Take Breaks from Facebook." CNN.com, 6 February 2013, http://www.cnn.com/2013/02/05/tech/social-media/facebook-breaks-pew.

———. "How to Post to Facebook, Twitter After You Die." CNN.com, 22 February 2013, http://www.cnn.com/2013/02/22/tech/social-media/death-and-social-media.

"How Teens Use Media." Nielsen, 2009.

Hutchison, C. "Watching SpongeBob SquarePants Makes Preschoolers Slower Thinkers, Study Finds." ABC NightTime News, 12 September 2011, http://abcnews.go.com/ Health/Wellness/watching-spongebob-makes-preschoolers-slower-thinkers-study-finds/ story?id=14482447#.UHr0kRj0RaE.

"Infants Do Not Appear to Learn Words from Educational DVDs." *Science Centric*, 15 March 2012, http://www.sciencecentric.com/news/10031578-infants-do-not-appear-learn-words-from-educational-dvds.html.

Ito, M., H. Horst, M. Bittanti, D. Boyd, B, Stephenson, P. Lange, C. J. Pascoe, and L. Laura Robinson. "Living and Learning with New Media: Summary of Findings from the Digital Youth Project." John D. and Catherine T. MacArthur Foundation Reports on Digital Media and Learning (Cambridge, MA: MIT Press, 2008).

Jackson, N. "More Kids Can Work Smartphones Than Can Tie Their Own Shoes." *The Atlantic*, 24 January 2011.

Jobs, Steve. "You've Got to Find What You Love." Stanford University Commencement Speech, 12 June 2005, http://news.stanford.edu/news/2005/june15/jobs-061505.html.

Johnson, C. "Researchers Study How Babies Think." *Boston Globe*, 28 March 2011, http://www.boston.com/lifestyle/health/articles/2011/03/28/researchers_study_how_babies_think/?rss_id=Boston.com+—+Latest+news.

Katz, J. *Tough Guise: Violence, Media and the Crisis in Masculinity*. Documentary, 1999, http://www.jacksonkatz.com/video2.html.

———. *The Macho Paradox: Why Some Men Hurt Women and How All Men Can Help*. Media Education Foundation, University of Massachusetts Amherst, www.mediaed.org.

———. *Spin the Bottle: Sex, Lies and Alcohol*. Educational video, 2004.

———. *Wrestling with Manhood*. Educational video, 2002.

Keim, B. "Is Multitasking Bad for Us?" NOVA scienceNOW, 4 October 2012, www.pbs.org/wgbh/nova/body/is-multitasking-bad.html.

Kessler, S. "Children's Consumption of Digital Media on the Rise." Mashable Social Media, 14 March 2011, http://mashable.com/2011/03/14/children-internet-stats/.

Kristof, N. "Blissfully Lost in the Woods." *New York Times*, 28 July 2012, http://www.nytimes.com/2012/07/29/opinion/sunday/kristof-blissfully-lost-in-the-woods.html?_r=0.

Kuhl, P. "The Linguistic Genius of Babies." TEDxRainier, October 2010.

Lagorio, C. "Resources: Marketing to Kids." CBS News, 11 February 2009.

Lee, C. "Media Gender Divide: Boys v. Girls (InfoGraphic)." NMR, 5 July 2012, http://newmediarockstars.com/2012/07/infographic-social-media-gender-divide-boys-v-girls/.

Lewin, T. "If Your Kids Are Awake, They're Probably Online." *New York Times*, 20 January 2010, http://www.nytimes.com/2010/01/20/education/20wired.html?_r=1&.

Maag, C. "When the Bullies Turned Faceless." *New York Times*, December 16, 2007, http://www.nytimes.com/2007/12/16/fashion/16meangirls.html?ref=meganmeier&pagewanted=print.

"Machine of the Year 1982: The Computer Moves In." *Time*, 5 October 1983, http://www.time.com/time/magazine/article/0,9171,952176,00.html.

McMillan, G. "Study: Women Better at Using Social Media to Keep in Touch." *Techland*, 30 September 2011, http://techland.time.com/2011/09/30/study-women-better-at-using-social-media-to-keep-in-touch.

Meisner, J. "Heads Up: Cell Phones Add to Risk When Crossing Street, Study Shows." *Mac News World*, 27 January 2009, http://www.macnewsworld.com/story/65963.html.

Merrill, M. "Play Again: What Are the Consequences of a Childhood Removed from Nature?" playagainfilm.com.

Meyer, P. "The Middle School Mess." Educationnext, Winter 2011, http://educationnext
.org/the-middle-school-mess/.

Miller, S. S. "Survey: Teens 3 Times Likely to Text and Drive." *Dayton Daily News*, 2 May
2012, http://www.daytondailynews.com/lifestyle/survey-teens–3-times-as-likely-to-text-
and-drive–1369411.html.

Moretz, P. "Traditional Books Provide More Positive Parent-Child Interaction According to
Temple, Erikson Researchers." *Temple Times*, 9 November 2006, http://www.temple.edu/
temple_times/november06/Traditionalbooks.html.

Moss, M. "The Extraordinary Science of Addictive Junk Food." *New York Times*, 20 February 2013, http://www.nytimes.com/2013/02/24/magazine/the-extraordinary-science-of-
junk-food.html?pagewanted=all.

"Most U.S. Drivers Engage in 'Distracted' Driving Behaviors." USAToday.com, 1 December 2011.

Muldoon, K. "Animal Nature: New Map Brims with Great Spots to Watch Oregon Wildlife;
Party Time at Bandon March." *The Oregonian*, 29 September 2011.

Murphy, K. "Cellphone Radiation May Alter Your Brain. Let's Talk." *New York Times*,
30 March 2011, http://www.nytimes.com/2011/03/31/technology/personaltech/31basics
.html?_r=1&gwh=52BDB3D2D3CF0EACA3E80E27C61CD719.

Nass, C. Interview with Clifford Nass, PBS, 1 December 1, 2009, www.pbs.org/wgbh/pag
es/frontline/digitalnation/interviews/nass.html.

"No Einstein in Your Crib? Get a Refund." *New York Times*, October 23, 2009, http://www
.nytimes.com/2009/10/24/education/24baby.html?_r=1.

O'Brien, K. "Privacy Advocates at Odds over Web Tracking." *New York Times*, 5 October 2012, http://www.nytimes.com/2012/10/05/technology/privacy-advocates-and-
advertisers-at-odds-over-web-tracking.html?gwh=25CC86CF71D6958540510F08915
5F000.

"Parents Taking New Steps to Protect Their Children from New Porn." www.netnanny.com.

Park, A. "Baby Wordsworth Babies: Not Exactly Wordy." *Time*, 2 March 2010. http://www
.time.com/time/health/article/0,8599,1968874,00.html.

Paton, G. "Twitter and Facebook 'Harming Children's Development.'" Independent.ie.com,
20 October 2012, http://www.independent.ie/business/technology/twitter-and-facebook-
harming-childrens-development–3266055.html.

Perlow, Leslie. "Why 'Work-Life Balance' Doesn't Work." *Washington Post*, 11
July 2012, http://www.washingtonpost.com/national/on-leadership/step-away-from-the-
smartphone/2012/07/11/gJQA3AhDdW_print.html.

Perna, G. "More Young Kids Can Use Technology Than Tie Shoes." *International Business
Times*, 20 January 2011, http://www.ibtimes.com/articles/103217/20110120/young-kids-
technology-study-play-computer-game-operate-a-smartphone.htm.

Play Again. Documentary about a childhood removed from nature, Portland, playagainfilm .com.

Ravichandran, P., and B. France de Bravo. "Children and Screen Time." ChildWise, 2010, http://boysdevelopmentproject.org.uk/index.html.

———. *Children and Screen Time (Television, DVD's, Computer).* 2010, boysdevelopment-project.org.uk.

———. "Pre-School Children." ChildWise, 2009, http://boysdevelopmentproject.org.uk/ index.html.

———. "Special Report Digital Lives." ChildWise, 2010, http://boysdevelopmentproject .org.uk/index.html.

Rich, M. Center on Media and Child Health, Children's Hospital, Boston, http://www .cmch.tv.

Richtel, M. "Attached to Technology and Paying the Price." *New York Times,* 6 June 2010, http://www.nytimes.com/2010/06/07/technology/07brain.html?pagewanted=all&_r=0.

———. "In Online Games, a Path to Young Consumers." *New York Times,* 20 April 2011, http://www.nytimes.com/2011/04/21/business/21marketing.html?pagewanted=all&_r=0.

Robins, M., and G. Wilson. "Porn-Induced Sexual Dysfunction Is a Growing Problem." *Psychology Today,* 11 July 2011, http://www.psychologytoday.com/blog/cupids-poisoned-arrow/201107/porn-induced-sexual-dysfunction-growing-problem.

Rochman, B. "Pediatricians Say Cell Phone Radiation Standards Need Another Look." *Time,* Healthland, 20 July 2012, http://healthland.time.com/2012/07/20/pediatricians-call-on-the-fcc-to-reconsider-cell-phone-radiation-standards/.

RTI International and Blue. "Prevention in Middle School Matters: A Summary of Find-ings of Teen Dating Violence Behaviors and Associated Risk Factors Among 7th-Grade Students." Robert Wood Johnson Foundation, 1 January 2011, www.rwjf.org/goto/middle schoolmatters.

Sadowski, M. "School Readiness Gap." *Harvard Education Letter,* 2006.

Schryver, Kelly. "Keeping Up Appearances," honors thesis, Brown University, 2011.

Schwartz, M. "The Trolls among Us." *New York Times,* 3 August 2008, http://www.ny times.com/2008/08/03/magazine/03trolls-t.html?ref=meganmeier&pagewanted=print.

"Secret of Facebook's Success: Sharing Gossip with Friends Is Addictive and Arousing." Womenist.net, http://www.womenist.net/en/p5704/technology/secret_of_facebooks_success .html.

Sedensky, M. "Dad's Techxecution Hits a Nerve." *Sunday Oregonian,* Associated Press, 19 February 2012.

Seligson, H. "When the Work-Life Scales Are Unequal." *New York Times,* 1 September 2012, http://www.nytimes.com/2012/09/02/business/straightening-out-the-work-life-balance .html?pagewanted=all.

Shellenbarger, S. "Single and Off the Fast Track: It's Not Just Working Parents Who Step Back to Reclaim a Life." *Wall Street Journal*, 23 May 2012, http://online.wsj.com/article/SB 10001424052702304791704577420130278948866.html#printMode.

———. "Tips for Improving Your Teens' Sleep Schedule." The Juggle: *Wall Street Journal* Blogs, 17 October 2012, http://blogs.wsj.com/juggle/2012/10/17/tips-for-improving-your-teens-sleep-schedule/.

———. "Understanding the Zombie Teen's Body Clock." *Wall Street Journal*, updated 16 October 2012, http://online.wsj.com/article/work_and_family.html.

———. "When Curious Parents See Math Grades in Real Time." *Wall Street Journal*, 2 October 2012, http://online.wsj.com/article/SB10000872396390444592404578032360255233782.html.

Shrivastava, A. "Is It Okay to Be a 'Helicopter Parent?'" The Juggle: *Wall Street Journal* Blogs, 12 October 2012, http://responsibility-project.libertymutual.com/articles/the-case-for-the-relaxed-parent#fbid=k_VjLUEe6bn.

Siebel Newsom, J. *Miss Representation*. Documentary, 2011, http://www.missrepresentation.org/the-film/.

Siegel, D. J. "Understanding Your Child's Attachment Style." Psychalive, http://www.psychalive.org/2011/10/identifying-your-childs-attachment-style/.

Singer, N. "Do Not Track? Advertisers Say 'Don't Tread on Us.'" *New York Times*, 13 October 2012, http://www.nytimes.com/2012/10/14/technology/do-not-track-movement-is-drawing-advertisers-fire.html.

Smith, A., L. Segall, and S. Cowley. "Facebook Reaches One Billion Users." CNNMoney, 4 October 2012, http://money.cnn.com/2012/10/04/technology/facebook-billion-users/index.html?hpt=hp_t3.

Smith, S., and A. Granados. "Gender and the Media." *National PTA Magazine*, December/January 2009–2010, http://www.pta.org/3736.htm#14a.

Sroufe, A. "Ritalin Gone Wrong." *New York Times*, 28 January 2012, http://www.nytimes.com/2012/01/29/opinion/sunday/childrens-add-drugs-dont-work-long-term.html?pagewanted=all.

Steinhauer, J. "Verdict in MySpace Suicide Case." *New York Times*, 27 November 2008.

Stober, D. "Multitasking May Harm the Social and Emotional Development of Tweenage Girls, but Face-to-Face Talks Could Save the Day, Say Stanford Researchers." *Stanford News*, 25 January 2012, news.stanford.edu/pr/2012/pr-tweenage-girls-multitasking-012512.html.

Stout, H. "Effort to Restore Children's Play Gains Momentum." *New York Times*, 5 January 2011, http://www.nytimes.com/2011/01/06/garden/06play.html?pagewanted=all&_r=0.

Taffel, R. "Decline and Fall of Parental Authority." *AlterNet*, 22 February 2012, http://www.alternet.org/story/154249/the_decline_and_fall_of_parental_authority?page=entire.

Takeuchi, L. "Families Matter: Designing Media for a Digital Age." Joan Ganz Cooney at Sesame Workshop, June 2011.

"Tech Addiction Symptoms Rife among Students." CBC News, 6 April 2011, http://www.cbc.ca/news/technology/story/2011/04/06/technology-addiction-students.html.

"Teens Warn Peers against Texting and Driving." Safety + Health, www.nsc.org/safety-health/Pages/teens_warn_peers_against_texting_and_driving.aspx#.UIHdmhjbAhc.

Thompson, D. "The Atlantic: Kids Are Changing, Neuroplasticity Is Real, and Education Needs a Revolution." Brain Power, 2 March 2012, http://www.brainpowerinitiative.com/2012/03/the-atlantic-kids-are-changing-neuroplasticity-is-real-and-education-will-need-to-change/.

Thompson, M. Raising Cain: Boys in Focus. PBS/Oregon Public Broadcasting and Powderhouse, 12 January 2006.

"TRUSTe Releases Survey Results of Parents and Teenagers on Social Networking Behaviors: National Poll Conducted in Partnership with Lightspeed Research Reveals Alignment between Parents and Teens in Desire for Privacy." TRUSTe, 18 October 2010, http://www.truste.com/about-TRUSTe/press-room/news_truste_2010_survey_snsprivacy.

Walsh, D. "Interactive Violence and Children." Testimony submitted to the Committee on Commerce, Science, and Transportation, United States Senate, 21 March 2000.

Weiss, D. "The New Normal? Youth Exposure to Online Pornography." Rock, 6 April 2011, http://www.myrocktoday.org/default.asp?q_areaprimaryid=7&q_areasecondaryid=74&q_areatertiaryid=0&q_articleid=861.

"A Window onto Family Facebook Use: TRUSTe Study." Net Family News.org, 18 October 2012, http://www.netfamilynews.org/a-window-onto-family-facebook-use-truste-study.

Worthen, B. "The Perils of Texting while Parenting." Wall Street Journal, 29 September 2012, C1, http://online.wsj.com/article.

———. "What Happens When Toddlers Zone Out with an iPad." Wall Street Journal, 22 May 2012, http://online.wsj.com/article/SB100014240527023043631045773918139618539838.html.

"'Your Baby Can Read': No, He Can't." Editorial, Boston Globe, 26 July 2012.

Professional Interviews

Craig Anderson, November 2011

Mark Bertin, October 2011

Ellen Birnbaum, August 2012

Tina Payne Bryson, October 2012

Dimitri Christakis, February 2012

Gene Cohen, December 2000

JoAnn Deak, January 2012

Ned Hallowell, March 2011

Mimi Ito, October, 2011

Jackson Katz, November 2011

Michael Langlois, September 2011

Madeleine Levine, January 2012

Liz Perle, April 2011

Denise Pope, April, 2011

Harvey L. Rich, January 2012

Michael Rich, April 2012

Kelly Schryver, April 2011

Nancy Schulman, August 2011

Robin Shapiro, January 2011

Daniel Siegel, February 2012

Lydia Soifer, November 2011

Michael G. Thompson, June 2012

Janice Toben, February 2012

Yalda Uhls, March 2012

Donna Wick, November 2011

Maryanne Wolf, March 2011

Index

addiction
  alcohol or substance abuse, 160, 214
  to pornography, 165
  to screen games, 151, 156, 157–58
  to tech, 3, 5–6, 25, 35, 38, 55, 81, 120,
    156, 158, 182–83, 219–20, 265–66
adolescents. *See* teens/adolescents
advertising and marketing, 18, 30, 90–94,
  170, 185
  screen learning programs, 83
  targeting children, 42, 53, 111–12
aggression, 38, 126
  addressing, 250
  in boys, 41, 42, 187
  media and model of masculinity, 112–13,
    143, 145
  media and normalizing of, 51, 146
  media violence and, 82, 113, 123, 139, 143,
    154, 157
  social (normal), 249
alcohol and substance abuse, 160, 214
American Academy of Pediatrics (AAP)
  cell-phone-radiation-emissions and, 5
  elimination of tech for baby's first 24
    months, recommendation, 88
  TV and tech time, minimizing and
    monitoring, 53
"Am I Ugly," 36
Anderson, Craig, 157
Angry Birds, 84, 107
Annenberg Center for the Digital Future,
  University of California, 24
anxiety, 36, 50, 160, 207, 214, 219, 231, 283
  media images and, 82, 126

presentation, teens and, 206, 207
social, 141, 283
tech and, 38, 74, 124, 158, 191, 197, 249
AO "always-on" generation, 58
apps, 53, 271, 295
  for babies or toddlers, 42, 60, 70, 77, 80,
    83, 86, 90, 92, 93, 100, 103
  imagination vs., 100, 101
  read-aloud with digital voice, 108
  as stimulant and overstimulation, 93
attentional problems, 56–60, 115, 126, 283
  eliminating risk factors, 123
  tech and, 80, 82, 104, 123, 139, 157, 182
attention deficit hyperactivity disorder
  (ADHD), 38, 82, 93, 116, 122–23, 182

babies and toddlers (birth to age two)
  advertising, marketing, and, 90–94
  apps for, 42, 60, 70, 77, 80, 83, 86, 90, 92,
    93, 100, 103
  bonding, attachment, and attunement,
    66–68, 69, 76, 77, 80, 94, 269
  brain of, 25, 66–98
  children's programming and, 87–88
  elimination of tech for first 24 months,
    68, 88, 92
  embodied connection as optimal for
    development, 80, 83
  family and development, 39
  first words, 95
  Gen Y mothers and toddlers' use of
    smartphones, 74
  human tone and touch for core lessons,
    84–86

babies and toddlers (*cont.*)
"instrumental parenting," 71–72
mirroring exchange, 69
overstimulation and, 93
parental interactions with, 68–75
parent's focus of attention as object of
desire, 71–72
physical activity and, 87
primacy of parental presence and, 71,
79–81, 88, 98
read-aloud time, 88–90
response to facial expressions, 70–71
response to mother's voice, 62–63, 68,
70, 80
room, as screen-free, 97–98
safe media for, 88
"secure attachment" and, 69
sensorium, 78–79
tech and expectation of instant gratification, 86–87
tech intrusion on parent-child relationship, 69–77, 94–97
tech's pervasiveness in life of, 83–84
Baby Einstein, 91, 107
Birnbaum, Ellen, 101
boys
aggression in, 41, 42, 49
conversation about sex, script for, 254–55
homophobia and, 142, 208–10
media and gender messages, 112–13, 140–47
media and violent messages, 112
model of masculinity, 112–13
pornography and, 165, 183–89, 196, 253
brain
ages three to five, 109
birth to age three, 25, 69, 77–83
development of empathy, 50–52
executive function, 37, 139, 287
growth of, 37, 78
language development and, 80–81
lifelong learning and, 65
mediated interaction vs. direct human-to-
human interaction, 27
patterns and "environmental input," 113
reading and, 79, 89
screens, media, and attention disorders,
58, 80, 82, 104, 123, 139, 157, 182

sensorium, 78–79, 109
social brain, 69
stimulation of development, 77
tech and dopamine, 5, 59, 151, 262, 264
tech's adverse effects on development,
37, 55, 78, 79, 81–82, 89–90, 113–14,
128, 139
texting as left-brain response, 22
unplugged downtime, importance of, 55
wiring of, screen games and, 158
Brain and Culture: Neurobiology, Ideology,
and Social Change (Wexler), 71
Bryson, Tina Payne, 158
bullying, 49–50, 51
dealing with, 246–50
online harassment, 129–31
psychiatric disorders and, 50
zero-tolerance policies and, 249–50

"Cat's in the Cradle" (Chapin), 28
cell phones, 5, 15. *See also* texting
disregard for privacy and, 10
reminders for child about, 272–73
rules for, 272
Chapin, Harry, 28
character or character development
accountability and building, 146–47
ages six to ten and, 134–35, 160, 166
author's work with parents and, 27–28, 35
creating opportunities for, 160
family as teacher of character, 40
mistakes as teachable moments, 251
online influences, 166–67
parenting and, 28, 121–22
real-life practice and, 139
in a school setting, 166, 249–50
SEL and, 249–50
teachable moments, example, 250–52
childhood
"achievement culture" of, 53
collapsed or speeded up, 129–61, 165, 166
digitalized, 25–27
early job of, sensorium and, 78
innocence, premature loss, 26, 41–44
"magic years," 102, 125–28
protection of, 18, 44–48, 65, 155–61, 263,
294

tech's adverse effects on, 79, 129–61
traditional canon for elementary school,
    134–35
value development during, 134
children. *See also* family; parenting; play;
    specific age groups
computer in child's room, 57–58
controlling media's access to, 263
descriptions of parents' tech habits by,
    11–16
"fast-twitch wiring" and generational
    characteristics of, 58, 59, 60, 62, 93,
    197, 273, 294
inappropriate online content accessible to,
    33–35, 36, 37, 43–44, 82, 132–33, 149,
    150–56, 167, 238 (*see also* pornogra-
    phy; violence)
message conveyed to by preoccupied
    parents, 17, 124–25
obesity in, 38, 53
off-line experiences, importance of, 55,
    291–94
overscheduling of, 35, 53
parenting message to convey to, 150
psychiatric disorders in, 50, 160–61, 214
secondhand tech exposure, 74–75, 82
sexualizing of, 41, 42, 43, 112, 140–47
television in child's room, 53, 74, 108
Christakis, Dimitri, 58, 154–55, 157,
    166–67
cognitive development
    ages six to ten, 131–33
    ages three to five, 99–103, 109
    birth to age three, 77–83, 89
    boredom, importance of, 55
    creativity and imagination, 53–56, 86,
        99–103, 107
    critical thinking, 107
    deep thinking, 55, 58, 59, 79, 90, 109
    "embodied cognition," 62–63
    "fast-twitch wiring" and, 58, 60
    higher-level thinking, 79, 102
    logical thinking, 102
    magical thinking to "real world" develop-
        ment, 102
    problem-solving abilities, 54, 55, 86, 102,
        105, 277, 279, 280

reading, 51, 52, 58, 77, 79, 89–90, 136,
    179, 200
reflection, 52, 55, 59, 65
solitude, importance of, 55, 63–65
sustained attention, 56–60
tech and expectation of instant gratifica-
    tion, 86–87
tech's adverse effects on, 51, 55, 58, 79,
    86–87, 90, 94
Cohen, Gene, 4
communication. *See also* texting
    attentive response and, 17
    conversation, values of, 64–65, 271,
        278–82
    dinnertime, 279–80
    e-mail, 19
    in family, 52, 271, 278–82
    family arguing, 280–81
    "fast-twitch wiring" and, 62, 197
    free play and, 53
    genuine connection and, 17
    human voice, importance of, 84–86
    human voice vs. texting, 21–22, 61–62,
        195, 200–201, 272–73
    manners and courtesy for, 273
    read-aloud time and, 89
    rules for, via tech, 272
    skills needed for effective, 61
    social conversation, 281–82
    tech and loss of direct human interac-
        tion, consequences of, 7, 52, 60–65,
        193–225
    tech as new psychological territory, 22–23
    tech redefining, 20
    "convergence culture," 19
    creativity and imagination
        absence of media and, 289–90
        boredom, importance of, 55
        family and nurturing, 289
        play and, 55, 99–102, 125–26, 275–76
    culture (dominant media and online),
        4, 36. *See also* pornography; texting;
        violence
    access to peers 24/7, 165
    achievement-focused, 53, 58
    breakdown in norms, 35
    commercialism, 35, 42, 91, 111, 112, 144

communication (*cont.*)
"convergence culture," 19
cynicism and crass entertainment in, 35, 37, 42, 147–48, 185, 112
dehumanization, 20, 263
disappearance of basic courtesy and civil conduct, 20, 150
empathy missing, 48–52
family as counterculture force, 263–64
family replaced by online relationships, 38–41, 165
gender messages, 43, 112–13, 141, 142, 146
instant gratification and stimulation, 93
loss of quiet respect, Arlington National Cemetery, 10
mealtimes, tech use during, 10
model of connection in, 18
multitasking and, 58–59
normalizing harmful and unethical behaviors, 30, 35, 41–42, 49, 51, 140, 146, 147, 150, 151–52, 155, 185, 198, 217, 250, 255, 257, 294
parental control in the home, 65
tech-driven (cyber), 24, 26, 27, 30, 36, 37, 38, 41–42, 50, 64, 76, 90, 91, 93, 142, 165, 173, 181, 192, 197, 217, 219, 230, 237, 258–59, 261, 290
ubiquitous sexual imagery and sexuality, 183–85, 186, 219, 254, 256
young adult and drinking, 277

Davidson, Cathy, 51
Deak, JoAnn, 78, 159
Denckla, Martha Bridge, 57
depression, 45, 50, 204, 130, 139, 160, 214, 219
Disappearance of Childhood, The (Postman), 26, 263
distractibility
injuries and, 59
media exposure and, 157
multitasking and, 56–60
reading comprehension and, 58
in school, 58–59
divorcing parents, 13, 22, 214, 220, 234
Driven to Distraction (Hallowell), 33

eating disorders, 38, 141, 145, 160, 207, 220
e-books, 89
elementary school children (ages six to ten), 129–61
character development, 134–35, 160, 166
childhood lost for, 140, 150–56
computer as new playground, 132
controlling tech and media use, 137, 156–59
emotional development, 139, 159–60
empathy and, 134, 138, 139
gender code and, 140–47
healthy growth, foundations, 134–35
impulsivity, 159
inappropriate online content accessible to, 132–33, 149, 150–56
inner critic and, 135–37
media and life online, negative effects of, 135–36, 137
media and tech curriculum in school, 131–32
moral development, 134, 135, 139, 160
online harassment and, 129–30
out-of-control circumstances caused by overexposure to TV, movies, and online content, 132–33
pace of development, consequences of accelerated, 137–59
parenting and, 136–37, 159–61
real-life practice and, 139
relationships and, 134–35
rules of social engagement, 138–39
social cruelty, 143, 145–50
social development, 138–39, 159–60
tech accessible to, beyond ability to manage use, 140, 156–59
tech-mediated environment and, 135
traditional childhood canon, 134–35
values and, 134, 135
e-mail, 19, 22, 59, 237
sexually harassing, 129–31, 150
emotional development
ages eleven to thirteen, 171
ages six to ten, 139, 159–60
ages three to five, 108–14
emotional intelligence, 109, 121–22
emotional literacy, 106

emotional sturdiness, 160
human tone and touch for core lessons,
    84–86
loss of intimacy capability, 62
parental cueing child's, 110
parents as foundation for child's, 72
resilience, 40, 85, 110, 135, 160
self-identity/identity issues, 174, 175–76,
    192, 196, 205–6, 220
self-made play and, 99–102
self-regulation, 85, 102, 109, 123
self-soothing, 85
social awareness and, 110
tech and teen addiction, 197
tech's adverse effects on, 25, 26, 37,
    61–62, 94, 105, 123
tech's inadequacy for, 101
teens, characteristics of, 196
texting and validating feelings, 63–65
empathy, 48–52
ages six to ten and, 134, 138, 139
family and, 52, 65
missing in school, 49
missing in social media, 49
social cruelty and, 148, 190–91
tech and decline in, 52
violent video games and, 153, 158
Erikson, Erik, 134

Facebook, 19–20, 183, 228, 261, 262, 264
as afterschool study group, 165–66
gaming and, 177
inappropriate contacts, 35, 170–71
personas and malleable identities, 196,
    204–8
pranks and incidents, 162–63, 171–72,
    209–10, 212, 215, 223
privacy issues, 170
social cruelty and, 165
underage users, 166
relationships and, 193, 210, 230, 240, 243
teens and time spent on, 206–7
tweens on, caution, 188–89
family, 260–95. See also family: practical
    guide to sustainability
adapting to tech culture, 264–65
arguing in, 280–81

author's son Daniel and tech, 9–10
capacity for connection and, 29
as center of human development, 1, 30,
    39–40, 65
children's definition of, 38–39
child's development of character and,
    40–41
child's development of self and, 39–40
communication and interaction in, 40–41,
    52, 60–63, 64–65
conflict over tech use in, 7–8
consequences learned in protected envi-
    ronment, 45, 52
convergence culture, impact of, 19–20
conversation in, 52, 271
as counterculture, 263–64
creating opportunities for, 122
creativity nurtured in, 289
dealing with change or crisis, 284–85
development steps and, 102
Digital Future report on Internet replac-
    ing family time, 24
early job of, sensorium and, 78
as empathic envelope, 52, 65
five elements for well-being and, 264
Gillian's story: reactivity of parents and
    restart, 240–48
house rules and tech contract, 270–71
infant development and, 39
life's lessons taught in, 264
loyalty, breaching of, 45
making meaning for children and, 40
manners and courtesy, 281
mealtimes/dinner together, 10, 65, 136,
    279–80
media and social networking, loss of fam-
    ily and personal privacy, 44–48
members feeling ignored in, 24
mentoring and, 40
modeling relationships, 40
online relationships vs., 165–66
parental attention/time and, 10–11,
    273–82
plugged in vs. unplugged, 265–69
primacy of, 29, 30, 38–41, 135, 262, 276,
    295
privacy, tech and loss of, 45

family (*cont.*)
  processing experiences through, 136
  protected childhood and, 65
  rituals and routines, 123, 136
  Skype and keeping connected, 111
  social conversation training, 281–82
  sustainable, 261–65
  tech, negative effects on, 7, 6–11, 13–14,
    17, 18, 27, 40–41, 262, 274
  television viewing and, 3, 82, 274–75
  values defined by, 40–41, 44, 135, 192,
    233, 245, 263, 276, 278, 288
  zones of interaction in, 40, 114–25
family: practical guide to sustainability,
    270–95
  encouraging play and playing together,
    273–78
  family philosophy about tech that reflects
    and supports the family's values and
    well-being, 269–73
  flexibility of, and as a work in progress, 291
  house rules and tech contract, 270–71
  nourishing meaningful connection and
    thoughtful conversation, 278–82
  parental authority and healthy disagree-
    ment, 285–87
  providing experiences off-line to experi-
    ence and cultivate an inner life,
    solitude, and connection with nature,
    291–94
  recognizing uniqueness of members; fos-
    tering their independence, 282–85
  stewardship and, 292–95
  tech-specific principles for, 271–73
  values, wisdom, links to past and future,
    and common language, 288–91
fear, 12, 44, 169, 191
  ages three to five, screen images and,
    126–28
  bullying and online harassment, 49–50,
    148, 249
  inappropriate online content and, 151, 152
  "magic years" and, 102
  media violence and, 82
  online mistakes and, 224
  parental, and reactivity, 149, 152, 166,
    231, 249, 250, 251

  sexual, 257
  teens and, 200, 204, 205, 214
Fine, Steven, 58–59
First Idea, The (Greenspan and Shanker),
    102
Flourish (Seligman), 264
Fraiberg, Selma, 39–40, 126
Friedlander, Michael, 139

Gallagher, Winifred, 29
gender codes, 112–13, 140–47
  school workshop to decode, 143–47
  stereotypes, 42, 43
  television and, 155
Gibran, Kahlil, 243
girls
  body image, 141, 174, 207
  media and gender messages 112–13,
    140–47
  pornography and unhealthy sexuality,
    187–88
  Rugrats' Angelica and, 113
  sex education conversation, script for,
    255–57
  sexualizing of, 41, 43, 112
  social cruelty and, 113
  television and self-image, 42
gorenography, 154
Gossip Girl (TV show), 42, 147, 215, 216
Grafman, Jordan, 139
Greenspan, Stanley, 102, 107

Hallowell, Ned, 33, 93
Harrison, Kristen, 142
*Homesick and Happy: How Time Away
    from Parents Can Help a Child Grow*
    (Thompson), 55–56
How Children Succeed (Tough), 249

impulsivity, 106
  ages six to ten, 138, 159
  media and boys, 112–13
  teens/adolescents and, 196, 197, 210, 216,
    219
instant gratification, 58, 84, 86–87
Internet. *See also* Facebook; social media
  addiction, 157

dangers to children, 24–25, 44–46
as external brain, 58
hunt-and-bump dynamic, 59
inappropriate online content accessible
    to children, 33–35, 36, 37, 43–44, 82,
    132–33, 149, 150–56, 167, 238
online harassment, 129–31, 217–21
online incidents and risks to teens, 198,
    204–5, 214–17, 221–25
online incidents and risks to tweens,
    162–63, 167–73
replacing family time, 24
social cruelty and, 143, 145–50
In the Moment: Celebrating the Everyday
    (Rich), 39
intimacy (emotional intimacy). *See also*
    sexuality or sexual behaviors
adolescents and, 196, 197, 200–201
family conversations about, 183
in human relationships, 21
real-life experiences necessary for, 197,
    200–201
sexual intimacy disconnected from, 186,
    198, 254, 257, 258
tech's negative impact on authentic, 18,
    62, 196, 197
texting and problems of, 62, 200–201
Ito, Mimi, 150–51, 173

Jenkins, Henry, 19
Jobs, Steve, 17

Kaiser Family Foundation surveys, 4, 57, 74
Katz, Jackson, 153, 157, 166–67, 186, 187,
    221
Kristof, Nicholas, 293, 294
Kuhl, Patricia, 27, 80

language development. *See also* communi-
    cation
ages three to five, 102, 109, 110
baby brain and mother's presence, 71,
    79–81
effective learning of, 80, 83
mediated vs. embodied connection and,
    80–81, 83
texting and loss of, 63

LEAP (Language Enrichment Arts Pro-
    gram), 107
learning. *See also* cognitive development;
    distractibility
advertising and marketing pitches, 90–94
baby brain and mother's presence, 71,
    79–81
core lessons and human tone/touch,
    84–86
dealing with children's mistakes, 246–52
discomfort as necessary for, 105–6
feeling safe, secure and, 69, 107, 269
instant gratification vs., 84, 86
mediated vs. embodied connection and,
    80–81
minimal results from infant media, 83
real-life practice or immersion-style learn-
    ing, 139
screen-based, 25, 59, 86–87, 90 (*see also*
    school)
screen vs. real-life experience and, 99–103
slow-paced hands-on practice as necessary
    for, 86–87
social-emotional learning (SEL), 121, 144,
    146–47, 249–50
tech's adverse effects on, 25, 26, 35,
    56–60, 105–6

Magic Years, The (Fraiberg), 1, 39, 126
Martha Speaks, 124
Martins, Nicole, 142
Maslow, Abraham, 67
moral development. *See also* empathy
ages six to ten, 134, 135, 139, 160
amorality of cyber culture, 37, 219
inappropriate online content accessible to
    children and, 35
need to internalize formative lessons, 164
parental concerns about tech, 35–36
SEL experiences and, 146–47
tech and media threats to, 164, 181
Morrongiello, Barbara, 6
motor skills, 53, 86
multitasking, 4, 15, 56–60, 178, 181, 200,
    266, 270
parenting and, 2, 15, 22, 73, 75–77, 98,
    115–16, 117

netiquette and digital citizenship, 132, 146–47, 167, 271
   online comments and posturing, 208–10
   online mistakes, 221–25
   parental violation of, 228
Norton, Bryan G., 261

obesity (childhood), 38, 53
101 Dalmatians (film), 127–28
optimism, 40, 41, 54, 85, 87, 278, 281, 282

panic disorder, 50, 283
parenting, 226–59. *See also* family
   abdicating of role, 237–39
   adolescents and, 226–59
   ages eleven to thirteen and, 164, 165–68, 189–92, 234–35
   ages six to ten and, 136–37, 159–61
   ages three to five and, 109–28
   birth to age three and, 66–68, 70, 71, 72, 73, 77
   amnesty policy, 229
   approachability, 226–32, 269
   basic message to convey to child, 150
   basic message to define limits on computer use, 191
   as behavioral model, 11, 16, 20, 27–31, 279
   character development and, 27–28, 121–22, 250
   children's cell phone use, helpful reminders, 272–73
   children competing with tech for attention, 11–16, 270
   children's need for parental time/attention, 10–11, 16–19, 70–71, 116–24
   children's "secondhand" exposure to tech and, 74, 82
   children's sense of self and self-esteem and, 136
   children's social-emotional toolkit and, 107, 110
   children's testing for reaction to something they did, 234–35
   communication via tech devices, 60–61
   communication with child, 165–66
   compared to gardening, 261

conflict resolution and mediation, 280–81, 286
conversation, values of, 64–65
distracted by tech, 1–3, 6, 16–19, 28, 29, 72–75, 124–25, 230, 266, 274
elimination of tech for baby's first 24 months, 68
everyday patterns of behavior that erode trust, 232–35
face-to-face words of love and, 270
Gen Y mothers and toddlers' use of smart phones, 74
"grandmother principle," 80
"instrumental parenting," 71–72, 91
intuitive sense, 77, 85, 97, 118
"mindsight," 72, 73
mistakes as teachable moments, 246–50, 251, 272
monitoring of inappropriate content accessible to children, 33–35, 36, 37
moral development and, 35
multitasking and, 73, 115–16, 117, 266
online incidents and parental response, 228
parent-child interactions and healthy development, 102
primacy of family and, 40–41, 135, 262, 295
primacy of presence and, 71, 79–81, 88, 98, 115–25, 160, 267
protecting childhood innocence, 41–44
protecting zones of interaction with children, 114–34
reacting with equanimity, 246, 258–59
reactivity: example of, and restart, 240–46
reactivity: scary, crazy, or clueless, 227–28, 231, 235–50, 286
"reflective parenting," 73–74
rules for communication with child, via tech, 272
setting limits and keeping authority, 285–87
sex education and conversation, 184, 186–89, 252–58
shifting from correcting to connecting, 287
spiritual dimension of, 97–98
supervision of computer use, 48, 164, 165–68, 189–92

teaching emotional intelligence, 109,
121–22
tech and loss of control, 230–31
tech and triangulation of parent-child
relationship, 23, 68–77, 94–97, 107
tech as de facto co-parent, 18, 231
tech habits and destructive messages to
child, 17, 74–75, 114–25
trust and, 225, 226–27, 228, 229, 232, 234,
240, 242, 258–59
as "the ultimate RPG," 28–29
unplugging, 90, 94–97, 120–21
value development and, 192, 229, 239,
241, 242, 263
Parenting by Heart (Taffel), 52
"peak experience," 67
Phillips Academy Andover, 253
physical development, 53, 86, 87, 107
play. See also video or screen games
blocks or other hands-on, 35
computer play and lack of moderation, 54
creativity, imagination, and, 55, 99–102,
125–26, 275–76
dress-up, 99–102
emotional literacy and, 106
family encouraging and playing together,
273–78
free play, 53, 54, 105
healthy, opportunities for, 123
losses to screen and digital diversions,
54–55
Magna Tiles, 104
nonscreen ideas, 276
parenting and providing richness of, 65
preschoolers, 103–8, 110
screen vs. real-life experience, 99–103
sleep-away camp and, 56
solitary, importance of, 275
tech and loss of self-generated, creative, at
every age, 53–56
tech's adverse effects on, 25, 35, 123
unstructured replaced by tech, 136–37
values of real-life play, 54–55
as window into child's experience, 276–77
pornography, 35, 42, 149, 151–52, 153, 154,
165, 219
teens and, 196, 210–13, 221, 253

tweens and, 165, 183–89
POS (Parents over Shoulder) screen, 57
Postman, Neil, 26, 263
posttraumatic stress disorder (PTSD), 82
Power Rangers, 42, 84, 112, 121
Practical Wisdom for Parents: Raising Self-
Confident Children in the Preschool Years
(Schulman and Birnbaum), 101
preschool to kindergarten (ages three to
five), 99–128
advertising and marketing to, 111–12
alternatives to tech, healthy options, 120
brain development, 109
child's school problems and lack of parental attention, 116–24
development, tech's adverse effects on,
99–103
emotional development and literacy,
108–14
fear reaction to film image, 126–28
gender messages from media, 112–13
magical thinking, 127–28
parental behaviors and, 233
parental tech habits and destructive messages, 114–25
percentage using a touch-screen device,
108
play, tech and changes in, 103–8
protecting the "magic years," 125–28
protecting zones of interaction with,
114–34
rituals and routines, 123, 124
school readiness, 106–8
screen vs. real-life experience, 99–103
sensorium experiences, 109
tech and media, immersion in, 102–3
tech's positive elements for, 111
privacy
adolescents, tech and, 224–25
cell phone use and disregarding, 10
sexting and threats to, 213
social media issues about, 47–48,
170–712, 223–24
tech trading away family and personal
privacy, 44–46, 48
Proust and the Squid: The Story and Science
of the Reading Brain (Wolf), 51

Raising Cain (PBS documentary), 184
Rapt: Attention and the Focused Life (Gallagher), 29
reading
  deep thinking and, 79
  distractibility and comprehension, 58
  e-books vs. printed page, 89
  editing of scary material and, 127
  neural connections and, 52, 79
  read-aloud apps, 108
  read-aloud time, 80, 88–90
  tech's adverse effects on, 79
relationships
  ages six to ten and, 134–35
  baby brain and, 69
  on-demand presence and relational fatigue, 23
  destructive role of tech in, 20–21, 204
  developing skills of, 110
  development of social brain and, 69
  family and modeling of, 40
  friendship, 8, 102, 110, 165
  intimacy skills and, 18, 21
  "intimate strangers" of the online world, 18
  parent-child, healthy, 10–11
  screen priority and ignoring others, 23–24
  sibling, 52, 156
  tech as coparent, shortcomings of, 18
  tech as third party in parent-child relationship, 23, 68–77, 94–97, 107
  tech redefining interactions, 18–19
  with tech vs. one another, 20–25
  teen, abusive, 217–21
  teens and mediated experience, 193–225
resilience, 40, 85, 110, 135, 160
Rich, Harvey, 39
Rich, Michael, 55, 127
Rugrats, 112
Runaway Bunny, The, 89

school
  academic cheating, 161
  addressing aggression, 250
  afternoon pickup scene, parents and tech use, 114
  bullying in, 49–50, 129–30
  character development and, 166, 249–50
  computer policies, contracts, and expectations for, 190–91
  distractibility of students, 58–59
  empathy lacking in culture of, 49–50
  Garage Band boys incident, handling of, 221–24, 250–52
  laptops and computer use in, 9, 130–31, 181, 189–91
  losses in child development and, 38
  moral education, 147
  out-of-control circumstances caused by overexposure to TV, movies, and online content, 132–33
  overscheduling of children, 53
  pressure for competitive success, 35
  problems and lack of parental attention, 116–24
  removal of screens, preschool, 107
  school readiness, 106–8
  screen-based learning, 59
  social-emotional learning (SEL), 144, 146, 249–50
  workshop on gender codes, 143–47
  zero-tolerance policies and, 249
school buses, 133
Schryver, Kelly, 206
Schulman, Nancy, 101
screen games. *See* video or screen games
self-esteem, 136, 142–43
Seligman, Martin, 264
Sesame Street Workshop study, 74
sexting, 43, 183, 185, 210–13
sexuality or sexual behaviors
  copying behaviors of media and online content, 138
  disconnect from healthy, 186, 197
  emotional intimacy disconnected from, 186, 198, 254, 257, 258
  exposure to sexualized images and gender messages, effect of, 43
  FWB (friends with benefits), 183, 185–86, 254–57
  Gillian's story: reactivity of parents and restart, 240–46
  high-risk, 160
  hookup culture, 185–86

inappropriate online content accessible to children, 35, 42, 43, 149, 151–52
online information about, 165
parental education of child, 252–58
parenting and, 184, 188–89
pornography, 35, 42, 149, 151–52, 153, 154, 165, 219
pornography and masturbation, 183–89
teens and, 196, 210–13, 221, 253
television portrayal of, 155
tweens and, 165, 183–89
value development and, 241, 253, 254, 257
sexualizing of children, 41, 42, 43, 112, 140–47
Shanker, Stuart, 102, 107
Shapiro, Robin, 107
Siegel, Dan, 50–51, 76
Skype, 29, 111, 201, 244
sleep-away camp, 56
sleepovers, 36
sleep problems
    exposure to media violence and, 123
    inappropriate online content, children viewing and, 44, 149
    overstimulation and, 74
    school bullying and, 49
    social cruelty and, 148–49
social cruelty, 49, 143, 145–50, 172–73, 174–75, 190–91, 208–10
social development
    ages eleven to thirteen, 163, 164–67, 171, 174–76
    ages six to ten, 138–39, 159–60
    ages three to five, 102, 104, 109
    self-made play and, 99–102
    tech-mediated correspondence and, 138
    tech's adverse effects on, 25, 26, 94, 104–5, 148
    tween and social media, 174–76
social-emotional learning (SEL), 121, 144
    character development and, 249–50
    core values and, 145–46, 166
    moral development and, 146–47
social media. See also Facebook
    abuse and harassment through, 49–50, 129–31, 217–25
    cyber bullying, 249
    damaging consequences of, 25, 45, 46–48
    gaming and, 177
    indelible digital footprint of, 46–48, 130, 207, 213
    online personas and, 175–76, 204–8
    posting on vs. real-life conversation, 61
    privacy issues, 170
    social cruelty and, 49, 172–73, 174–75, 190–91, 208–10
    teens and, 204–8
    tween social development and, 174–76
Soifer, Lydia, 80
spirituality
    as dimension of parenting, 97
    as mediated experience, 195
    sustainable families and cultivating in nature, 291–94SpongeBob SquarePants, 42, 121
Strong, Jake, 46–47
Strudel Theory, 159
suicide, 50, 219–20

Taffel, Ron, 52
tech (electronic devices and digital communication). See also children; family; parenting; play; specific forms of technology
    addiction of, 3, 5–6, 25, 35, 38, 55, 81, 120, 156, 158, 182–83, 219–20, 265–66
    adverse effects on internal processes of childhood, 25, 26–27, 41, 58, 230, 264
    average time spent on daily, children eight to eighteen, 4
    "BlackBerry nod," 15
    brain and cognition, adverse effects on development, 37, 51, 55, 58, 78, 79, 81–82, 86–87, 89–90, 94, 113–14, 128, 139
    capacity for sustained attention eroded by, 56–60
    character development, online influencing of, 166–67
    childhood, premature loss of, 41–44, 129–61, 165, 166
    children's "secondhand" exposure, 74–75, 82
    communication and loss of direct human interaction, consequences of, 7, 52, 60–65, 193–225

tech (*cont.*)

dehumanizing effects of, 20, 64, 263

on-demand presence and relational fatigue, 23

developmental risks to children, 36–38, 45–46, 78–79

as dominating daily life, 4, 18, 19

emotional development, adverse effects on, 25, 26, 37, 61–62, 94, 105, 123

empathy, adverse effects on, 52, 148, 153, 158, 190–91

expanded job demands and, 24

family values displaced by, 40–41, 112, 135, 145, 173

hyperconnectivity, 63–65

inappropriate online content accessible to children, 33–35, 36, 37, 43–44, 82, 132–33, 149, 150–56, 167, 238

infinite access and, 24

intimacy (authentic emotional), adverse effects on, 18, 62, 196, 197, 200–201

"junk values" and, 111, 181

as "left-brain" activity, 76

loss of boundaries and, 29

media and social networking, loss of family and personal privacy, 44–48

as model of connection, 18–19

moral development, adverse effects on, 35–36, 37, 164, 181, 219 (*see also* empathy)

parental distraction and, 1–3, 6, 11–19, 94–96

play skills, adverse effects on, 53–56, 103–8

positive attributes of, 36

as redefining communication, 20

redefining interactions, 18–19

social development, adverse effects on, 25, 26, 94, 104–5, 148

time using, as eroding time spent with others, 23–24

teens/adolescents

brain of, executive functioning, 37

case of phantom boyfriend, 193–95

characteristics of, 196, 198

dating violence and abuse, 185

drinking culture of, 277

family play and, 277–78

Garage Band boys incident, 221–24, 229, 250–52

intimacy issues 196, 197, 200–201

mediated experience vs. real-life, 193–225

online incidents and risks, 198, 204–5, 214–17, 221–25

privacy issues and tech, 47–48, 223–25

pornography and unhealthy sexuality, 196

real-life experiences lacking, 200–201

self-identity/identity issues, 196, 198, 204–8, 220

self-injury or addiction, 196, 219–20

sexting and, 210–13

suicide and serious psychological issues, 219–20

tech and risk-taking behavior, 196–97

tech and teen drama, 197

texting, 193–95, 199–203

as unsupervised online, 48

values of, 221

wounded psyche and tech exposure, 213–17

television, 3

as babysitter, 83

children's programming and, 87–88

in child's room, 53, 74, 108

cynicism and crass entertainment, 42, 147–48, 185

damaging content, 82

family viewing, 3, 274–75

gender messages on, 155

influencing of teen behavior, 213–17

lack of family interaction and, 82

lack of morality and, 181

loss of protected childhood and, 26

normalizing harmful and unethical behaviors, 51

programs not recommended, 42, 113, 121, 147

programs recommended, 54

racial and sexual stereotypes on, 155

sexual behavior on, 155, 183–84

social-emotional learning (SEL) and, 121

viewing by children under age two, 74

violent content, 155

texting, 19

as crippling to mature relationships, 204

dehumanizing effects of, 64, 199–200

distraction and injury, 6, 59
driving and, 38, 59–60, 232–33
emotional content and confusion, 201–2, 203
emotional discharge increased, 21
gender differences in styles, 203
human voice vs., 21–22
impulsivity and addiction to, 64
impulsivity encouraged by, 22
indelible digital footprint of, 47–48
as left-brain response, 22
loss of conversational skills and, 61
negative impact on emotional intimacy, 62
parental distraction and, 6
pseudointimacy promoted by, 200–201
reactivity and, 22
reflection absent, 202
speed of exchanges, 21
teens and, 193–95, 199–203
validation of feelings and, 63–65
young children, loss of training language and speech centers, 63
Thompson, Michael, 55–56, 184–85, 252–53
Toben, Janice, 99
Tough, Paul, 249
transitions/transitioning skills, 86, 106, 117
tweens/preadolescents (ages eleven to thirteen), 162–92
collapsed or speeded-up childhood, 165, 166
coming of age in a cyber culture, 162–63, 167–73
coping mechanisms for, 177, 182
emotional development and cyber culture, 171
family time for, 192
"geeking out," or screen addictions, 162, 174, 177–83
"hanging out" online, 173–74
Internet and loss of parental control, 164, 165–67
"messing around" online, 174
online incidents and risks, 162–63, 167–73
parental supervision of computer use, 164, 165–68, 189–92
pornography available to, 184–88

rites of passage for, 166
screen games and, 167, 176–77
self-identity/identity issues, 174, 175–76, 192
sexual development and cyber culture, 183–89
social cruelty and, 172–73, 174–75
social development and cyber culture, 171, 174–76
testing parents for reaction to something they did, 234–35
as time of transition, 163, 164
value development and, 192
tweets, 47–48

values. See also culture
adolescents and, 221
ages six to ten and, 134
defined by family, 40–41, 44, 135, 192, 233, 245, 263, 276, 278, 288
family philosophy about tech use and, 269–73
humanistic, as grounding, 294
media and tech displacing family's influence, 40–41, 112, 135, 145, 173
overcoming impact of negative tech content and, 276
parenting and developing, 192, 229, 239, 241, 242, 263
SEL experiences and, 145–46, 166
sexual behavior and, 186, 253, 254, 257
tech and "junk values," 111, 181
tweens/preadolescents and, 192
video or screen games, 9
addiction of, 157–58
aggression and, 49, 157
ARG (alternative reality games), 174
"boy code" and, 42
Facebook and, 177
gaming addiction, 35, 55, 151
Gangstar, 152
gender codes and, 143
Grand Theft Auto, 33–34
inappropriate M or AO available to children, 167
lack of moderation in time spent on, 54
positive effects of prosocial, 54, 264

video or screen games (*cont.*)
  preschoolers imitating, 35
  recommended games, 54
  violence in, 33–34, 152, 153, 157, 158
violence
  aggressive behavior and media violence,
    82, 113, 123, 139, 143, 154, 157
  desensitization to, 82, 153, 157
  exposure to media images of, 43
  "gorenography," 154
  guns and mass shootings, 154
  loss of empathy and, 153
  in media, effect on children, 82, 152–53
  media and screen games, gender coding
    for boys and, 112, 142
  media ideals and being "cool," 145
  normalizing, and damage to childhood,
    140, 151
  parental guidance and, 155–59
  real-world mimicking screen violence, 154

  sexual, 160
  troubled "outcomes" and, 160
  video game addiction and, 151, 156
  in video games, 33–34, 152, 153
  watching vs. imagining, 127

Warren, Robert, 189–91
Wexler, Bruce, 71–72
Whole-Brain Child, The (Bryson), 158
Wick, Donna, 73–74
Wii, 156
Wolf, Maryanne, 51–52, 79, 80, 90

YouTube videos, 43–44
  antisocial themes in, 216
  "four guys in a shower" clip, 152, 238
  indelible digital footprint of, 47–48
  sexual content of, 151–52, 184, 238

Zuckerberg, Mark, 26

About the Authors

Dr. Catherine Steiner-Adair is an internationally recognized clinical psychologist, consultant, and educator. She divides her time working as a therapist with children, couples, and families, and as a school consultant, researcher, and speaker. Her current work focuses on strengthening students' social and emotional resilience to unhealthy cultural values, helping schools develop curricula and programs to strengthen students' confidence and competence as emerging leaders, and nourishing healthy relationships in the age of technology. She is a clinical instructor in the Department of Psychiatry at Harvard Medical School, and an associate psychologist at McLean Hospital. She speaks to numerous audiences, including educators, health professionals, PTAs, religious groups, and corporate and nonprofit organizations. Dr. Steiner-Adair lives in Chestnut Hill, Massachusetts, with her husband, Fred Adair, a leadership consultant. They have a son and a daughter, both young adults, with whom they happily text, talk, and connect.

Visit her web site at www.catherinesteineradair.com.

Teresa H. Barker is a journalist, mother of three digital natives, and book cowriter whose previous collaborations include *Raising Cain: Protecting the Emotional Life of Boys*, *The Pressured Child: Helping Your Child Find Success in School and Life*, and *Girls Will Be Girls: Raising Confident, Courageous Daughters*. Barker and her husband live in Portland, Oregon.